HANDBOOK OF EVAPORATION TECHNOLOGY

HANDBOOK OF EVAPORATION TECHNOLOGY

by

Paul E. Minton

Union Carbide Corporation
South Charleston, West Virginia

NOYES PUBLICATIONS
Park Ridge, New Jersey, U.S.A.

CHEMISTRY

Copyright © 1986 by Paul E. Minton
No part of this book may be reproduced in any form
without permission in writing from the Publisher.
Library of Congress Catalog Card Number 86-17978
ISBN: 0-8155-1097-7
Printed in the United States

Published in the United States of America by
Noyes Publications
Mill Road, Park Ridge, New Jersey 07656

10 9 8 7 6 5 4 3 2 1

Library of Congress Cataloging-in-Publication Data

Minton, Paul E.
  Handbook of evaporation technology.

  Bibliography: p.
  Includes index.
  1. Evaporation--Handbooks, manuals, etc.
2. Evaporators--Handbooks, manuals, etc. I. Title.
TP363.M56 1986   660.2'8426   86-17978
ISBN 0-8155-1097-7

Preface

This book results from an evaporation technology course I have taught for some time. Evaporation is one of the oldest unit operations; it is also an area in which much has changed in the last quarter century. This book is my attempt to present evaporation technology as it is generally practiced today. Although there are other methods of separation which can be considered, evaporation will remain the best separation process for many applications. However, all factors must be properly evaluated in order to select the best evaporator type.

Evaporation technology has often been proprietary to a few companies who design evaporation systems. This situation has benefits, but it also has drawbacks to users of evaporation equipment. Evaporation does not need to be considered an art; good engineering can result in efficient evaporation systems which operate reliably and easily. However, some experience in evaporator design is certainly an advantage in understanding the many problems that can and do occur in evaporation processes.

Much of what is said in this book has been said before. There have, however, been few attempts to combine all this information into one location. I am indebted to the many people who have pioneered evaporation processes and have shared their experiences.

I would like to thank Charlie Gilmour for his mould upon my engineering career. He encouraged me and proved that heat transfer is the most rewarding engineering discipline. I would like to acknowledge the assistance of Howard Freese in the area of mechanically-aided, thin-film evaporation as well as his encouragement in the writing of this book.

South Charleston, West Virginia　　　　　　　　　　　　　　　　　　Paul E. Minton
October 1986

NOTICE

To the best of the Publisher's knowledge the information contained in this publication is accurate; however, the Publisher assumes no liability for any consequences arising from the use of the information contained herein. Final determination of the suitability of any information or product for use contemplated by any user, and the manner of that use, is the sole responsibility of the user.

Contents

1. INTRODUCTION 1
2. EVAPORATION 2
3. WHAT AN EVAPORATOR DOES 3
4. EVAPORATOR ELEMENTS 5
5. LIQUID CHARACTERISTICS 6
   Concentration 6
   Foaming 6
   Temperature Sensitivity 6
   Salting 7
   Scaling 7
   Fouling 7
   Corrosion 7
   Product Quality 7
   Other Fluid Properties 7
6. IMPROVEMENTS IN EVAPORATORS 8
7. HEAT TRANSFER IN EVAPORATORS 9
   Modes of Heat Transfer 10
   Types of Heat Transfer Operations 10
      Sensible Heat Transfer Inside Tubes 11
         Turbulent Flow 11
         Laminar Flow 12
         Transition Region 12
         Helical Coils 12

      Sensible Heat Transfer Outside Tubes 13
      Air-Cooled Heat Exchangers 17
      Agitated Vessels 17
      Condensation 18
          Vertical Tubes 19
          Inside Horizontal Tubes 21
          Outside Horizontal Tubes 21
              Lowfin Tubes. 23
          Sloped Bundles. 23
          Immiscible Condensates 24
          Condensate Subcooling. 24
          Enhanced Condensing Surfaces 24
          Condensation of Vapors Containing Noncondensable Gases. . . . 25
          Desuperheating. 26
          Fog Formation 26
      Falling Film Heat Transfer 27
      Boiling 29
          Natural Convection 33
          Nucleate Boiling 33
          Convective Boiling 34
          Improved Boiling Surfaces. 37
  Physical Properties. 38

8. PRESSURE DROP IN EVAPORATORS 39
  Flow Inside Tubes 39
      U-Bend Tubes. 40
      Helical Coils 41
      Condensing Vapors 41
  Flow Across the Tube Banks 42
      Tubes Omitted from Baffle Windows. 45
      Lowfin Tubes. 45
      Air-Cooled Heat Exchangers 46
      Two-Phase Flow 46
      Condensing Vapors 46

9. FLOW-INDUCED VIBRATION. 48
  Mechanisms 49
  Vortex Shedding. 50
  Turbulent Buffeting. 51
  Fluid-Elastic Whirling. 52
  Parallel Flow Eddy Formation 53
  Acoustic Vibration 53
  Recommendations. 55
  Design Criteria 56
  Fixing Vibration Problems in the Field 58
  Proprietary Designs to Reduce Vibration 59

10. NATURAL CIRCULATION CALANDRIAS 60
- Operation 60
- Surging 64
- Flow Instabilities 65
- Internal Calandrias 67
- Feed Location 69
- Summary 69

11. EVAPORATOR TYPES AND APPLICATIONS 70
- Jacketed Vessels 71
- Coils 73
- Horizontal Tube Evaporators 74
  - Horizontal Spray-Film Evaporators 76
- Short Tube Vertical Evaporators 77
  - Basket Type Evaporators 78
  - Inclined Tube Evaporators 79
  - Propeller Calandrias 80
  - Applications 81
- Long Tube Vertical Evaporators 81
  - Rising or Climbing Film Evaporators 82
  - Falling Film Evaporators 83
  - Rising-Falling Film Evaporators 84
  - Applications 84
- Forced Circulation Evaporators 84
  - Circulation Pumps 86
  - Applications 87
- Plate Evaporators 87
  - Spiral-Plate Evaporators 87
  - Gasketed-Plate Evaporators 88
  - Patterned-Plate Evaporators 89
- Mechanically-Aided Evaporators 90
  - Agitated Vessels 90
  - Scraped-Surface Evaporators 90
    - Low-Speed Units 92
    - High-Speed Units 92
  - Mechanically Agitated Thin-Film Evaporators 92
    - Application for Thin-Film Evaporators 92
    - Process Considerations and Performance 97
    - Maintenance 100
- Submerged Combustion Evaporators 100
  - Applications 103
- Flash Evaporators 103
  - Flash/Fluidized Bed Evaporators 105
  - Special Application 105
- Special Evaporator Types 106
  - Vertical Tube Foaming Evaporator 106
  - Direct-Contact Multiple-Effect Evaporator 107
  - Grainer 107

x   Contents

    Disk or Cascade Evaporator............................107
    Refrigerant Heated Evaporators........................108
    Ball Mills..108
    Fired Heaters..108
    Electrical Heaters....................................108

12. FOULING..113
    Cost of Fouling......................................114
    Classification of Fouling..............................114
    Net Rate of Fouling..................................115
    Sequential Events in Fouling..........................116
        Initiation..116
        Transport..116
        Attachment......................................117
        Removal...117
        Aging..117
    Precipitation Fouling.................................118
        Industrial Systems................................118
            Cooling Water Systems.......................119
            Steam Generating Systems....................119
            Saline Water Distillation.....................119
            Geothermal Brine............................119
            Potable Water Supply........................120
            Evaporation and Crystallization................120
    Particulate Fouling...................................120
    Chemical Reaction Fouling............................121
        Foodstuffs.......................................121
    Corrosion Fouling....................................123
    Biofouling...124
    Solidification Fouling.................................125
    Fouling in Evaporation................................125
        Optimum Cleaning Cycles..........................126
    Design Considerations................................127
    Fouling: Philosophy of Design........................127
        Tubeside Velocities...............................129
        Shellside Design..................................130
        Shellside Velocities...............................131
        Tube Wall Temperatures..........................131
        Conclusions......................................131

13. EVAPORATOR PERFORMANCE.........................133
    Venting..133
        Effect of Noncondensables on Heat Transfer.........134
        Design Precautions................................134
        Tubeside Condensation............................135
        Shellside Condensation............................135
            Small Units..................................135
            Intermediate Units...........................135

Contents   xi

       Large Units 138
       Water-Cooled Equipment 139
  Time/Temperature Relation 139
  Pressure Versus Vacuum Operation. 140
  Energy Economy. 140
       Steam 141
       Cooling water 141
       Tempered Water Systems 141
       Heat Exchange 142
       Evaporative-Cooled Condensers 143
       Air-Cooled Condensers 145
       Pumping Systems 147
       High Temperature Heat Transfer Media 147
       Heat Pumps 148
       Multiple-Effect Evaporators. 148
       Thermal Engine Cycles 149
  Steam Condensate Recovery 150
       Problem Areas 151
           Contamination 151
           Return Line Corrosion 151
           Steam Traps 151
           Improper Steam and Heat Balance 151
       Instrumentation Used for Monitoring Condensate Quality. 151
           Conductivity 152
           pH 152
           Carbon and Hydrogen Analyzers 152
       Summary. 152

14. VAPOR-LIQUID SEPARATION 153
  Entrainment. 153
  Flash Tanks 155
  Wire Mesh Separators. 155
  Vane Impingement Separators 157
  Centrifugal Separators 159
  Cyclones 159
  Other Separators. 162
  Comparison 162
  Solids Deposition 162
  Falling Film Evaporators. 164
  Flashing. 164
  Splashing. 164
  Foaming 164

15. MULTIPLE-EFFECT EVAPORATORS. 166
  Forward Feed. 167
  Backward Feed. 168
  Mixed Feed 169
  Parallel Feed. 169

xii Contents

    Staging 169
    Heat Recovery Systems 170
    Calculations 170
    Optimization 170

16. HEAT PUMPS 172
    Conventional Heat Pump 172
    Overhead Vapor Compression 172
    Calandria Liquid Flashing 172

17. COMPRESSION EVAPORATION 175

18. THERMAL COMPRESSION 176
    Thermocompressor Operation 177
    Thermocompressor Characteristics 178
    Thermocompressor Types 179
        Single Fixed Nozzle 179
        Multiple Nozzle 180
        Single Nozzle, Spindle Operated 180
    Estimating Data 180
    Control 184
    Application 184

19. MECHANICAL VAPOR COMPRESSION 186
    Thermodynamics 187
    Factors Affecting Costs 188
    Compressor Selection 189
        Positive-Displacement Compressors 189
        Axial-Flow Centrifugal Compressors 189
        Radial-Flow Centrifugal Compressors 190
    Factors Influencing Design 190
    Drive Systems 192
    Centrifugal Compressor Characteristics 192
        Constant Speed 192
        Capacity Limitations 193
        Variable Speed 195
        Factors Affecting Control 195
    System Characteristics 199
    Reliability 203
    Evaporator Design 203
    Application 204
    Summary 204
    Economics 204

20. DESALINATION 206
    Startup and Operability 206
    Complexity 206
    Maintenance 207

Contents   *xiii*

    Energy Efficiency 207
    Capital Cost 208
    Operating Temperature. 208
    Materials of Construction 209
    Pretreatment 209
    Chemicals and Auxiliary Energy. 209

21. EVAPORATOR ACCESSORIES 210

22. CONDENSERS. 211
    Direct Contact Condensers 211
        Design of Direct Contact Condensers. 212
    Surface Condensers 213
        Shell-and-Tube Condensers 214
        Updraft Versus Downdraft Condensers 215
        Flooding in Horizontal Tubes 217
        Shellside Flooding in Horizontal Condensers. 218
        Integral Condensers 218
        Condensate Connections. 221

23. VACUUM PRODUCING EQUIPMENT 222
    Jet Ejectors 222
        Multistage Jet Ejectors 225
        Intercondensers. 225
        Precondensers. 225
        Aftercondensers 227
        Barometric Legs and Hotwells 228
        Ejector Stage Characteristics 230
        Ejector Efficiency 230
        Basic Performance Curve–Critical Ejector 231
        Unstable Ejectors 232
        Multistage Ejector Characteristics. 233
        Troubleshooting Steam Jet Ejectors 234
            Test Procedure 235
    Mechanical Pumps. 236
        Liquid-Ring Pumps 237
            Application 239
    Vacuum System Reliability/Maintenance. 240
        Summary. 240
    Multistage Combinations. 240
    Sizing Information. 241
        Specifying the Load. 242
        Air Leakage 242
        Condensable Load. 243
        Free Dry Air Equivalent 244
        Safety Factors 245
    Estimating Energy Requirements 246
        Steam Jets 247

xiv Contents

      Mechanical Pumps............................251
   Initial System Evacuation............................251
   Control of Vacuum Systems............................252
      Suction Throttling............................253
      Load Gas............................253
      Combination Suction Throttling and Load Gas............255
      Variable Speed Drive............................255
      Spindle Jets............................256
   Costs of Vacuum Systems............................256
      Steam Jets............................256
      Liquid-Ring Pumps............................257
   Comparisons............................257
   Energy Conservation............................258

24. CONDENSATE REMOVAL............................259
   Liquid Level Control............................259
   Steam Traps............................261
      Operating Principles............................262
   Mechanical Traps............................263
   Thermostatic Traps............................263
   Thermodynamic Traps............................263
   Steam Trap Specification............................263
   Common Trap Problems............................264
   Selection of Steam Traps............................264
      Process Steam Traps............................265
   Installation............................265
      Double Trapping............................265
      Multiple Trapping............................265
      Part-Load Operation............................266
      Steam Locking............................266
      Inlet Piping............................266
      Discharge Piping............................266
      Large Loads............................266
      Piping Details............................267
      Condensate Return Headers............................267
   Effect of Carbon Dioxide............................268
   Steam Trap Maintenance............................269

25. PROCESS PUMPS............................270
   General Types of Pump Designs............................270
   Net Positive Suction Head (NPSH)............................271
   Cavitation............................272
   Principles of Pumps and Pumping Systems............................272
      Centrifugal Pumps............................273
         Safety Factors for Performance............................274
         Net Positive Suction Head (NPSH) Available............275
      Propeller Pumps............................275
   Avoiding Common Errors............................277

Contents xv

    Temperature Rise and Minimum Flow................... 277
    Oversizing Pumps 277

26. PROCESS PIPING.. 279
    Designing Drain Piping 280
    Compressible Fluids.................................. 281
    Two-Phase Flow 281
        Lockhart-Martinelli 281
        Dukler Homogeneous Model 282
        Flashing Liquids 283
        Elevation Effects................................. 283
        Two-Phase Flow Valve and Fitting Losses 283
        Siphon Flashing 284
    Slurry Flow 284
    Piping Layout....................................... 285
        Process Requirements............................. 286
        Transmission of Stresses and Vibration 286
        Economy....................................... 286
        Accessibility.................................... 286
        Maintenance and Replacement..................... 287
        Piping Stresses 287
            Flexibility Through Layout..................... 287
            Expansion Joints............................. 287
            Cold Springing 287

27. THERMAL INSULATION 288

28. PIPELINE AND EQUIPMENT HEAT TRACING 290

29. PROCESS VESSELS 292

30. REFRIGERATION...................................... 294
    Mechanical Refrigeration 295
    Steam Jet Refrigeration 295
    Absorption Refrigeration 296

31. CONTROL... 297
    Manual Control..................................... 297
    Evaporator Control Systems 298
        Feedback Control 299
        Multielement Control............................. 299
            Cascade Control 299
            Ratio Control................................ 300
            Override Control............................. 300
        Feedforward Control 300
    Control of Evaporators............................... 302
    Auto-Select Control System 304
    Product Concentration............................... 304

Boiling Point Rise 305
  Conductivity 305
  Differential Pressure 306
  Gamma Gage 306
  U-Tube Densitometer 306
  Buoyancy Floats 306
 Condenser Control 306
  Constant-Pressure Vent Systems 306
  Condensate Flooding 307
  Direct-Contact Condensers 308
  Water-Cooled Condensers 308
  Air-Cooled Condensers 308
  Refrigerated Condensers 310
 Calandria Control 310
  Kettle-Type Reboilers 310
  Falling Film Vaporizers 311
  Steam Heated Calandrias 311
  Liquid Heated Calandrias 312
 Evaporator Base Sections and Accumulators 312
 Guidelines for Instruments 313
  Display Information 313
  Measurement Locations 313
  Control Valve Stations 313
  Primary Flow Elements 314
  Primary Level Measurements 314
  Primary Pressure Measurements 315
  Pressure Regulators 316
  Control System Utilities 316
 Process Computers 316

32. THERMAL DESIGN CONSIDERATIONS 318
  Tube Size and Arrangement 318
   Tube Diameter 318
   Tube Length 318
   Arrangement 318
  Extended Surfaces 319
  Shellside Impingement Protection 319
  Flow Distribution 321

33. INSTALLATION 322
  Venting 322
  Siphons in Cooling Water Piping 322
  U-Bend Exchangers 323
  Equipment Layout 323
  Piping 323

34. DESIGN PRACTICES FOR MAINTENANCE 324
  Standard Practices 325

       Repair Features..325
       Chemical Cleaning Equipment...........................325
       Mechanical Cleaning Equipment.........................325
       Backwashing...325
       Air Injection...326

35. MECHANICAL DESIGN.......................................327
       Maximum Allowable Working Pressure and Temperature....327
       Upset Conditions......................................328
       Thermal Expansion.....................................328
       Tube-to-Tubesheet Joints..............................329
       Double Tubesheets.....................................329
               Conventional Design...........................330
               Integral Design...............................330
               Summary.......................................330
       Inspection Techniques.................................331

36. SAFETY..332
       Common Errors...333
       Safety Relief...334
               Pressure Relief...............................334

37. MATERIALS OF CONSTRUCTION...............................336
       Basic Questions.......................................336
       Selection...338

38. TESTING EVAPORATORS.....................................339
       Planning the Test.....................................339
       Causes of Poor Performance............................340
               Low Steam Economy.............................341
               Low Rates of Heat Transfer....................341
               Excessive Entrainment.........................341
               Short Cleaning Cycles.........................342

39. TROUBLESHOOTING...343
       Calandrias..344
       Condensers..345
       Vacuum Fails to Build.................................346
               Steam Jet Vacuum Systems......................346
               Mechanical Pumps..............................347
       No Vacuum in Steam Chest..............................347
       Vacuum Builds Slowly..................................347
       Foaming...348
       Inadequate Circulation................................348
       Sudden Loss of Vacuum.................................348
       Vacuum Fluctuates.....................................348
       Water Surge in Tail Pipe..............................349
       Barometric Condenser Flooding.........................349

40. UPGRADING EXISTING EVAPORATORS..................350
Areas for Upgrading Existing Evaporators 351
Fine Tuning Existing Systems 351
Venting Rates.................................352
Water Leakage 352
Operating Pressures 352
Fouling......................................352
Separator Efficiencies........................352
Heat Losses 352
Improper Cleaning...........................353
Modifying Auxiliary Hardware.......................353
Heat Recovery 353
Condensate Reuse 353
Instrumentation and Control..................354
Major Hardware Modifications.......................354
Vapor Compression 354
Thermal Compression....................354
Mechanical Compression................355
Installing Additional Effects 356
Combinations...............................356
Economic Effects of Improvements 356
Guidelines for Upgrading Program 357
Evaluating Existing Evaporators.....................357
Determining Appropriate Methods 358
Economic Analysis 358
Implementation 358

41. ENERGY CONSERVATION...............................359

42. SPECIFYING EVAPORATORS.............................360
Comparing Vendors' Offerings...........................361

43. NEW TECHNOLOGY 363

44. NOMENCLATURE 365
Greek 371
Subscripts 371

BIBLIOGRAPHY 372
Evaporation 372
Heat Transfer 373
Boiling 374
Heat Exchangers.................................374
Flow-Induced Vibration 375
Fouling 375
Direct Contact Heat Transfer......................375
Energy Conservation 376
Vapor Compression Evaporation 376

Vacuum Systems. 377
Steam Traps. 377
Control 378
Pumps. 378
Process Piping and Fluid Flow 379
Separators 379
Thermal Insulation 379
Troubleshooting 380
Venting. 380
Air-Cooled Heat Exchangers 380
Heat Transfer Fluids 381
Testing 381
Electrical Heating 381
Steam Tracing. 381
Jacketed Vessels 382
Turbines 382
Mechanical Design 382
Materials of Construction 382
Desalination 382
Evaporators 383

INDEX 384

1
Introduction

The industrial society in which we live has depended during recent decades upon the earth's supply of oil and gas as its principal source of energy. These resources are dwindling, and most knowledgeable observers expect them to attain peak production on a worldwide basis during the next quarter-century. Possibly, the most important problem we face in the years immediately ahead is the timely development of alternate energy sources in sufficient quantity to avert serious economic and social disruption. Efficient utilization of the energy resources currently available will extend the time period during which new energy sources can be developed.

Approximately 25% of the cost of products is the cost of energy to operate plants. Energy is the fastest growing element of manufacturing cost.

Proper specification, design, and operation of evaporator systems will help to reduce the cost of producing a product by evaporation. Upgrading of existing evaporator systems is a fruitful area for achieving reduced energy requirements.

2
Evaporation

Evaporation is the removal of solvent as vapor from a solution or slurry. For the overwhelming majority of evaporation systems the solvent is water. The objective is usually to concentrate a solution; hence, the vapor is not the desired product and may or may not be recovered depending on its value. Therefore, evaporation usually is achieved by vaporizing a portion of the solvent producing a concentrated solution, thick liquor, or slurry.

Evaporation often encroaches upon the operations known as distillation, drying, and crystallization. In evaporation, no attempt is made to separate components of the vapor. This distinguishes evaporation from distillation. Evaporation is distinguished from drying in that the residue is always a liquid. The desired product may be a solid, but the heat must be transferred in the evaporator to a solution or a suspension of the solid in a liquid. The liquid may be highly viscous or a slurry. Evaporation differs from crystallization in that evaporation is concerned with concentrating a solution rather than producing or building crystals.

This discussion will be concerned only with evaporation. Distillation, drying, or crystallization will not be emphasized.

3
What an Evaporator Does

As stated above, the object of evaporation may be to concentrate a solution containing the desired product or to recover the solvent. Sometimes both may be accomplished. Evaporator design consists of three principal elements: heat transfer, vapor-liquid separation, and efficient utilization of energy.

In most cases the solvent is water, heat is supplied by condensing steam, and the heat is transferred by indirect heat transfer across metallic surfaces. For evaporators to be efficient, the equipment selected and used must be able to accomplish several things:

(1) Transfer large amounts of heat to the solution with a minimum amount of metallic surface area. This requirement, more than all other factors, determines the type, size, and cost of the evaporator system.

(2) Achieve the specified separation of liquid and vapor and do it with the simplest devices available. Separation may be important for several reasons: value of the product otherwise lost; pollution; fouling of the equipment downstream with which the vapor is contacted; corrosion of this same downstream equipment. Inadequate separation may also result in pumping problems or inefficient operation due to unwanted recirculation.

(3) Make efficient use of the available energy. This may take several forms. Evaporator performance often is rated on the basis of steam economy—pounds of solvent evaporated per pound of steam used. Heat is required to raise the feed temperature from its initial value to that of the boiling liquid, to provide the energy required to separate liquid solvent from the feed, and to vaporize the solvent. The greatest increase in energy economy is achieved by reusing the vaporized solvent as a heating medium. This can be accomplished in several ways to be discussed later. Energy efficiency may be

increased by exchanging heat between the entering feed and the leaving residue or condensate.

(4) Meet the conditions imposed by the liquid being evaporated or by the solution being concentrated. Factors that must be considered include product quality, salting and scaling, corrosion, foaming, product degradation, holdup, and the need for special types of construction.

Today many types of evaporators are in use in a great variety of applications. There is no set rule regarding the selection of evaporator types. In many fields several types are used satisfactorily for identical services. The ultimate selection and design may often result from tradition or past experience. The wide variation in solution characteristics expand evaporator operation and design from simple heat transfer to a separate art.

4
Evaporator Elements

Three principal elements are of concern in evaporator design: heat transfer, vapor-liquid separation, and efficient energy consumption. The units in which heat transfer takes place are called heating units or calandrias. The vapor-liquid separators are called bodies, vapor heads, or flash chambers. The term body is also employed to label the basic building module of an evaporator, comprising one heating element and one flash chamber. An effect is one or more bodies boiling at the same pressure. A multiple-effect evaporator is an evaporator system in which the vapor from one effect is used as the heating medium for a subsequent effect boiling at a lower pressure. Effects can be staged when concentrations of the liquids in the effects permits; staging is two or more sections operating at different concentrations in a single effect. The term evaporator denotes the entire system of effects, not necessarily one body or one effect.

5
Liquid Characteristics

The practical application of evaporator technology is profoundly affected by the properties and characteristics of the solution to be concentrated. Some of the most important properties of evaporating liquids are discussed below.

CONCENTRATION

The properties of the feed to an evaporator may exhibit no unusual problems. However, as the liquor is concentrated, the solution properties may drastically change. The density and viscosity may increase with solid content until the heat transfer performance is reduced or the solution becomes saturated. Continued boiling of a saturated solution may cause crystals to form which often must be removed to prevent plugging or fouling of the heat transfer surface. The boiling point of a solution also rises considerably as it is concentrated.

FOAMING

Some materials may foam during vaporization. Stable foams may cause excessive entrainment. Foaming may be caused by dissolved gases in the liquor, by an air leak below the liquid level, and by the presence of surface-active agents or finely divided particles in the liquor. Many antifoaming agents can be used effectively. Foams may be suppressed by operating at low liquid levels, by mechanical methods, or by hydraulic methods.

TEMPERATURE SENSITIVITY

Many chemicals are degraded when heated to moderate temperatures for relatively short times. When evaporating such materials special techniques are needed to control the time/temperature characteristics of the evaporator system.

SALTING

Salting refers to the growth on evaporator surfaces of a material having a solubility that increases with an increase of temperature. It can be reduced or eliminated by keeping the evaporating liquid in close or frequent contact with a large surface area of crystallized solid.

SCALING

Scaling is the growth or deposition on heating surfaces of a material which is either insoluble or has a solubility that decreases with an increase in temperature. It may also result from a chemical reaction in the evaporator. Both scaling and salting liquids are usually best handled in an evaporator that does not rely upon boiling for operation.

FOULING

Fouling is the formation of deposits other than salt or scale. They may be due to corrosion, solid matter entering with the feed, or deposits formed on the heating medium side.

CORROSION

Corrosion may influence the selection of evaporator type since expensive materials of construction indicate evaporators affording high rates of heat transfer. Corrosion and erosion are frequently more severe in evaporators than in other types of equipment because of the high liquid and vapor velocities, the frequent presence of suspended solids, and the concentrations required.

PRODUCT QUALITY

Product quality may require low holdup and low temperatures. Low-holdup-time requirements may eliminate application of some evaporator types. Product quality may also dictate special materials of construction.

OTHER FLUID PROPERTIES

Other fluid properties must also be considered. These include: heat of solution, toxicity, explosion hazards, radioactivity, and ease of cleaning. Salting, scaling, and fouling result in steadily diminishing heat transfer rates, until the evaporator must be shut down and cleaned. Some deposits may be difficult and expensive to remove.

6

Improvements in Evaporators

Many improvements have been made in evaporator technology in the last half-century. The improvements have taken many forms but have served to effect the following:

(1) Greater evaporation capacity through better understanding of the heat transfer mechanisms.

(2) Better economy through more efficient use of evaporator types.

(3) Longer cycles between cleaning because of better understanding of salting, scaling, and fouling.

(4) Cheaper unit costs by modern fabrication techniques and larger unit size.

(5) Lower maintenance costs and improved product quality by use of better materials of construction as a result of better understanding of corrosion.

(6) More logical application of evaporator types to specific services.

(7) Better understanding and application of control techniques and improved instrumentation has resulted in improved product quality and reduced energy consumption.

(8) Greater efficiency resulting from enhanced heat transfer surfaces and better energy economy.

(9) Compressor technology and availability has permitted the application of mechanical vapor compression.

7
Heat Transfer in Evaporators

Whenever a temperature gradient exists within a system, or when two systems at different temperatures are brought into contact, energy is transferred. The process by which the energy transport takes place is known as heat transfer. The thing in transit, called heat, cannot be measured or observed directly, but the effects it produces are amenable to observations and measurement.

The branch of science which deals with the relation between heat and other forms of energy is called thermodynamics. Its principles, like all laws of nature, are based on observations and have been generalized into laws which are believed to hold for all processes occurring in nature, because no exceptions have ever been detected. The first of these principles, the first law of thermodynamics, states that energy can be neither created nor destroyed but only changed from one form to another. It governs all energy transformations quantitatively but places no restrictions on the direction of the transformation. It is known, however, from experience that no process is possible whose sole result is the net transfer of heat from a region of lower temperature to a region of higher temperature. This statement of experimental truth is known as the second law of thermodynamics.

All heat-transfer processes involve the transfer and conversion of energy. They must therefore obey the first as well as the second law of thermodynamics. From a thermodynamic viewpoint, the amount of heat transferred during a process simply equals the difference between the energy change of the system and the work done. It is evident that this type of analysis considers neither the mechanism of heat flow nor the time required to transfer the heat.

From an engineering viewpoint, the determination of the rate of heat transfer at a specified temperature difference is the key problem. The size and cost of heat transfer equipment depend not only on the amount of heat to be transferred, but also on the rate at which the heat is to be transferred under given conditions.

MODES OF HEAT TRANSFER

The literature of heat transfer generally recognizes three distinct modes of heat transfer: conduction, radiation, and convection. Strictly speaking, only conduction and radiation should be classified as heat-transfer processes, because only these two mechanisms depend for their operation on the mere existence of a temperature difference. The last of thre three, convection, does not strictly comply with the definition of heat transfer because it depends for its operation on mechanical mass transport also. But since convection also accomplishes transmission of energy from regions of high temperature to regions of lower temperature, the term "heat transfer by convection" has become generally accepted.

In most situations heat flows not by one, but by several of these mechanisms simultaneously.

Conduction is the transfer of heat from one part of a body to another part of the same body, or from one body to another in physical contact with it, without appreciable displacement of the particles of the body. Conduction can occur in solids, liquids, or gases.

Radiation is the transfer of heat from one body to another, not in contact with it, by means of electromagnetic wave motion through space, even when a vacuum exists between them.

Convection is the transfer of heat from one point to another within a fluid, gas or liquid, by the mixing of one portion of the fluid with another. In natural convection, the motion of the fluid is entirely the result of differences in density resulting from temperature differences; in forced convection, the motion is produced by mechanical means. When the forced velocity is relatively low, it should be realized that "free-convection" factors, such as density and temperature difference, may have an important influence.

In the solution of heat-transfer problems, it is necessary not only to recognize the modes of heat transfer which play a role, but also to determine whether a process is *Steady* or *Unsteady*. When the rate of heat flow in a system does not vary with time—when it is constant—the temperature at any point does not change and steady-state conditions prevail. Under steady-state conditions, the rate of heat input at any point of the system must be exactly equal to the rate of heat output, and no change in internal energy can take place. The majority of engineering heat-transfer problems are concerned with steady-state systems.

The heat flow in a system is transient, or unsteady, when the temperatures at various points in the system change with time. Since a change in temperature indicates a change in internal energy, we conclude that energy storage is associated with unsteady heat flow. Unsteady-heat-flow problems are more complex than are those of steady state and can often only be solved by approximate methods.

TYPES OF HEAT TRANSFER OPERATIONS

There are two types of heat transfer operation: sensible heat and change of phase. Sensible heat operations involve heating or cooling of a fluid in which the heat transfer results only in a temperature change of the fluid. Change-of-phase heat transfer results in a liquid being changed into a vapor or a vapor being changed into a liquid. *Boiling* or vaporization is the convection process involving a change in phase from liquid to vapor. *Condensation* is the convection process involving a change in phase from vapor to liquid. Many applications involve both sensible heat and change-of-phase heat transfer.

Sensible Heat Transfer Inside Tubes

Sensible heat transfer in most applications involves forced convection inside tubes or ducts or forced convection over exterior surfaces.

The heating and cooling of fluids flowing inside conduits are among the most important heat-transfer processes in engineering. The flow of fluids inside conduits may be broken down into three flow regimes. These flow regimes are measured by a ratio called the Reynolds number which is an indication of the turbulence of the flow inside the conduit. The three regimes are:

Laminar Flow	Reynolds numbers less than 2,100
Transition Flow	Reynolds numbers between 2,100 and 10,000
Turbulent Flow	Reynolds numbers greater than 10,000

Figure 7-1 indicates the shape of the curve correlating Reynolds number with a heat transfer parameter.

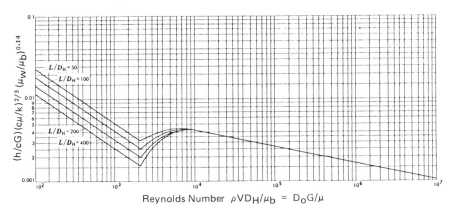

Figure 7-1: Recommended curves for determining heat-transfer coefficient in the transition regime. (Reprinted from *Industrial and Engineering Chemistry*, Vol. 28, p. 1429, December 1936, with permission of the copyright owner, The American Chemical Society).

Turbulent Flow: For engineering purposes, semi-empirical equations are generally used to describe heat transfer in turbulent flow. These correlations adequately predict heat transfer in this region. (Nomenclature is presented in Chapter 44.)

$$h/cG = 0.023(c\mu/k)^{-2/3}(D_i G/\mu)^{-0.2}(\mu_b/\mu_w)^{0.14} \quad (7.1)$$

For short tube lengths, the equation above should be corrected to reflect entrance effects as given below:

$$h_L/h = 1 + (D_i/L)^{0.7} \tag{7.2}$$

where h_L = average heat transfer coefficient for finite length L
$\quad$ h = heat transfer coefficient for tube of infinite length calculated using Equation 7.1
$\quad D_i$ = inside tube diameter
$\quad$ L = tube length

Laminar Flow: Although heat-transfer coefficients for laminar flow are considerably smaller than for turbulent flow, it is sometimes necessary to accept lower heat transfer in order to reduce pumping costs. The heat-flow mechanism in purely laminar flow is conduction. The rate of heat flow between the walls of the conduit and the fluid flowing in it can be obtained analytically. But to obtain a solution it is necessary to know or assume the velocity distribution in the conduit. In fully developed laminar flow without heat transfer, the velocity distribution at any cross section has the shape of a parabola. The velocity profile in laminar flow usually becomes fully established much more rapidly than the temperature profile. Heat-transfer equations based on the assumption of a parabolic velocity distribution will therefore not introduce serious errors for viscous fluids flowing in long ducts, if they are modified to account for effects caused by the variation of the viscosity due to the temperature gradient. The equation below can be used to predict heat transfer in laminar flow.

$$h/cG = 1.86(D_i G/\mu)^{-2/3}(c\mu/k)^{-2/3}(L/D_i)^{-1/3}(\mu_b/\mu_w)^{0.14} \tag{7.3}$$

For extremely viscous fluids (viscosity greater than 1,000 centipoise), this correlation is not adequate, especially for non-Newtonian fluids. The effects of viscous shear must be considered; more rigorous approaches are usually required.

Transition Region: The mechanism of heat transfer and fluid flow in the transition region varies considerably from system to system. In this region the flow may be unstable and fluctuations in pressure drop and heat transfer have been observed. There exists a large uncertainty in the basic heat transfer and flow friction performance, and consequently the designer is advised to design equipment, if possible, to operate outside this region. The equation below can be used to predict heat transfer in the transition region.

$$h/cG = 0.116 \left[\frac{(D_i G/\mu)^{2/3} - 125}{(D_i G/\mu)} \right] \left[\frac{1 + (D_i/L)^{2/3}}{(c\mu/k)^{2/3}(\mu_w/\mu_b)^{0.14}} \right] \tag{7.4}$$

Helical Coils: Heat transfer coefficients for fluids flowing inside helical coils can be calculated with modifications of the equations for straight tubes. The equations for straight tubes should be corrected as below:

$$h_h = h_s [1 + 3.5(D_i/D_c)] \tag{7.5}$$

where h_h = heat transfer for helical coil
$\quad h_s$ = heat transfer for straight tube

D_i = inside tube diameter
D_c = diameter of helix or coil

For laminar flow, the ratio of length to diameter should be calculated as below:

$$(L/D_i)^{1/3} = (D_c/D_i)^{1/6} \tag{7.6}$$

The Reynolds number required for turbulent flow can be determined as below:

$$Re_c = 2100\,[1 + 12(D_i/D_c)^{1/2}] \tag{7.7}$$

where Re_c = critical Reynolds number above which turbulent flow exists

Sensible Heat Transfer Outside Tubes

The heat-transfer phenomena for forced convection over exterior surfaces are closely related to the nature of the flow. The heat transfer in flow over tube bundles depends largely on the flow pattern and the degree of turbulence, which in turn are functions of the velocity of the fluid and the size and arrangement of the tubes. The equations available for the calculation of heat transfer coefficients in flow over tube banks are based entirely on experimental data because the flow is too complex to be treated analytically. Experiments have shown that, in flow over staggered tube banks, the transition from laminar to turbulent flow is more gradual than in flow through a pipe, whereas for in-line tube bundles the transition phenomena resemble those observed in pipe flow. In either case the transition from laminar to turbulent flow begins at a Reynolds number based on the velocity in the minimum flow area of about 100, and the flow becomes fully turbulent at a Reynolds number of about 3,000. The equation below can be used to predict heat transfer for flow across ideal tube banks.

$$h/cG = a(D_o G/\mu)^{-m}(c\mu/k)^{-2/3}(\mu_b/\mu_w)^{0.14} \tag{7.8}$$

$D_o G/\mu$	Tube Pitch	m	a
above 200,000	staggered	0.300	0.168
above 200,000	in-line	0.300	0.124
300 to 200,000	staggered	0.365	0.273
300 to 200,000	in-line	0.349	0.211
below 300	staggered	0.640	1.309
below 300	in-line	0.569	0.742

In the equation above, G is the mass velocity defined as below:

$G = W/a_c$ where W is the mass flow rate and a_c is given below.

For triangular and square tube patterns:

$$a_c = D_s P_b (S_t - D_o)/S_t \tag{7.9}$$

For rotated triangular tube patterns:

$$a_c = 1.55 D_s P_b (S_t - D_o)/S_t \quad \text{when } S_t \text{ is less than } 3.73 D_o \tag{7.10}$$

$$a_c = D_s P_b [1 - 0.577(D_o/S_t)] \text{ when } S_t \text{ is greater than } 3.73 D_o \quad (7.11)$$

For rotated square tube patterns:

$$a_c = 1.414 D_s P_b (S_t - D_o)/S_t \text{ when } S_t \text{ is less than } 1.71 D_o \quad (7.12)$$

$$a_c = D_s P_b [1 - 0.707(D_o/S_t)] \text{ when } S_t \text{ is greater than } 1.71 D_o \quad (7.13)$$

The values of "a" for Equation 7.8 are for ideal tube banks with no by-passing or leakage. For well-built tube bundles fabricated to industry-wide accepted standards for clearances (discussed in Chapter 12), the heat transfer coefficients obtained using these values of "a" are normally multiplied by a factor of 0.7 to account for unavoidable bypassing and leakage. However, more precise results can be obtained by using correction factors for the actual conditions as given below. This method assumes that:

(1) all clearances and tolerances are in accordance with TEMA
(2) one sealing device is provided for every five tube rows
(3) baffle cuts are 20% of the shell diameter
(4) fouled heat exchangers result in plugging of clearances between tubes and baffles.

Under these conditions:

$$h_o = h F_l F_r F_c F_b \quad (7.14)$$

where h_o = actual heat transfer coefficient
h = heat transfer coefficient for ideal tube bank
F_c = 1.1
F_b = 0.9
$F_l = 0.8(P_b/D_s)^{1/6}$ for fouled bundles (7.15)
$F_l = 0.8(P_b/D_s)^{1/4}$ for clean bundles (7.16)
F_r = 1.0 when $D_o G/\mu$ is greater than 100
$F_r = 0.2(D_o G/\mu)^{1/3}$ when $D_o G/\mu$ is less than 100 (7.17)

For banks of lowfin tubes, the mass velocity is lower because of the greater flow area afforded by the fins. The equations above for tube bundles can be used for lowfin tubes with the following adjustments:

(1) root diameters should be used instead of outside tube diameter

$$d_r'' = d_o'' - 1/8'' \quad (7.18)$$

where d_r'' = root diameter, inches
d_o'' = outside tube diameter, inches

(2) mass velocity calculated as already outlined must be reduced by the following ratio:

$$G(s_t'' - d_o'')/(s_t'' - d_o'' + 0.09'') \tag{7.19}$$

where G = mass velocity for tube bundles of plain tubes
s_t'' = tube center-to-center spacing, inches
d_o'' = outside tube diameter, inches

The fintube outside heat transfer coefficient can then be referred to the inside tube surface as below:

$$h_{fi} = h_o E_w (A_t/A_i) \tag{7.20}$$

where h_{fi} = effective inside heat transfer coefficient for the fintube
h_o = outside heat transfer coefficient
E_w = weighted fin efficiency (from Figures 7-2 and 7-3)
A_t = total outside surface of the tube
A_i = inside surface of the tube

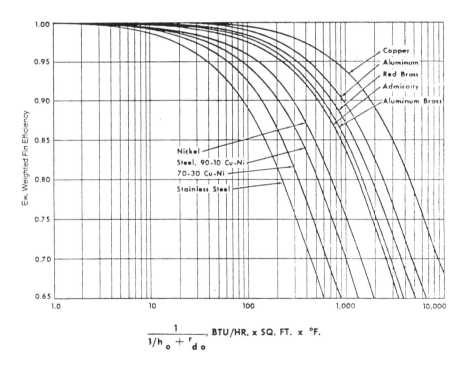

Figure 7-2: Weighted fin efficiencies 16-19 fins per inch.

16 Handbook of Evaporation Technology

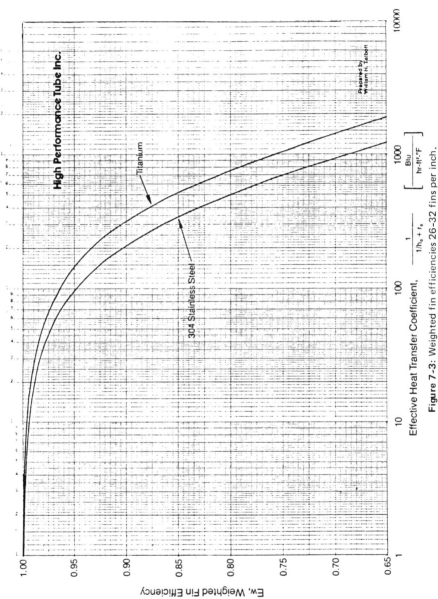

Figure 7-3: Weighted fin efficiencies 26-32 fins per inch.

Air-Cooled Heat Exchangers

Air-cooled exchangers are generally designed with standard tube geometries: plain tubes with outside diameters of 1 inch with 5/8 inch high aluminum fins spaced at 8 or 10 fins per inch. Standard tube arrangements are listed in the table below. Typical face velocities used for design are also tabulated below. These values result in air-cooled heat exchangers which approach an optimum cost. The total cost of an air-cooled exchanger must include the purchase cost, the cost for installation, and the cost of power to drive the fans. The optimum will vary for each user of air-cooled equipment, but generally the optimum cost is not much less than the designs on either side of the optimum point. Each designer may wish to establish his own values of typical design face velocities; they should not vary greatly from those tabulated.

Design Face Velocities

Number of Tube Rows	Face Velocity, Feet Per Minute		
	8 Fins/Inch 2-3/8" Pitch	10 Fins/Inch 2-3/8" Pitch	10 Fins/Inch 2-1/2" Pitch
3	650	625	700
4	615	600	660
5	585	575	625
6	560	550	600

For air-cooled equipment, the tube spacing is normally determined by the relative values of the inside heat-transfer coefficient. For inside coefficients much greater than the air-side coefficient, tubes spaced on 2-1/2" centers are normally justified. For low values of the inside coefficient, tubes spaced on 2-3/8" centers are normally justified.

The air-side heat-transfer coefficient is frequently calculated based on the outside surface of the bare tube. The equations for air-cooled heat exchangers can be simplified as below:

$$h_a = 8(FV)^{1/2} \text{ for 10 fins per inch} \quad (7.21)$$

$$h_a = 6.75(FV)^{1/2} \text{ for 8 fins per inch} \quad (7.22)$$

where h_a = air-side heat-transfer coefficient, Btu/(hr)(sq ft)(°F)
FV = face velocity of air, feet per minute

Agitated Vessels

Sensible heat transfer coefficients for agitated vessels can be predicted using the following correlation:

$$hD_j/k_l = a(L^2 N \rho_l/\mu_l)^{2/3} (c_l \mu_l/k_l)^{1/3} (\mu_b/\mu_w)^{0.14} \quad (7.23)$$

where D_j = diameter of the agitated vessel
L = diameter of the agitator
N = speed of the agitator, revolutions per hour

The term "a" has the values below:

Agitator Type	Surface	a
Turbine	Jacket	0.62
Turbine	Coil	1.50
Paddle	Jacket	0.36
Paddle	Coil	0.87
Propeller	Jacket	0.54
Propeller	Coil	0.83
Anchor	Jacket	0.46

Condensation

When a saturated vapor comes in contact with a surface at a lower temperature condensation occurs. Condensation is the convection heat transfer process involving a change in phase from vapor to liquid. Condensate forms on the cool surface and under the influence of gravity will flow down the surface. If the liquid wets the surface, a smooth film is formed, and the process is called *film condensation*. If the liquid does not wet the surface, droplets are formed which fall down the surface in some random fashion. This process is called *dropwise condensation*. In the film condensation process the surface is blanketed by the film, which grows in thickness as it moves down the surface. A temperature gradient exists in the film, and the film represents a resistance to heat transfer. In dropwise condensation a large portion of the surface is directly exposed to the vapor; there is no film barrier to heat flow; and higher heat transfer rates are experienced. In fact, heat transfer rates in dropwise condensation may be many times higher than in film condensation.

Because of the higher heat transfer rates, dropwise condensation would be preferred to film condensation, but it is extremely difficult to maintain since most surfaces become "wetted" after exposure to a condensing vapor over an extended period of time. Various surface coatings and vapor additives have been used in attempts to maintain dropwise condensation, but these methods have not met with general success to date.

Under normal conditions a continuous flow of liquid is formed over the surface and the condensate flows downward under the influence of gravity. Unless the velocity is very high or the liquid film relatively thick, the motion of the condensate is laminar and heat is transferred from the vapor-liquid interface to the surface merely by conduction. The rate of heat flow depends on the rate at which vapor is condensed and the rate at which the condensate is removed. On a vertical surface the film thickness increases continuously from top to bottom. As the surface is inclined from the vertical, the drainage rate decreases and the liquid film becomes thicker. This causes a decrease in the rate of heat transfer.

However, even at relatively low film Reynolds numbers, the assumption that the condensate layer is in laminar flow is open to some question. Experiments have shown that the surface of the film exhibits considerable waviness (turbulence). This waviness causes increased heat transfer rates. Better heat transfer correlations for vertical condensation were presented by Dukler in 1960. He obtained velocity distributions in the liquid film as a function of the interfacial shear (due to the vapor velocity) and film thickness. From the integration of the velocity and temperature profiles, liquid film thickness and point heat-transfer coefficients were computed. According to the Dukler development, there is no

Heat Transfer in Evaporators

definite transition Reynolds number and deviation from laminar theory is less at low Reynolds numbers.

Vertical Tubes: Heat transfer coefficients for condensation on vertical tubes may be calculated from laminar theory as given below:

$$h = 0.925 k_l (\rho_l^2 g/\mu_l \Gamma)^{1/3} \quad (7.24)$$

$$\Gamma = W_l/n\pi D_i \text{ when condensing inside tubes} \quad (7.25)$$

$$\Gamma = W_l/n\pi D_o \text{ when condensing outside tubes} \quad (7.26)$$

where W_l is the amount of liquid condensed
n is the total number of tubes
D_i is the inside tube diameter
D_o is the outside tube diameter

When vapor density is high, the term ρ_l^2 should be replaced by the term $\rho_l(\rho_l - \rho_v)$. If ρ_v is small in comparison to ρ_l, the latter term reduces to the former.

The Dukler theory assumes that three fixed factors must be known to establish the value of the average heat transfer coefficient for condensing for vertical tubes. These are the terminal Reynolds number $(4\Gamma/\mu_l)$, the Prandtl number $(c_l\mu_l/k_l)$ of the condensed phase, and a dimensionless group designated by A_d and defined as follows:

$$A_d = \frac{0.250 \mu_l^{1.173} \mu_v^{0.16}}{g^{2/3} D^2 \rho_l^{0.553} \rho_v^{0.78}} \quad (7.27)$$

Figure 7-4 presents the heat transfer data for values of $A_d = 0$, no interfacial shear. Figure 7-5 can be used to predict condensing heat transfer when interfacial shear is not negligible.

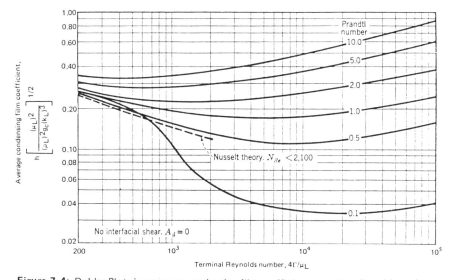

Figure 7-4: Dukler Plot gives average condensing-film coefficients at various Prandtl numbers.

20 Handbook of Evaporation Technology

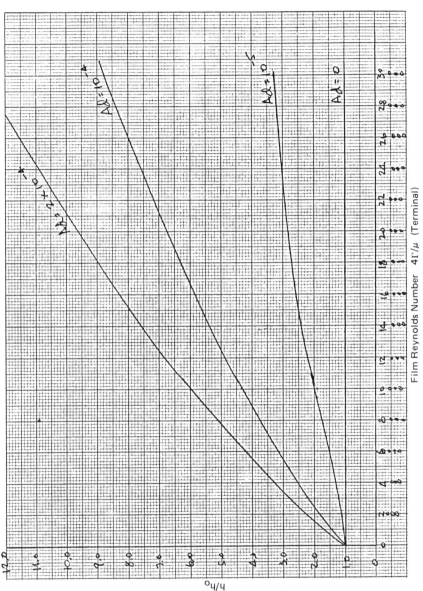

Figure 7-5.

Inside Horizontal Tubes: Condensation inside horizontal tubes can be predicted assuming two mechanisms: laminar film condensation and vapor shear dominated condensation in which the two-phase flow is in the annular region. For laminar film condensation the further assumption is made that the rate of condensation on the stratified layer of liquid running along the bottom of the tube is negligible. Consequently, this layer of liquid must not exceed values assumed without being approximately accounted for.

The actual condensing heat transfer coefficient should be taken as the higher of the values calculated using the two mechanisms.

For stratified flow:

$$h_c = 0.767 k_l (\rho_l^2 g n L/\mu_l W_l)^{1/3} \qquad (7.28)$$

This equation assumes a certain condensate level on the bottom of the tube. This should be evaluated and corrected using Figure 7-6 after calculating the value of $W_l/n\rho_l D_i^{2.56}$.

For annular flow:

$$h_s = h_{LO}[1 + x[(\rho_l/\rho_v) - 1]]^{1/2} \qquad (7.29)$$

where h_s = condensing heat transfer coefficient when vapor shear dominates
  h_{LO} = sensible heat transfer coefficient calculated from Equation 7.1 assuming that the total fluid is flowing with condensate properties
  x = vapor phase mass flow fraction (quality)
  ρ_l = liquid density
  ρ_v = vapor density

Outside Horizontal Tubes: Condensation on the outside of banks of horizontal tubes can be predicted assuming two mechanisms: laminar condensate flow and vapor shear dominated heat transfer.

For laminar film condensation:

$$h_c = a k_l (\rho_l^2 g n L/\mu_l W_l)^{1/3} (1/N_r)^{1/6} \qquad (7.30)$$

  a = 0.951 for triangular tube patterns
  a = 0.904 for rotated square tube patterns
  a = 0.856 for square tube patterns
  N_r = is defined by Equation 7.33

For vapor shear dominated condensation:

$$h_s D_o/k_l = b(D_o \rho_l V_v/\mu_l)^{1/2}(1/N_r)^{1/2} \qquad (7.31)$$

  b = 0.42 for triangular tube patterns
  b = 0.39 for square tube patterns
  b = 0.43 for rotated square tube patterns

22 Handbook of Evaporation Technology

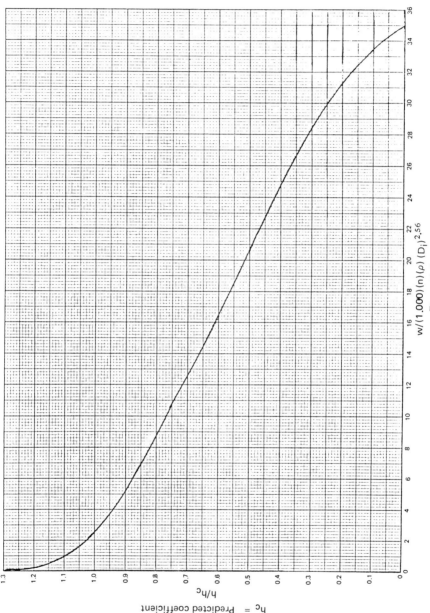

Figure 7-6.

$$V_v = W_v/\rho_v a_c \tag{7.32}$$

where W_v is the mass flow rate of the vapor and a_c is defined in an earlier section.

$$N_r = mD_s/S_t \tag{7.33}$$

m = 1.155 for triangular tube patterns
m = 1.0 for square tube patterns
m = 0.707 for rotated square tube patterns

The recommended procedure is to calculate the heat transfer coefficient using both mechanisms and select the higher value as the effective heat transfer coefficient (h). For baffled condensers, the vapor shear effects vary for each typical baffle section. The condenser should be calculated in increments with the average vapor velocity (V_v) for each increment used to calculate vapor shear heat transfer coefficients. When the heat transfer coefficients for laminar flow and for vapor shear are nearly equal, the effective heat transfer coefficient (h) is increased above the higher of the two values. The table below permits the increase to be approximated:

h_s/h_c	h/h_c
0.5	1.05
0.75	1.125
1.0	1.20
1.25	1.125
1.5	1.05

Lowfin Tubes — For lowfin tubes, the laminar condensing coefficient can be calculated by applying an appropriate correction factor, F, to the value calculated using Equation 7.30 above. The factor F is defined below:

$$F = [E_w(A_t/A_i)(D_i/D_r)]^{1/4} \tag{7.34}$$

where F = correction factor for Equation 7.30
E_w = weighted fin efficiency (Figures 7-2 and 7-3)
A_t = total outside surface of the fintube
A_i = inside surface of the fintube
D_i = inside tube diameter
D_r = root diameter; outside diameter minus twice the fin height

$$\text{Then } h = Fh_c \tag{7.35}$$

Sloped Bundles: For bundles with slight slopes, the following correction should be applied to the condensing heat transfer coefficient:

$$h = h_o(\cos \alpha)^{1/3} \text{ for L/D greater than } 1.8 \tan \alpha \tag{7.36}$$

where h = coefficient for sloped tube
h_o = coefficient for horizontal tube
α = angle from horizontal, degrees

Immiscible Condensates: Condensation of mixed vapors of immiscible liquids is not well understood. The conservative approach is to assume that two condensate films are present and all the heat must be transferred through both films in series. Another approach is to use a mass fraction average thermal conductivity and calculate the heat transfer coefficient using the viscosity of the film-forming component (the organic component for water-organic mixtures). The recommended approach is to use a shared-surface model and calculate the effective heat transfer coefficient as:

$$h = V_A h_A + (1 - V_A) h_B \qquad (7.37)$$

where h is the effective heat transfer coefficient
 h_A is the condensing heat transfer coefficient for liquid A assuming it only is present
 h_B is the condensing heat transfer coefficient for liquid B assuming it only is present
 V_A is the volume fraction of liquid A in the condensate

Condensate Subcooling: *For vertical condensers,* condensate can be readily subcooled if required. The subcooling occurs as falling-film heat transfer. Heat transfer coefficients can be calculated as presented in a later section.

For horizontal tubeside condensers, no good methods are available for predicting heat transfer coefficients when appreciable subcooling of the condensate is required. A conservative approach is to calculate a superficial mass velocity assuming the condensate fills the entire tube and use the equations presented previously for a single phase sensible heat transfer inside tubes

This method is less conservative for higher condensate loads.

For horizontal shellside condensers, condensate subcooling can be accomplished in two ways. The first method requires a condensate level in the shellside; heat transfer can then be calculated using the appropriate single phase sensible heat transfer correlations presented in an earlier section. The second method requires that the vapor make a single pass across the bundle in a vertical downflow direction. Subcooling heat transfer can then be calculated using falling-film correlations.

Enhanced Condensing Surfaces: Various devices have been used to improve condensing heat transfer by taking advantage of the surface tension forces exhibited by the condensate. One such device is the fluted condensing surface, first presented by Gregorig. The fluted condensing surface has a profile similar to that shown in Figure 7-7. Surface tension of the curved liquid-vapor interface produces a large excess pressure in the condensate film adjacent to the crests of the flutes. This causes a thinning of the film in that region, resulting in very high local heat transfer. The surface tension mechanism causes the condensate to accumulate in the troughs. Condensate is removed by flowing vertically downward in the troughs.

Enough condensate accumulates in the troughs within a short distance from the top of the tube to make heat transfer in the troughs negligible. Thus, heat transfer, averaged around the circumference of the tube, is essentially independent of the tube length as long as the troughs are not flooded.

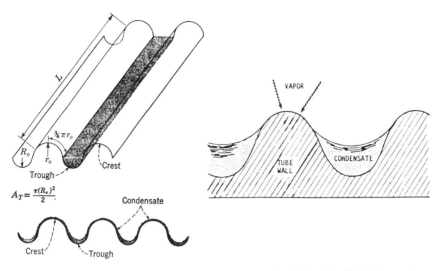

Figure 7-7: (Left) Fluted condensing surface of Gregorig's. (Right) Profile of the condensing surface developed by Gregorig.

Although only a portion of the tube surface is available for condensation, the mean heat-transfer coefficient over the total surface is 5 to 10 times that of a smooth tube of equivalent length. Heat transfer on a fluted tube approaches that of dropwise condensation. The flute profile is determined by the properties of the condensate. Fluids with lower surface tensions require flutes with a more pronounced curvature. The trough geometry is also determined by the amount of fluid to be removed and the fluid properties.

Fluted surfaces can also be designed for horizontal applications. However, this application has not been as extensively tested as have flutes for vertical tubes. Roped or swaged tubes are also used.

Condensation of Vapors Containing Noncondensable Gases: The presence of noncondensable gases in a vapor decreases the rate of heat transfer. This reduction depends upon the relative values of the gas-cooling heat load and the total cooling/condensing heat duty.

One mechanism for the analysis of condenser/coolers assumes that the latent heat of condensation is transferred only through the condensate film, whereas the gas-cooling duty is transferred through both the gas and condensate films. This mechanism yields good results for cases in which the condensate and cooling vapor-gas mixture are in substantial equilibrium. This mechanism applies only when the condensing surface is colder than the dew point of the mixture.

The following equations may be written:

$$\frac{Q_g}{h_g} + \frac{Q_T}{h_c} = A\Delta T \qquad (7.38)$$

$$h_{cg} = Q_T/A\Delta T \qquad (7.39)$$

From these equations we obtain:

$$h_{cg} = 1/[(Q_g/Q_T)(1/h_g) + (1/h_c)] \tag{7.40}$$

Some comments on these equations must be made:

(1) Sensible heat, Q_g, includes the sensible heat for cooling the non-condensable gases, plus the vapors that remain in the vent stream and are subsequently subcooled.

(2) Total heat, Q_T, includes the gas-phase sensible heat, Q_g, plus the latent heat of condensation of the vapor that condenses, plus the sensible heat of the liquid phase subcooling.

(3) The gas-film coefficient, h_g, is calculated by using a mean mass velocity based on the entering and exit gas-vapor rates. Better accuracy is obtained by calculating smaller increments and integrating the results.

(4) The condensing-film coefficient, h_c, is calculated for the loading rate represented by the condensate film.

(5) Value of the net-effective coefficient, h_{cg}, will generally be between the values for the gas cooling and condensing coefficients. However, this is not always true if h_g approaches the value of h_c.

(6) The greater the ratio of sensible heat to total heat, the more closely the net-effective coefficient approaches the gas-film efficient.

(7) The gas-film coefficient is usually controlling. Therefore, higher gas velocities result in higher net-effective coefficients.

When the condensing-cooling curves are other than straight lines, the mean temperature difference will vary from the logarithmic-mean temperature difference. For such cases, the condensing curve must be known, and the variance between the actual and log-mean temperature differences calculated. In most cases, the actual condensing curve will yield temperature differences that are greater than the log-mean obtained with straight-line condensing.

Desuperheating: Provided the condensing surface is cooler than the saturation temperature of the vapor, superheat is removed by a mechanism of condensing and re-evaporation. This mechanism maintains the condensing-surface temperature at essentially the same level as that obtained with saturated vapor. For superheated vapors (within limits at constant heat level), condensate loading is lower and therefore the condensing coefficient is often slightly higher. Desuperheating occurs at the expense of the temperature difference available between the superheat temperature and the saturation temperature. External desuperheating is seldom required or justified.

Fog Formation: A vapor or vapor-gas mixture may be cooled below the dew point. If cooled sufficiently, droplets of condensate may form in the bulk vapor stream. The droplet formation process is known as homogeneous nucleation and significant nucleation will result in a noticeable fog in the vapor. Fog droplets are small (1 to 20 microns in diameter); smaller droplets tend not to collect on the condensing surfaces and may be lost through the condenser vent unless special precautions are taken for removal. Usual demisters will not capture fog droplets.

Fog can be formed in a condenser when the rates of noncondensable to condensable vapor is high and the temperature differential is high. Supersaturation is measured by the ratio of the actual vapor pressure of the condensable vapor to the equilibrium vapor pressure at the given temperature. The tendency to form supersaturated vapor generally increases with increasing vapor molecular weight, increasing temperature difference, and with reduced inlet superheat. External nuclei in the inlet vapor will enhance the fog formation. The following are sources of foreign nuclei:

(1) solids in air;

(2) ions

(3) solids produced by combustion in upstream processing

(4) entrained liquids

(5) upstream reactions

Fogging may be prevented or reduced by reducing the temperature difference or increasing the inlet superheat. Inlet gases can be filtered to remove foreign nuclei. The gas stream can be seeded with condensation nuclei to produce drops that will be captured by conventional demisters. The vent can be heated to reduce fog. Most of these methods are costly or impractical; an alternative to preventing fog is to allow it to form and then remove it with special demisters which first coalesce the droplets before removal.

Falling Film Heat Transfer

Falling film heat transfer can be predicted using the correlation presented in Figure 7-8. Falling film vaporization occurs as a result of sensible heat transfer and subsequent flashing at the surface of the falling liquid film. Falling films generally are highly turbulent. In falling film vaporization, no boiling occurs on the heat transfer surface; however, low temperature differences (less than 25°C) are required to maintain this condition.

A minimum flowrate is required to induce a film for falling film equivalent. This minimum rate can be determined as below:

$$\Gamma_{min} = 19.5(\mu_l s_l \sigma^3)^{1/5} \tag{7.41}$$

where Γ_{min} = minimum tube loading, lb/(hr)(ft)
μ_l = liquid viscosity, centipoise
s_l = liquid specific gravity, referred to water
σ = surface tension, dynes/cm

Once the film has been formed, a lower terminal flow rate can be achieved without destroying the film. This rate can be determined as below:

28 Handbook of Evaporation Technology

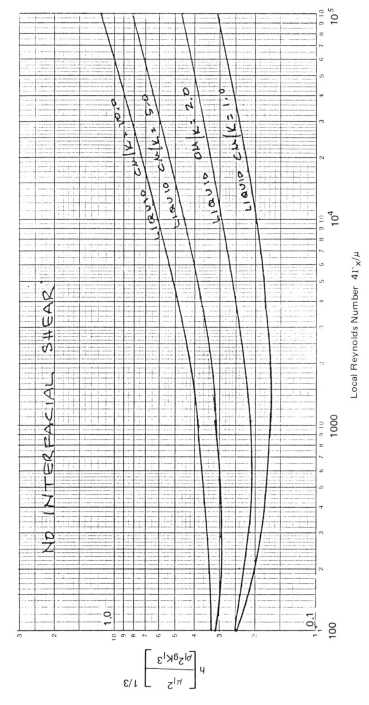

Figure 7-8: Falling film heat transfer. No interfacial shear ($A_d = 0$).

$$\Gamma_T = 2.4(\mu_{lsl}\sigma^3)^{1/5} \tag{7.42}$$

where Γ_T is the terminal flow rate, lb/(hr)(ft), and the other terms are as defined for Equation 7.41.

If the minimum rate is not achieved a film cannot be formed. If the terminal rate is less than that predicted by Equation 7.42, the film will break and form rivulets. Part of the tube surface will not be wetted and the result will be reduced heat transfer and increased fouling potential.

The distribution of liquid is critical to the performance of falling film units. Several methods are used to achieve distribution. The most common is an orifice plate located above the top tubesheet. Such a distribution system is illustrated in Figures 7-9 and 7-10.

Falling film units must be plumb. Tilted units will achieve lower heat transfer rates and offer higher fouling potential. Units with long tubes must be rigidly supported to counter the effects of wind and transmitted vibrations.

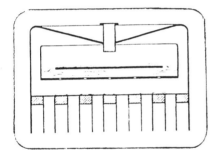 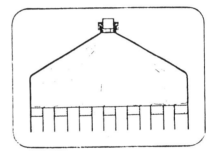

Figure 7-9: (Left) Liquid distribution by means of a trough (Example of a static distribution). (Right) Liquid distribution by means of a full cone nozzle (Example of a dynamic distribution).

Boiling

Heat transfer to boiling liquids is a convection process involving a change of phase from liquid to vapor. The phenomena of boiling heat transfer are considerably more complex than those of convection without phase change.

When a surface is exposed to a liquid and is maintained at a temperature above the saturation temperature of the liquid, boiling may occur, and the heat flux will depend on the difference between the surface and the saturation temperature. When the heated surface is submerged below a free surface of liquid, the process is referred to as *pool boiling*. If the temperature of the liquid is below the saturation temperature, the process is called *subcooled*, or *local boiling*. If the liquid is maintained at saturation temperature, the process is known as *saturated*, or *bulk boiling*.

30 Handbook of Evaporation Technology

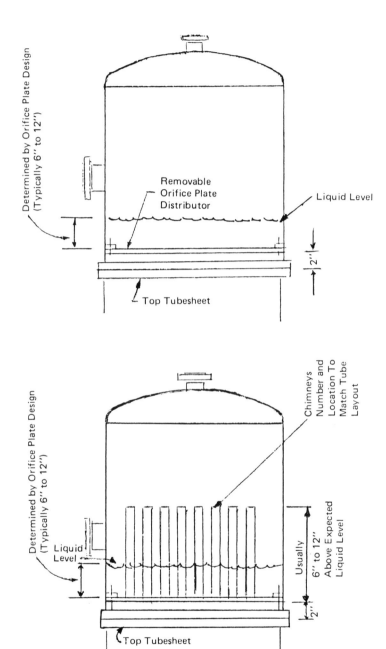

Figure 7-10A.

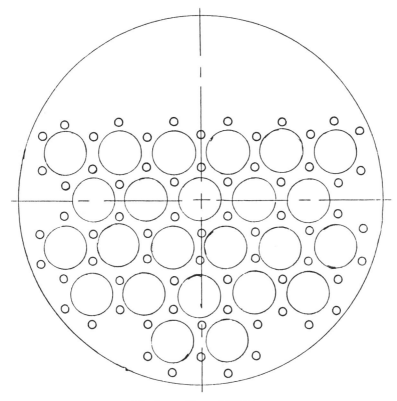

Distributor Plate – 3/8" Thick
68 – 1/4" Diameter Holes
24 – 2" O.D. Tubes

Figure 7-10B.

There are different regimes of boiling as indicated in Figure 7-11. In region AB free-convection currents are responsible for motion of the fluid near the surface. In this region the liquid near the heated surface is superheated slightly and it subsequently evaporates when it rises to the surface of the liquid pool. In region BB' bubbles begin to form on the heat-transfer surface and are dissipated by condensing in the liquid after breaking away from the surface. This regime indicates the onset of *nucleate boiling*. As the temperature difference is increased further, bubbles form more rapidly and rise to the surface of the liquid where they are dissipated. This is indicated in region B'C, the nucleate boiling region. Eventually bubbles are formed so rapidly that they blanket the heating surface and restrict a fresh liquid supply from reaching the surface. Point D marks the upper limit of nucleate boiling. The transition boiling region (DE) is characterized by the release of large patches of vapor at more or less regular intervals. The thermal resistance of the vapor film causes a reduction in heat transfer as indicated by this region which represents a transition from nucleate boiling to film boiling and is highly unstable. A stable vapor film eventually covers the surface in the film boiling region (EF) and vapor is released from the film periodically in the form of regularly spaced bubbles. Heat transfer is accomplished principally by conduction and convection through the vapor film with radiation becoming significant as the surface temperature is increased.

In nucleate boiling, bubbles are created by the expansion of entrapped gas or vapor at small cavities in the surface. The bubble grows to a certain size, depending on the surface tension at the liquid-vapor interface and the temperature and pressure. Depending on the temperature excess, the bubbles may collapse on the surface to be dissipated in the body of the liquid, or at sufficiently high temperatures may rise to the surface of the liquid before being dissipated. When local boiling conditions are observed, the primary mechanism of heat transfer is thought to be the intense agitation at the heat transfer surface which creates the high heat-transfer rates observed in boiling. In saturated, or bulk, boiling the bubbles may break away from the surface and move into the body of the fluid. In this case the heat-transfer rate is influenced by both the agitation caused by the bubbles and the vapor transport of energy into the body of the liquid.

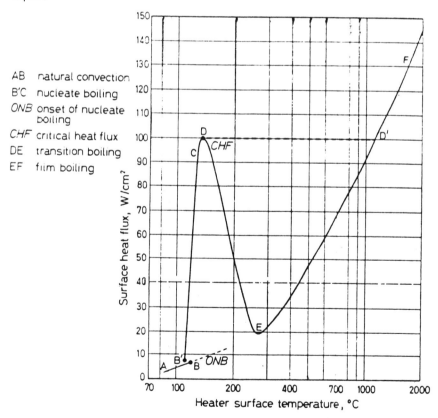

AB natural convection
B'C nucleate boiling
ONB onset of nucleate boiling
CHF critical heat flux
DE transition boiling
EF film boiling

Figure 7-11: Pool boiling curve for water at atmospheric pressure.

Natural Convection: Natural, or free, convection heat transfer occurs whenever a body is placed in a fluid at a higher or lower temperature than that of the body. As a result of temperature difference, heat flows between the fluid and the body and causes a change in the density of the fluid layers in the vicinity of the surface. The difference in density leads to downward flow of the heavier fluid and upward flow of the lighter. If the motion of the fluid is caused solely by differences in density resulting from temperature gradients, without the aid of a pump or fan, the associated heat-transfer mechanism is called natural or free convection. Free-convection currents transfer internal energy stored in the fluid in essentially the same manner as forced-convection currents. However, the intensity of the mixing motion is generally less in free convection, and consequently the heat-transfer coefficients are lower than in forced convection.

Heat transfer for natural convection systems may be represented by the following equations:

$$(h/cG)(c\mu/k)^m = a(LG/\mu)^{-n} \text{ for vertical surfaces} \quad (7.43)$$

$$(h/cG)(c\mu/k)^m = b(DG/\mu)^{-n} \text{ for horizontal cylinders} \quad (7.44)$$

$$G = (g\beta\Delta T\rho^2 L)^{1/2} \text{ for vertical surfaces} \quad (7.45)$$

$$G = (g\beta\Delta T\rho^2 D)^{1/2} \text{ for horizontal cylinders} \quad (7.46)$$

Film Reynolds Number	a	b	m	n
Greater than 10,000	0.13	0.13	2/3	1/3
100 to 10,000	0.59	0.53	3/4	1/2
Less than 100	1.36	1.09	5/6	2/3

Nucleate Boiling: High rates of heat transfer are obtained under nucleate boiling. The mechanism has not yet been clearly established, but, as a result of considerable activity in the field of nucleate boiling, there are available several expressions from which reasonable values of the film coefficients may be obtained. These expressions do not yield exactly the same numerical results even though the correlations were based upon much of the same data. There is thus neither a prominent nor unique equation for nucleate boiling. Either convenience or familiarity usually governs the user's selection of the available equations.

The following equation has been widely used to predict nucleate boiling heat transfer. The equation includes a term to treat variation of coefficient with pressure. A special Reynolds number is also defined.

$$(h/c_l G)(c_l \mu_l/k_l)^{0.6}(\rho_l \sigma g/P^2 g_c)^{0.425} = \phi(DG/\mu_l)^{0.3} \quad (7.47)$$

$$G = (W_v/A)(\rho_l/\rho_v) \quad (7.48)$$

(Note that A is the surface area, not a cross-sectional area.) ϕ, which is a numerical correlating factor, varies with the nature (material of construction) of the surface. Recommended values are:

0.001 for commercial copper and steel surfaces
0.00059 for stainless steel or chromium and nickel alloys
0.0004 for polished surfaces

Surface conditions have a profound effect on boiling phenomena. The above numerical factors have been obtained from plots of existing data. Extreme accuracy cannot be claimed for these values, because of the variable conditions of the surfaces in these tests.

Other equations for predicting nucleate boiling include the following:

$$hD/k_l = 0.225(DQ/A\lambda\mu_l)^{0.69}(PD/\sigma)^{0.31}[(\rho_l/\rho_v) - 1]^{0.33}(c_l\mu_l/k_l)^{0.69} \quad (7.49)$$

$$c_l(T_w - T_s)/\lambda = C_{sf}[(Q/A\mu_l\lambda)[\sigma/g(\rho_l - \rho_v)]^{1/2}]^n(c_l\mu_l/k_l)^{m+1} \quad (7.50)$$

C_{sf} is specific to various liquid-surface conditions, but a value of $C_{sf} = 0.013$ may be used as a first approximation.

$$n = 0.33$$
$$m = 0.7 \text{ except for water where } m = 0$$
$$T_w = \text{surface temperature}$$
$$T_s = \text{saturation temperature}$$

$$h_B = 0.00122 \left[\frac{k_l^{0.79} c_l^{0.45} \rho_l^{0.49}}{\sigma^{0.5} \mu_l^{0.29} (\lambda\rho_v)^{0.24}} \right] (T_w - T_s)^{0.24}(p_w - p_s)^{0.75} \quad (7.51)$$

p_w and p_s are the saturation pressures corresponding to T_w and T_s respectively.

There is both an upper and lower limit of applicability of Equations 7.47, 7.49, 7.50, and 7.51. At the upper limit, nucleation is diminished because of the insulating effects of a vapor film. At the lower limit, nucleation is inhibited because of natural convection effects.

For horizontal tubes, the criterion for determining the upper limit is the maximum allowable value of vapor velocity. Vapor bubbles can escape at a velocity determined by the pressure of the system. When vapor is generated faster than it can escape, all or a portion of the tube or tube bundle will become blanketed by a film of vapor. The maximum vapor velocity can be determined from the following equation:

$$v = 0.0825(\rho_l - \rho_v)^{1/4} \sigma^{1/4} (\rho_v)^{-1/2} \quad (7.52)$$

where v is velocity, feet per second; ρ_l is liquid density, lb/ft^3; ρ_v is vapor density, lb/ft^3; σ is surface tension, dynes/cm. For a tube bundle, this velocity is the superficial vapor velocity and must be based on the area projected above the bundle.

The best method to determine the lower limit of the equations for nucleate boiling is to plot two curves: one of h versus ΔT for natural convection; the other of h versus ΔT for nucleate boiling. The intersection of these two curves may be considered the lower limit of applicability.

Convective Boiling: The mechanisms occurring in convective boiling are quite different from those occurring in pool boiling. The following conditions occur in different parts of the tube as shown in Figure 7-12:

(1) subcooled single-phase liquid region
(2) subcooled boiling region
(3) saturated boiling region
(4) dry wall region.

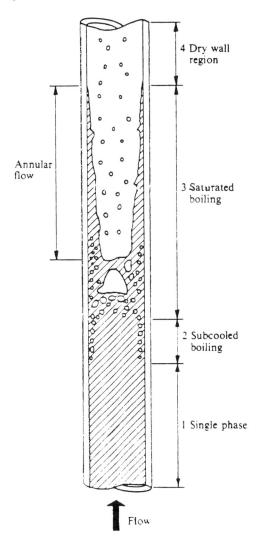

Figure 7-12: Heat-transfer mechanisms in convective boiling in a tube.

The saturated boiling region is the most important region encountered in design. Chen has shown that the coefficient in this region is made up of two components: a nucleate-boiling component and a convective component:

$$h = h_B + h_c \qquad (7.53)$$

where h = convective boiling heat transfer coefficient
h_B = nucleate boiling heat transfer coefficient
h_c = convective heat transfer coefficient

The convective component, h_c, can be estimated by modifying the sensible heat transfer coefficient:

$$h_c = F_c h_1 \tag{7.54}$$

where h_1 is calculated using Equation 7.1 if the Reynolds number is calculated on the assumption that the liquid alone is flowing in the tube:

$$Re_l = (1 - x)GD_i/\mu_l \tag{7.55}$$

where G is the total mass velocity in the tube and x is the mass fraction vapor. The two-phase convective correction factor F_c is given in Figure 7-13, where $1/X_{tt}$, is a two-phase parameter given by:

$$1/X_{tt} = [x/(1-x)]^{0.9}(\rho_l/\rho_v)^{0.5}(\mu_v/\mu_l)^{0.1} \tag{7.56}$$

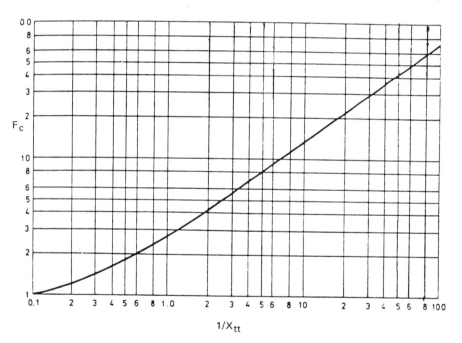

Figure 7-13: Convective boiling factor, F_c.

Equation 7.54 gives an increase in coefficient with vapor fraction, except for vapor fractions greater than 0.8.

The nucleate-boiling component, h_B, is given by a modified form of Equation 7.51:

$$h_B = S_c h_2 \tag{7.57}$$

where h_2 is the coefficient given by Equation 7.51. S_c is known as the suppression factor and accounts for the fact that nucleate boiling is more difficult in convective conditions than in pool boiling. The factor S_c is given in Figure 7-14. With increasing quality, h_B decreases but this decrease is compensated by the increase in S_c.

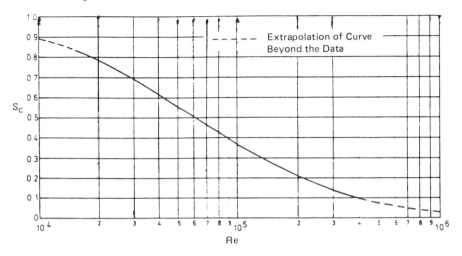

Figure 7-14: Nucleate boiling suppression factor, S_c.

Improved Boiling Surfaces: It has long been known that nucleate boiling heat transfer can be increased by artificially roughening the surface with techniques such as scratching, sand blasting, and etching. Union Carbide Corporation, Linde Division, has developed a special boiling surface which has wide application.

Fluted boiling surfaces have also been used. Boiling on fluted tubes is a thin-film process that takes place on each crest. The performance of fluted boiling surfaces is relatively poor when compared to the results achieved when condensing. The reason for this is not clear but is probably due to the inability of maintaining thin films on the crests of the flutes. The crests may run dry; they may be flooded by liquid displaced from the troughs by axial flow of the accumulated vapor; or nucleate boiling may occur in the troughs.

The Linde surface consists of a thin layer of porous metal bonded to the heat transfer substrate. (See Figure 7-15.) Bubble nucleation and growth are promoted within a porous layer that provides a large number of stable nucleation sites of a predesigned shape and size.

Microscopic vapor nuclei in the form of bubbles entrapped on the heat-transfer surface must exist in order for nucleate boiling to occur. Surface tension at the vapor-liquid interface of the bubbles exerts a pressure above that of the liquid. This excess pressure requires that the liquid be superheated in order for the bubble to exist and grow. The porous surface substantially reduces the superheat required to generate vapor. The entrances to the many nucleation sites are restricted in order to retain part of the vapor in the form of a bubble and to prevent flooding of the site when liquid replaces the escaping bubble.

38 Handbook of Evaporation Technology

Many individual sites are also interconnected so that fresh liquid is continually supplied.

Boiling heat-transfer coefficients, which are functions of pore size, fluid properties, and heat flux, are 10 to 50 times greater than smooth-surface values at a given temperature difference. Boiling coefficients are stable and relatively independent of convection effects. Nucleate boiling exists at much lower temperature differences than required for smooth surfaces.

The Linde surface can be used in conjunction with fluted tubes or other enhanced surfaces to greatly increase heat transfer rates.

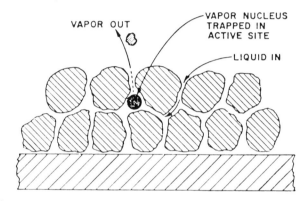

Figure 7-15: Linde porous boiling surface—mechanism of operation.

PHYSICAL PROPERTIES

To calculate heat-transfer rates, physical-property data for the fluids being treated must be available. Physical property data should be as accurate as possible, especially as more accurate heat-transfer correlations become available. However, most physical properties of mixtures must be calculated or estimated; consequently there is little need to attempt to determine true film temperatures. Physical-property data at the average bulk fluid temperature are generally sufficient.

The following physical properties are usually required in order to obtain satisfactory calculated heat transfer rates:

(1) viscosity

(2) thermal conductivity

(3) density

(4) specific heat

(5) latent heat

(6) surface tension

When condensing or vaporizing over a temperature range, a curve representing heat load as a function of temperature should also be available. In addition, any concentration effects should be known.

8
Pressure Drop in Evaporators

Pressure drop is one of the process variables which have the greatest effect on the size (cost) of an evaporator. Selection of optimum pressure drops involves consideration of the overall process. While it is true that higher pressure drops result in smaller equipment, investment savings are realized only at the expense of operating costs. Only by considering the relationship between operating costs and investment can the most economic pressure drop be determined.

Pressure drop correlations are available which enable pressure drop to be confidently predicted.

FLOW INSIDE TUBES

The total pressure drop for fluids flowing inside straight tubes results from frictional pressure drop as the fluid flows along the tube, from pressure drop as the fluid enters and leaves the tube side heads or channels, and from pressure drop as the fluid enters and leaves the tubes from the heads or channels. The frictional pressure drop can be calculated from the equation:

$$\Delta P_f = 4fG^2 LN_p/2(144)g\rho D_i \qquad (8.1)$$

where ΔP_f = pressure drop, psi (pounds per square inch)
   f = friction factor
   G = mass velocity, lb/(hr)(sq ft)
   L = tube length, feet
   N_p = number of tubeside passes
   g = gravitational constant (4.18×10^8 ft/hr^2)
   ρ = density, lb/cu ft
   D_i = inside tube diameter, feet

For fluids with temperature dependent viscosities, the pressure drop must be corrected by the ratio:

$$\phi = (\mu_w/\mu_b)^n \tag{8.2}$$

where μ_w = viscosity at the surface (wall) temperature
μ_b = viscosity at the bulk fluid temperature
n = 0.14 for turbulent flow (Re greater than 2100)
n = 0.25 for laminar flow (Re less than 2100)

The friction factor, f, can be calculated from the equations:

$f = 16/(D_iG/\mu)$ for D_iG/μ less than 2100 (8.3)

$f = 0.046/(D_iG/\mu)^{0.2}$ for D_iG/μ greater than 2100 for tubing (8.4)

$f = 0.054/(D_iG/\mu)^{0.2}$ for D_iG/μ greater than 2100 for commercial pipe (8.5)

The pressure drop as the fluid enters and leaves a radial nozzle at the heads or channels can be calculated as below:

$$\Delta P_n = K\rho v_n^2/9266 \tag{8.6}$$

where ΔP_n = pressure drop, psi (pounds per square inch)
ρ = density, lb/cu ft
v_n = velocity in the nozzle, ft/sec
K = 0 for inlet nozzles
K = 1.25 for outlet nozzles

The pressure drop associated with inlet and outlet nozzles can be reduced by selection of other channel types, but the expense is seldom warranted except for situations in which the pressure drop is critical or costly.

The pressure drop as a result of entry and exit from the tubes can be calculated as below:

$$\Delta P_e = \rho K N_p v_t^2/9266 \tag{8.7}$$

where ΔP_e = pressure drop, psi (pounds per square inch)
N_p = number of tubeside passes
v_t = velocity in the tube, ft/sec
ρ = density, lb/cu ft
K = 1.3 if D_iG/μ is greater than 2100
K = 2.3 if D_iG/μ is less than 2100

The total pressure drop is the sum of all these components:

$$\Delta P_t = \Delta P_f + \Delta P_n + \Delta P_e \tag{8.8}$$

U-Bend Tubes

For U-bend tubes the following modifications are required:

Frictional pressure drop, ΔP_f:

Substitute $L_o + R$ for L

where L_o is the straight tube length, feet (tube end to bend tangent)
R is the maximum bend radius, feet

Entry/exit pressure drop, ΔP_e:

K = 0.8 for $D_i G/\mu$ greater than 2100
K = 1.3 for $D_i G/\mu$ less than 2100

Helical Coils

The procedure used for straight tubes can be used to calculate the pressure drop in helical coils. For turbulent flow, a friction factor for curved flow is substituted for the friction factor for straight tubes. For laminar flow, the friction loss for a curved tube is expressed as an equivalent length of straight tube and the friction factor for straight tubes is used. The Reynolds number required for turbulent flow is:

$$Re_c = 2100[1 + 12(D_i/D_c)^{1/2}] \tag{8.9}$$

where Re_c = Reynolds number required for turbulent flow
D_i = inside tube diameter
D_c = diameter of coil or helix

The friction factor for turbulent flow is calculated as below:

$$f_c(D_c/D_i)^{1/2} = 0.0073 + 0.076[(D_i G/\mu)(D_i/D_c)^2]^{-1/4} \tag{8.10}$$

for values of $(D_i G/\mu)(D_i/D_c)^2$ between 0.034 and 300.

fc is equal to the friction factor for curved flow.
For values of $(D_i G/\mu)(D_i/D_c)^2$ less than 0.034, the friction factor for curved flow is practically the same as that for straight tubes.
For laminar flow, the equivalent length (L_e) can be predicted as below:

For $(D_i G/\mu)(D_i/D_c)^{1/2}$ between 150 and 2000,

$$L_e/L = 0.23[(D_i G/\mu)(D_i/D_c)^{1/2}]^{0.4} \tag{8.11}$$

For $(D_i G/\mu)(D_i/D_c)^{1/2}$ between 10 and 150,

$$L_e/L = 0.63[(D_i G/\mu)(D_i/D_c)^{1/2}]^{0.2} \tag{8.12}$$

For $(D_i G/\mu)(D_i/D_c)^{1/2}$ less than 10,

$$L_e/L = 1 \tag{8.13}$$

where L = straight length
L_e = equivalent length of a curved tube in laminar flow

Condensing Vapors

The frictional pressure drop for vapors condensing inside tubes can be predicted as below:

$$\Delta P_c = (1/2)[1 + (v_2/v_1)]\Delta P_1 \tag{8.14}$$

where ΔP_c = condensing frictional pressure drop
ΔP_1 = pressure drop based on inlet flow conditions
v_1 = vapor velocity at tube inlet
v_2 = vapor velocity at tube outlet

For a total condenser, Equation 8.14 reduces to:

$$\Delta P_c = (1/2)\Delta P_1 \qquad (8.15)$$

FLOW ACROSS THE TUBE BANKS

The following equation may be used to predict pressure drop for fluids flowing across banks of tubes:

$$\Delta P = F_s(20.34)(W/1000)^2 L/(\rho D_s{}'' P_b{}''^3) \qquad (8.16)$$

F_s has the values tabulated below:

Tube O.D. (inch)	Triangular Pitch Tube Spacing/Tube O.D.				Square Pitch Tube Spacing/Tube O.D.			
	1.20	1.25	1.30	1.35	1.20	1.25	1.30	1.35
5/8	2.4	1.6	1.0	0.6	1.8	1.2	0.75	0.45
3/4	1.8	1.2	0.85	0.65	1.4	0.9	0.65	0.5
1	1.7	0.95	0.7	0.5	1.1	0.65	0.5	0.4
1-1/4	1.1	0.75	0.5	0.35	0.85	0.5	0.4	0.3
1-1/2	1.0	0.6	0.45	0.3	0.75	0.45	0.35	0.25
2	0.8	0.45	0.3	0.25	0.6	0.35	0.25	0.2

Equation 8.16 applies only when the shellside Reynolds number ($D_o G/\mu$) is greater than 300.

Equation 8.16 is a simplified equation which is based upon equipment with a certain amount of fouling present on the shell side. Consequently, it may predict values of pressure drop higher than actually present for certain applications. A more rigorous method for calculating pressure drop across banks of tubes is presented here. The pressure drop for fluids flowing across the tube banks may be determined by calculating the following components:

(1) inlet nozzle pressure drop

(2) outlet nozzle pressure drop

(3) frictional pressure drop for inlet and outlet baffle sections

(4) frictional pressure drop for intermediate baffle sections

(5) pressure drop for flow through the baffle windows.

Heat exchangers with well constructed shell sides will have a certain amount of bypassing and leakage which will reduce the pressure drop that would be experienced with an ideal tube bundle (one with no fluid bypassing or leakage). The amount of bypassing for a clean heat exchanger is more than that for a fouled heat exchanger. The leakage factors below are based upon data for operating heat exchangers and include the typical effects of fouling on pressure drop. The method assumes that:

(1) all clearances and tolerances are in accordance with TEMA

(2) one sealing device is provided for every five tube rows

(3) baffle cuts are 20% of the shell diameter

(4) fouled heat exchangers result in plugging of some of the clearances and bypass lanes.

Pressure drop for nozzles may be calculated as below:

$$\Delta P_n = K\rho v_n^2/9266 \tag{8.17}$$

where ΔP_n = pressure drop, psi
ρ = density, lb/cu ft
v_n = nozzle velocity, ft/sec
K = 0 for inlet nozzle with no impingement plate
K = 1.0 for inlet nozzle with impingement plate
K = 1.25 for outlet nozzle

Frictional pressure drop for tube bundles may be calculated as below:

For intermediate baffle sections:

$$\Delta P_f = 4fG^2 N_r (N_b - 1) R_l R_b \phi/2(144)g\rho \tag{8.18}$$

For inlet and outlet baffle sections, combined:

$$\Delta P_{fi} = 4(2.66)fG^2 N_r R_b \phi/2(144)g\rho \tag{8.19}$$

$$R_l = 0.6(P_b/D_s)^{1/2} \text{ for clean bundles} \tag{8.20}$$

$$R_l = 0.75(P_b D_{/s})^{1/3} \text{ for bundle with assumed fouling} \tag{8.21}$$

$$R_b = 0.80(D_s)^{0.08} \text{ for clean bundles} \tag{8.22}$$

$$R_b = 0.85(D_s)^{0.08} \text{ for bundle with assumed fouling} \tag{8.23}$$

$$\phi = (\phi_w/\phi_b)^n \text{ where n = 0.14 for } D_o G/\mu \text{ greater than 300} \tag{8.24}$$
where n = 0.25 for $D_o G/\mu$ less than 300

$$N_r = bD_s/S_t \quad b = 0.7 \text{ for triangular tube patterns} \tag{8.25}$$
b = 0.6 for square tube patterns
b = 0.85 for rotated square tube patterns

The friction factor, f, can be calculated:

$$f = z/(D_g G/\mu)^{0.25} \text{ for } D_g G/\mu \text{ greater than 100} \tag{8.26}$$

where z = 1.0 for square and triangular tube patterns
z = 0.75 for rotated square tube patterns

$$f = r/(D_g G/\mu)^{0.725} \text{ for } D_g G/\mu \text{ less than 100} \tag{8.27}$$

where r = 10 for triangular patterns
r = 5.7 for square and rotated square tube patterns.

D_g is defined as the gap between the tubes:

$$D_g = S_t - D_o \tag{8.28}$$

Pressure drop for baffle windows may be calculated:

44 Handbook of Evaporation Technology

For $D_o G/\mu$ greater than 100,

$$\Delta P_w = [G^2/2(144)\rho\phi](a_c/a_w)(2 + 0.6N_w)N_b R_l$$

The factor $(2 + 0.6N_w)$ can be approximated with the following term:

$$2 + 0.6N_w = m(D_s)^{5/8} \qquad (8.29)$$

where m = 3.5 for triangular tube patterns
m = 3.2 for square tube patterns
m = 3.9 for rotated square tube patterns

For $D_o G/\mu$ less than 100:

$$\Delta P_w = (26\mu G/\rho g)(a_c/a_w)^{1/2}\left[\frac{N_w}{S_t - D_o} + \frac{P_b}{D_e^2}\right] + (2G^2/2g\rho)(a_c/a_w) \qquad (8.30)$$

For baffle cuts which are 20% of the shell diameter, the following values can be used:

Tube Geometry		$N_w/(S_t-D_o)$	D_e (feet)
D_o =	5/8 inch; S_t = 13/16 inch triangular pitch	$173 D_s$	0.059
D_o =	5/8 inch; S_t = 7/8 inch square pitch	$105 D_s$	0.090
D_o =	3/4 inch; S_t = 15/16 inch triangular pitch	$151 D_s$	0.063
D_o =	3/4 inch; S_t = 1 inch square pitch	$92 D_s$	0.097
D_o =	1 inch; S_t = 1¼ inches triangular pitch	$85 D_s$	0.083
D_o =	1 inch; S_t = 1¼ inches square pitch	$74 D_s$	0.104

$$a_c = D_s P_b (S_t - D_o)/S_t \text{ for triangular and square tube patterns} \qquad (8.31)$$

$$a_c = 1.414 D_s P_b (S_t - D_o)/S_t \text{ for square tube patterns} \qquad (8.32)$$

$$a_w = 0.055 D_s^2 \text{ for trianglular tube patterns} \qquad (8.33)$$

$$a_w = 0.066 D_s^2 \text{ for square and rotated square tube patterns} \qquad (8.34)$$

ΔP_f = total pressure drop for intermediate baffle sections, psi
ΔP_{fi} = total pressure drop for inlet and outlet baffle sections, psi
ΔP_w = total pressure drop for baffle windows, psi
f = friction factor
G = mass velocity, lb/(hr) (sq ft), = W/a_c where W = flow, lb/hr
g = gravitational constant, 4.18×10^8 ft/hr^2
ρ = density, lb/cu ft
N_r = number of tube rows crossed
N_b = number of cross baffles
a_c = minimum flow area for cross flow, sq ft
a_w = flow area in baffle window, sq ft

D_s = shell diameter, feet
S_t = tube center to center spacing, feet
P_b = baffle spacing, feet
D_o = outside tube diameter, feet
D_g = gap between tubes, feet = $S_t - D_o$
μ = viscosity, lb/ft hr
μ_w = viscosity at wall temperature, lb/ft hr
μ_b = viscosity at bulk fluid temperature, lb/ft hr
R_l = correction factor for baffle leakage
R_b = correction factor for bundle bypass
D_e = equivalent diameter of baffle window, feet

The total pressure drop is the sum of all these components:

$$\Delta P = \Delta P_n + \Delta P_f + \Delta P_{fi} + \Delta P_w \qquad (8.35)$$

Tubes Omitted from Baffle Windows

Frequently, tubes are omitted from the baffle window areas. For this configuration, maldistribution of the fluid as it flows across the bank of tubes may occur as a result of the momentum of the fluid as it flows through the baffle window. For this reason baffle cuts less than 20% of the shell diameter should be used only with caution. Maldistribution will normally be minimized if the fluid velocity in the baffle window is equal or less than the fluid velocity in crossflow across the bundle. This frequently requires baffle cuts greater than 20% of the shell diameter. For such cases, the number of tube rows in crossflow will be less than that assumed in the methods above; consequently a correction is required. In addition, the pressure drop for the first and last baffle sections assumes tubes in the baffle windows; the factor 2.66 in Equation 8.19 should be reduced to 2.0 for the case of 20% baffle cuts.

For baffles with no tubes in the baffle windows, the pressure drop for the window section can be calculated as below:

$$\Delta P_w = 1.8 \rho v_w^2 N_b / 9266 \qquad (8.36)$$

where ΔP_w = pressure drop, psi
ρ = density, lb/cu ft
v_w = velocity in the baffle window, ft/sec
N_b = number of baffles

For baffles with no tubes in the baffle windows, the flow area may be calculated as below:

$$a_w = 0.11 D_s^2 \text{ for 20\% baffle cuts} \qquad (8.37)$$

$$a_w = 0.15 D_s^2 \text{ for 25\% baffle cuts} \qquad (8.38)$$

where D_s is the shell diameter.

Lowfin Tubes

The method presented above can be used to predict pressure drop for banks of lowfin tubes. For lowfin tubes, the pressure drop is calculated assuming that the tubes are bare. The mass velocity and tube diameter used for calculation are those for a bare tube with the same diameter as the fins of the lowfin tubes.

Air-Cooled Heat Exchangers

For air-cooled heat exchangers, the following relations can be used to predict air-side pressure for fin geometries discussed in Chapter 7:

$$\Delta P_a = aN_r(FV/100)^{1.8} \tag{8.39}$$

- a = 0.0047 for 10 fins per inch; 2-3/8 inches spacing
- a = 0.0044 for 8 fins per inch; 2-3/8 inches spacing
- a = 0.0037 for 10 fins per inch; 2-1/2 inches spacing
- ΔP_a = air-side pressure drop, inches of water column
- N_r = number of tube rows
- FV = face velocity, feet per minute

Two-Phase Flow

Two-phase flow pressure drop for tube bundles may be calculated using the equation:

$$\Delta P_{tp}/\Delta P_{LO} = 1 + (K^2 - 1)[Bx^{(2-n)/n}(1-x)^{(2-n)/n} + x^{2-n}] \tag{8.40}$$

where ΔP_{tp} is the two-phase pressure drop
ΔP_{LO} is the pressure drop for the total mass flowing as liquid
x is the mass fraction vapor

$$K = (\Delta P_{GO}/\Delta P_{LO})^{1/2} \tag{8.41}$$

where ΔP_{GO} is the pressure drop for the total mass flowing as vapor
B has the values below for flow across tube bundles:

- B = 1.0 for vertical up-and-down flow
- B = 0.75 for horizontal flow other than stratified flow
- B = 0.25 for horizontal stratified flow

The value of n can be calculated from the relation:

$$K = (\Delta P_{GO}/\Delta P_{LO})^{1/2} = (\rho_l/\rho_v)^{1/2} (\mu_v/\mu_l)^{n/2} \tag{8.42}$$

For the baffle windows, n = 0 and Equation 8.40 becomes:

$$\Delta P_{tp}/\Delta P_{LO} = 1 + (K^2 - 1) [Bx(1 - x) + x^2] \tag{8.43}$$

where B = $(\rho_{ns}/\rho_l)^{1/4}$ for vertical up-and-down flow
ρ_{ns} is the no-slip density defined in Chapter 26
B = $2/(K - 1)$ for horizontal flow $\tag{8.44}$

Condensing Vapors

The frictional pressure drop for shellside condensation can be calculated:

$$\Delta P_c = (1/2)[1 + (v_2/v_1)]\Delta P_1 \tag{8.45}$$

where ΔP_c = condensing frictional pressure drop
  ΔP_1 = pressure drop based upon the inlet conditions of flow rate, density, and viscosity
  v_1 = vapor velocity at the inlet
  v_2 = vapor velocity at the outlet.

For a total condenser, this becomes:

$$\Delta P_c = \Delta P_1/2 \tag{8.46}$$

9
Flow-Induced Vibration

The concern for flow-induced tube vibration has become a serious consideration in the design of shell-and-tube equipment. These problems can lead to tubes and tubejoints that leak, increased shellside pressure drop, and intolerably loud noises. The result is that equipment must be removed from service for repair and modification.

Flow-induced vibrations can damage tubes in evaporators. All tubes vibrate under all flow conditions! However, we are concerned with vibrations which cause significant tube damage. As larger evaporators, greater flowrates and higher shellside velocities become more prevalent, damaging tube vibrations are more likely to occur. No evaporator design is complete without considering the possibility of damage as a result of flow-induced vibrations.

Damage is more likely to occur with gases or vapors on the shellside than with liquids. Flow-induced vibrations also occur with liquids on the shellside, but the damage is often limited to localized areas of relatively high velocity. In severe cases, tubes can leak within a few days or even in a few hours after the equipment has been placed in service. More often, damage will appear a year or so after startup. Additional tube damage will develop after the initial damage has been repaired, but the number and frequency of further damages will decrease with time.

In a number of cases, heat exchanger tube failures attributed to flow-induced vibration have resulted in consequential damage to other equipment within a plant. Failures of this nature have proven to be the most destructive, most costly, and have required the longest plant shutdowns for rectification.

Currently available methods for predicting flow-induced vibration damage are inadequate for predicting failures. At best, they identify the equipment that are susceptible to damage. The primary reason for this lack of precision is that flow-induced vibrations are extremely complicated. Much has been learned, but the probability of its occurrence is still not known. However, the cost penalty for equipment designed to completely avoid damaging vibration is modest and is almost always easily justified.

Flow-Induced Vibration 49

Some of the problem areas concerned with prediction of vibration include:

(1) the complex pattern of flow through a tube bundle
(2) the complicated fluid mechanics of a bank of vibrating tubes
(3) the role of damping
(4) the rates of wear and fatigue.

Nevertheless, it is possible to develop design criteria, especially when tempered with experience, to ensure that equipment will be safe from vibration damage.

Flow-induced vibrations problems in tubular equipment are commonly thought of as consisting entirely of mechanical failure of the tubes. However, the vibration can increase the shellside pressure drop, sometimes as much as double. Further, an acoustically vibrating unit can produce an intolerably high noise level. With an increasing emphasis on noise control, acoustic vibration must be an important consideration in design of tubular equipment.

MECHANISMS

Induced vibration of any system involves the coupling of some exciting forces with an elastic structure. In the case of flow-induced vibration, the exciting forces result from the flow of the shellside fluid and the elastic structure in the bundle of tubes. The exciting forces fluctuate at characteristic frequencies which increase continuously with increasing flow rate. The tubes vibrate only at unique responding frequencies called their natural frequencies. Coupling occurs when the exciting frequencies match the responding frequencies and tube vibration results.

The natural frequency of tubes depends primarily on their geometry and material of construction. The intensity of vibration is evidenced by the amount of periodic movement; the extent of this peak-to-peak movement about the at-rest centerline is termed the amplitude of vibration. Energy must be available to excite the tubes into vibration. The energy of vibration is dissipated by internal and external damping. The exciting force could be the result of:

(1) fluid dynamic mechanisms as a result of flow parallel to or across the tubes
(2) pulsations of a compressor or pump
(3) mechanical vibrations transmitted through a structure.

Unless amplified by resonant phenomena, the flow forces normally enountered in equipment are not sufficient to cause damage. Resonance, which can increase the tube deflection by orders of magnitude, occurs when the frequency of a cyclic exciting force coincides with the natural frequency of the tube.

In order to predict the occurrence of flow-induced vibration, the phenomena that produces the exciting forces and the dynamic response by the tubes must be understood. The determination of tube natural frequencies is relatively straightforward. However, determination of the exciting forces created by the shellside fluid flow is extremely more difficult. The shellside flow in a heat exchanger follows a complex flow path. It is subjected to changes of direction, acceleration, and deceleration. At times, the flow is either perpendicular to the tubes (crossflow), axially along the tubes (parallel flow), or at any angle in between. Flow phenomena in crossflow include vortex shedding, turbulent buffeting, and fluid-elastic whirling. The flow phenomena found in parallel flow includes axial-flow eddy formation.

VORTEX SHEDDING

Flow across a tube produces a series of vortices in the downstream wake formed as the flow separates alternately from the opposite sides of the tube. This alternate shedding of vortices produces alternating forces which occur more frequently as the velocity of flow increases. For a single cylinder the tube diameter, the flow velocity, and the frequency of vortex shedding can be described by the dimensionless Strouhal number:

$$St = f_{vs} D_o / v_c \qquad (9.1)$$

where St = Strouhal number
f_{vs} = vortex shedding frequency, Hz
D_o = outside tube diameter, ft
v_c = crossflow velocity, ft/sec

For single cylinders the vortex shedding Strouhal number is a constant with a value of about 0.2. Vortex shedding occurs in the range of Reynolds numbers 100 to 5×10^5 and greater than 2×10^6. The gap is due to a shift of the flow separation point of the vortices in this intermediate transcritical Reynolds number range.

Vortex shedding also occurs for flow across tube banks. The Strouhal number is no longer a constant, but varies with the arrangement and spacing of the tubes. There is less certainty that there is a gap in the vortex shedding Reynolds number. Vortex shedding is fluid-mechanical in nature and does not depend upon any movement of the tubes. For a given arrangement and tube size, the frequency of the vortex shedding increases as the velocity increases. The vortex shedding frequency can be an exciting frequency when it matches the natural frequency of the tube and vibration results. With tube motion the flow areas between the tubes are being expanded and contracted in concert with the frequency of vibration. This in turn changes the flow velocity which controls the frequency of the vortex shedding. Since tubes vibrate only at unique frequencies, the vortex shedding frequency can become "locked in" with a natural frequency.

Strouhal numbers for banks of tubes can be estimated using information from Figure 9-1.

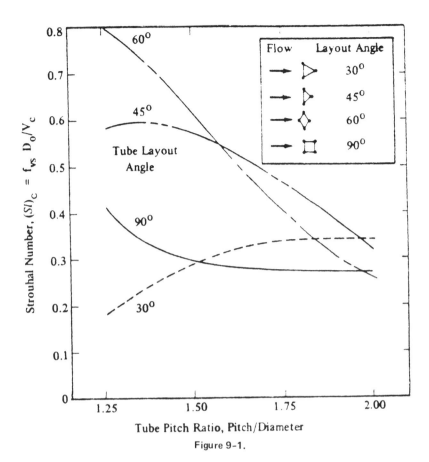

Figure 9-1.

TURBULENT BUFFETING

Turbulent buffeting is the name given to the fluctuating forces acting on tubes due to extremely turbulent flow of the shellside fluid. The turbulence has a wide spectrum of frequencies distributed around a central dominant frequency which increases as the crossflow velocity increases. This turbulence buffets the tubes which selectively extract energy from the turbulence at their natural frequencies from the spectrum of frequencies present. This is an extremely complex form of excitation. Turbulent buffeting frequencies can be predicted as below:

$$f_{tb} = (v_c D_o / S_t S_l)[3.05[1 - (D_o/S_t)^2] + 0.28] \qquad (9.2)$$

where f_{tb} = dominant turbulent buffeting frequency, Hz
v_c = crossflow velocity, ft/sec
D_o = outside tube diameter, ft
S_t = transverse tube spacing, ft
S_l = longitudinal tube spacing, ft

FLUID-ELASTIC WHIRLING

Fluid-elastic whirling is used to describe a phenomenon which is evidenced by the tubes vibrating in an orbital motion. This is produced by flow across the tubes causing a combination of lift and drag displacements of the tube at their natural frequencies. Typically, once fluid-elastic whirling begins, it can lead to a "run away" condition if the energy fed to the tubes exceeds that which can be dissipated by damping.

The prediction depends upon the natural frequency of the tubes and the damping characteristics of the system. Both natural frequency and system damping can be experimentally measured for existing equipment. However, to date the estimation of system damping is uncertain.

The equation below can be used to predict fluid-elastic whirling:

$$v_{cr}/f_n D_o = K(w\delta/\rho_s D_o^2)^{1/2} \qquad (9.3)$$

where v_{cr} = critical crossflow velocity, ft/sec
f_n = tube natural frequency, Hz
D_o = outside tube diameter, ft
K = instability threshold constant
w = virtual mass per unit length, lb/ft
δ = log decrement of the tube bundle in the shellside fluid under static conditions (recommended value is 0.03)
ρ_s = shellside fluid density, lb/cu ft

The instability factor is a function of tube arrangement and spacing and can be estimated from Figure 9-2.

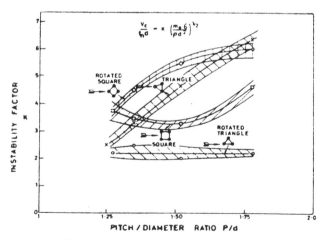

Figure 9-2: Instability factors for bundles tested.

PARALLEL FLOW EDDY FORMATION

Vibration due to axial or parallel flow results from the development of eddies along the tube. Axial or parallel flow eddy formation occurs when there are long, unsupported tube spans, relatively narrow shellside flow passages, and very high axial flows. Vibration due to this phenomenon rarely occurs in industrial heat exchangers.

ACOUSTIC VIBRATION

Another type of vibration that can occur is due to acoustic resonance when the shellside fluid is a vapor or a gas and in some two-phase flow situations. It does not occur when the shellside fluid is a liquid because liquids have extremely high sonic velocities. The acoustic frequencies of an exchanger can be excited by either vortex shedding or turbulent buffeting. When the exciting frequencies coincide with a natural acoustic frequency of the bundle a coupling occurs, and kinetic energy in the flow stream is converted into acoustic pressure waves. Acoustic resonance is due to a gas column oscillation which is molecular in nature and excited usually by phased vortex shedding. The oscillation occurs perpendicular to both the flow direction and the tube axis, creating an acoustic vibration of a standing wave type similar to the excitation of the air column in an organ pipe. The wave frequency and the tone produced are a function of the appropriate cavity or shell dimension. Acoustic resonance requires that the bounding walls in the characteristic dimension be nearly parallel, or have nearly parallel tangents. Otherwise the incident pressure wave cannot reflect back upon itself, providing the reinforcement and amplification necessary for a resonant condition.

The sound field generated within the shell does not affect the tube bundle unless the acoustic resonance frequency approaches the tube natural frequency. However, very loud, low frequency noise can be emitted. Also, fluctuating forces are developed that can potentially destroy the shell, anchor bolts, and reinforced concrete foundations, and can cause severe vibration of connecting piping. Yet another effect sometimes noticed is an increase in pressure losses with the onset of resonance.

Vortex shedding and acoustic resonance can have a "tuning" effect, forcing their frequencies into coincidence as they near each other. "Detuning" baffles are sometimes required in tube bundles. These baffles are designed and located so as to modify the characteristic cavity dimension without significantly affecting the shellside flow pattern.

The acoustic frequency can be predicted by the following equation:

$$f_a = mS/2Y \qquad (9.4)$$

where f_a = acoustic frequency, Hz
m = mode number, dimensionless integer
S = velocity of sound in shellside fluid, ft/sec
Y = characteristic length for acoustic vibration, ft
(usually the shell diameter for heat exchangers)

54 Handbook of Evaporation Technology

The lowest acoustic frequency is achieved when m = 1 and the characteristic length is the shell diameter. This is called the fundamental tone and the higher overtones vibrate at acoustic frequencies 2, 3, or 4 times the fundamental (m = 2, 3, or 4). These are illustrated in Figure 9-3. Reports of higher overtones in heat exchangers are rare. The length of the sides of a square inscribed inside the shell circumference is also a characteristic length and leads to the structure shown in Figure 9-4.

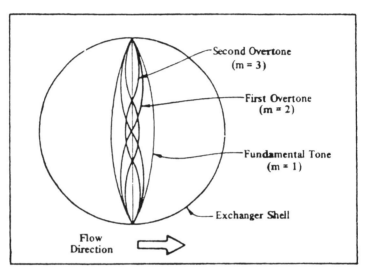

Figure 9-3: Most probable location of standing acoustic waves for fundamental and first two higher overtones.

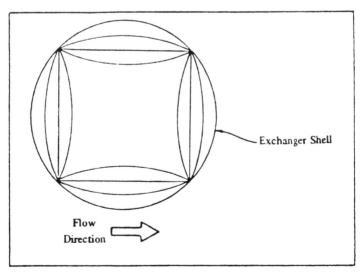

Figure 9-4: Possible location of standing acoustic waves.

The velocity of sound in a gas is given by the equation below:

$$S = (z\gamma gRT/M)^{1/2} \tag{9.5}$$

where S = velocity of sound in gas, ft/sec
        z = compressibility factor, dimensionless
        γ = specific heat ratio, C_p/C_v
        g = gravitational constant, 32.2 ft/sec^2
        R = gas constant, 1544 lb ft/(mol)°R
        T = absolute temperature, °R
        M = molecular weight, lb/mol

When the exciting frequencies are within 20% of an acoustic frequency, a loud noise is produced. This acoustic vibration can become destructive when it is in resonance with some component of the equipment. Good practice insures that the natural frequencies of the tubes differ from the acoustic frequencies of the shell.

RECOMMENDATIONS

The following recommendations can be made for design of tubular equipment:

(1) Avoid vortex shedding resonance. The fundamental frequency of the tube must be higher than the exciting frequency. The following ratio is suggested, assuming that f_n has been evaluated including the effects of axial loads:

    f_{vs}/f_n should be less than 0.75

(2) Avoid fluid-elastic instability. Maximum intertube velocity must be less than that calculated using the criterion:

    v_c/v_{cr} should be less than 1.0

Calculate v_{cr} using = 0.03 and K from Figure 9-2.

(3) Avoid vibration as a result of turbulent buffeting. The following ratio is suggested:

    f_{tb}/f_n should be less than 0.7

(4) Check for acoustic vibration. The following criteria can be used to predict acoustic vibration:

    f_{vs}/f_a and f_{tb}/f_a between 0.8 and 1.2 will result in acoustic resonance.

DESIGN CRITERIA

When designing equipment in which tube vibrations are likely to be a problem, the following design practices are recommended:

(1) Design the shellside so that the flow pattern is predictable. Avoid large (greater than 30%) and small (less than 15%) baffle cuts, because both conditions provide poor velocity distribution.

(2) Block any bypass flowpaths between the bundle and the shell, as well as through the pass-rib lanes; high local velocities in these areas can cause local damage in an otherwise sound design.

(3) In cases where the only troublesome velocity is that near the entrance or exit nozzle, it may be questionable if this local velocity can produce damage to downstream spans. Three methods are available for reducing these effects:

  (a) install a nozzle-velocity reducing device such as a distributor belt or an impingement plate.

  (b) install a tube-support baffle directly under the nozzle. This puts the exciting force at a node, and significantly diminishes the amplitude of the vibration that can be produced in the other spans.

  (c) roll the tubes into the first baffle in the vicinity of the nozzle. This partially isolates the exciting force from the rest of the tube, and lessens the amplitude of the vibration.

Extra-thick baffles reduce the rate of wear. They do not, however, increase the natural frequency unless the holes are precision-machined to very close tolerances. The additional cost for machined baffles is not usually justified, but the cost for extra-thick baffles may be.

The heat exchanger design most resistant to tube-vibration damage is the segmental-baffle type with no tubes in the baffle windows, as shown in Figure 9-5. There are two advantages to such a design:

(1) The most troublesome tubes, those that are supported only at every other baffle, are eliminated.

(2) Intermediate tube supports can be installed between baffles as necessary to increase the natural frequency beyond any exciting frequency.

Impingement plates should be used at the shell inlet nozzle.

Vibration problems are best prevented rather than corrected after the fact. A careful analysis at the design stage can greatly reduce future vibration problems.

Detuning baffles can be provided to prevent acoustic vibration. Detuning baffles, as shown in Figure 9-6, decrease the characteristic length.

Flow-Induced Vibration 57

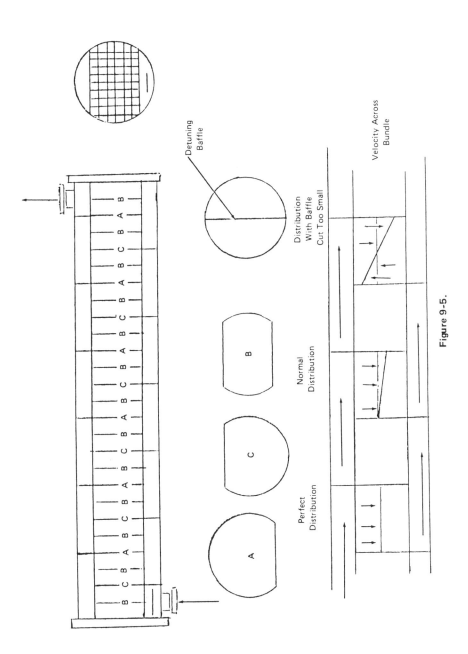

Figure 9-5.

58 Handbook of Evaporation Technology

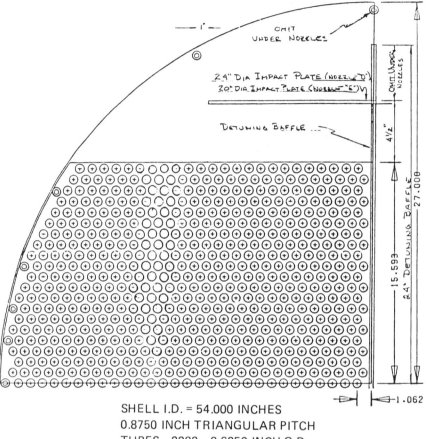

SHELL I.D. = 54.000 INCHES
0.8750 INCH TRIANGULAR PITCH
TUBES - 2282 - 0.6250 INCH O.D.
SPACERS - 24 - 5/8 INCH O.D.
TIE RODS - 24 - 3/8 INCH O.D.

Measured. Not design. Shell I.D. minus baffle O.D. to be 0.225 inch maximum.

Figure 9-6

FIXING VIBRATION PROBLEMS IN THE FIELD

When vibration problems are observed in the field, it should first be established that the problem is due to flow-induced vibration or from some transmitted source. If it has been determined to be flow-induced, several solutions can be considered:

(1) plug leaking tubes

(2) reduce shellside flow rate

(3) insert bars, wires, or wedges to stiffen the bundle

(4) roll tubes in baffles near nozzles
(5) remove tubes to create bypass lanes
(6) replace the tube bundle.

PROPRIETARY DESIGNS TO REDUCE VIBRATION

Two proprietary designs have been developed to combat tube vibration:

(1) Phillips "Rod Baffle" units
(2) Ecolaire "NESTS" systems.

In some cases, improved heat transfer and reduced pressure drop are achieved as additional benefits.

10
Natural Circulation Calandrias

OPERATION

Natural circulation calandrias (or thermosiphon reboilers) depend upon density differences to produce required flow rates. Vaporization creates an aerated liquid with a density less than that of the liquid in the system. The hydraulic head resulting from this density difference causes the fluid in the system to circulate. Circulation rates are high with liquid-to-vapor ratios ranging from 1 to 50.

Boiling on the tubeside has many advantages over boiling on the shellside. Since calandrias are often used in fouling services, the tubeside can be more readily cleaned. Tubeside distribution is more uniform, and pressure drop is lower. Consequently, heat transfer is higher because of the resulting higher circulation rates. Performance of horizontal thermosiphon reboilers with boiling on the shellside is difficult to predict. Normally, several shellside nozzles are required to effect adequate distribution. Stagnant areas are difficult to prevent. Residence time also is greater.

Vertical units provide more hydraulic head and higher circulation rates. However, such units have boiling point elevations due to the hydrostatic head. To offset this, they can be inclined, but not less than 15 degrees from horizontal. Inclined units result in different flow patterns and velocity in various tubes because elevation is not constant, especially for large units. Horizontal units frequently have poor distribution and vapor binding, especially when boiling on the shellside. The vertical boiling-in-tube unit is generally the most economical choice, both in initial investment and operating costs. The greater boiling-point elevation is seldom a serious penalty when all the advantages are weighed.

Liquid levels in the evaporator are generally maintained at high values. Greater capacity can be obtained by lowering the liquid level—to a point. Beyond this point, the top portion of the calandria operates free of liquid and the heat tranferred is in superheating the vapor. The increase in capacity obtained by lowering the liquid level is modest but the decrease at levels beyond this point is rather drastic. Figure 10-1 indicates a typical performance curve. Lower levels may require surge capacity in the base of the evaporator. Lower levels also reduce circulation rates increasing fouling and maintenance, especially if the top portion of the calandria runs dry.

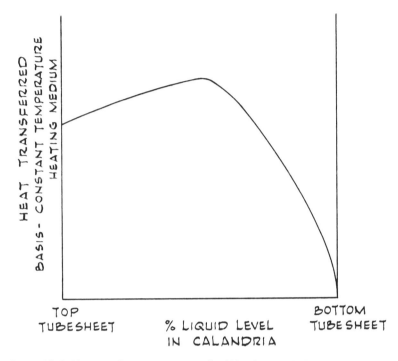

Figure 10-1: Heat transfer versus common liquid level—natural circulation calandria.

The optimum liquid level depends upon the system. The optimum can be predicted with confidence, but the final point must be established by observation since two-phase pressure drops, product composition, and piping arrangements cannot always be precisely calculated. For most systems the optimum point results when the evaporator liquid level is approximately half the distance between the two tubesheets.

The mechanism of operation of natural circulation calandrias is quite complicated. Figures 10-2, 10-3, and 10-4 will be used to illustrate the operation of calandrias. Because of the hydrostatic head, the liquid entering the calandria is not at its boiling point. Consequently, the bottom portion of the calandria is a sensible heat zone in which the liquid is heated to its boiling point. The length of this sensible zone depends upon the hydrostatic head as well as the liquid circulation rate. Once the liquid is heated to its boiling point, boiling occurs. Generally, boiling heat transfer rates are much higher than sensible heat transfer rates. The design of the calandria should be such that most of the tube is operating in the boiling zone.

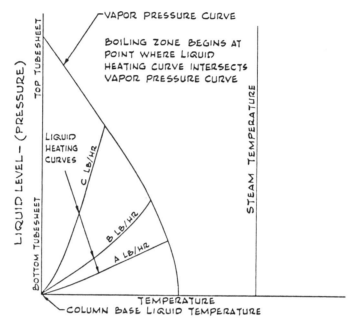

Figure 10-2: Natural circulation calandria—mechanism of operation—sensible heat zone and boiling zone.

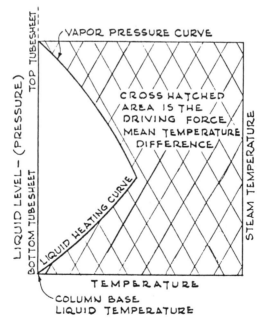

Figure 10-3: Natural circulation calandria—mechanism of operation—mean temperature differences.

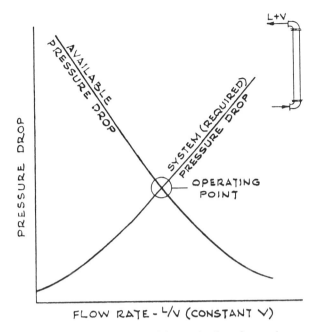

Figure 10-4: Natural circulation calandria—mechanism of operation—pressure drop.

Boiling point elevation as a result of the hydrostatic heat results in a temperature profile along the tube. The temperature of the liquid is increased in the sensible zone until it reaches the boiling point. From that point on the temperature decreases as the boiling pressure decreases progressively up the tube. This temperature profile must be known in order to calculate the driving force (temperature difference).

Lowering the liquid level in the evaporator reduces the boiling point elevation. However, it also reduces the liquid circulation rate because there is less available pressure difference to drive the system. Generally, lowering the liquid level will result in less of the tube being required for the sensible zone and more of the tube available for the boiling zone. This results not only in higher heat transfer coefficients but also a higher driving force. Consequently, heat transfer is increased as liquid level is dropped to the optimum point. At the optimum point, the liquid entering the calandria is entirely vaporized so that saturated vapor exits the tube. A further drop in liquid level will result in superheating of the vapor—a very poor heat transfer mechanism. Therefore, at liquid levels below the optimum heat transfer of the system is drastically reduced.

Normally, calandrias are not designed to operate at the optimum liquid level. The design point for liquid level is usually at the top tubesheet because this gives the highest circulation rates. High circulation rates reduce chances of fouling; we are usually justified in sacrificing heat transfer area to achieve reduced maintenance costs. For systems where fouling does not occur, the optimum liquid level is utilized in equipment design.

The amount of liquid circulated is a function of several things:

(1) the amount of vaporization,
(2) the liquid level in the evaporator,
(3) the temperature difference between the process and utility fluids,
(4) the flow characteristics of the calandria itself,
(5) the piping size and configuration associated with the calandria, and
(6) the physical properties of the fluid, especially its vapor pressure.

The calandria will operate at the condition such that the pressure drop through the piping and equipment is exactly equal to the pressure drop available for the system (Figure 10-4). The available pressure drop is a function of the liquid level in the evaporator and the difference in density between the saturated liquid and the aerated liquid. For a constant vapor rate, increasing liquid rates will reduce the available pressure drop while increasing the pressure drop through the system. Lowering the liquid level in the evaporator reduces the available pressure drop; therefore the liquid circulation rate will decrease.

As stated previously, the mechanism of operation of natural circulation calandrias is quite complicated. The final design is based on the net effect of several interdependent variables. Because of the nature of these variables, natural circulation calandrias are not advantageous for vacuum service. Because of the pronounced rise in boiling point due to hydrostatic head, the temperature difference is low, the sensible heat area (low coefficient) is large, the boiling area is small, and the circulation rate is low. If an attempt is made to reduce the effect of hydrostatic head by reducing tube length, the result is an uneconomical design.

SURGING

Surging of natural circulation calandrias can be experienced in certain circumstances. Surging most often occurs when the composition of the liquid at the base of the system is lean with respect to the fluid component which is actually being boiled. Circulation rates must be sufficient to ensure an adequate supply of the component being boiled. If the liquid is deficient in this component, the boiling point will be elevated. This results in inadequate heat transfer, reducing the vapor loading. At low vapor loadings, the feed will enrich the base liquid with the desired component. Now the calandria will function properly until all the component is boiled away from the base liquid. The cycle will repeat resulting in an unstable system. This situation can be corrected only by maintaining the concentration such that the calandria will receive an adequate supply of the boiling component. The base liquid may have to be controlled at a different concentration point. In some cases, higher temperature differences between the process and utility fluids may solve the problem.

Surging can also result if the base section has insufficient liquid volume to maintain circulation through the calandria system. Surging can also occur when the liquid level is maintained such that the calandria vapor nozzle is wholly or partially covered with liquid.

Surging can also occur when two or more calandrias are installed on the same body. Proper piping design will prevent such surging; therefore, calandrias mounted in parallel are not to be avoided. However, proper attention should be given to such an installation. Parallel calandrias should not be connected to a common liquid supply line. Each calandria should be mounted independently to assure that the fluid circulation in one calandria will not interfere with the circulation in the other calandrias. Figure 10-5 indicates the proper piping arrangement for parallel calandrias.

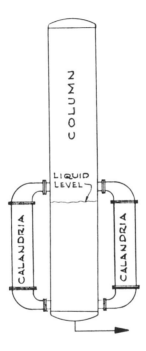

Figure 10-5: Piping arrangement—parallel natural circulation calandria.

FLOW INSTABILITIES

Two-phase fluid flows are susceptible to instability and this is often a very important parameter in design. The variation of total pressure drop as a function of flow rate may take the form illustrated in Figure 10-6. For imposed pressure drop ΔP_1 or ΔP_4 there is only one possible value of flow. However, at ΔP_3 there are two possible values and at ΔP_2, three values. Cases with more than one flow rate for a given pressure drop are susceptible to flow excursion from one operating condition to another. Similar excursions are possible when the imposed pressure drop is not constant but follows a pump characteristic such as shown by line PQ.

66 Handbook of Evaporation Technology

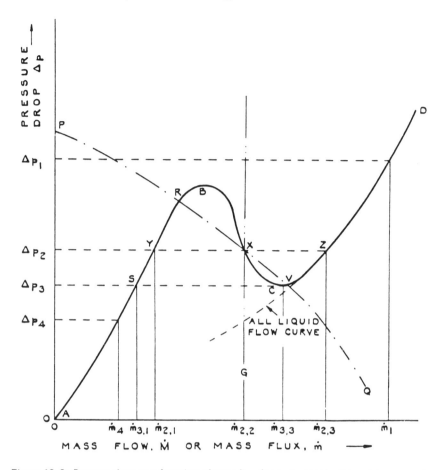

Figure 10-6: Pressure drop as a function of mass flux for a constant heat input in an evaporating channel.

Chugging instability occurs when the liquid is heated well above its saturation temperature. When vapor formation eventually occurs, it does so very rapidly and may expel liquid in both directions. The vapor formed can thus enter a region of subcooled liquid where it can collapse violently. This situation can normally occur only at rather low heat fluxes.

Oscillatory instability is the most important form and occurs because of feed-back effects due to time lags in the system. If the inlet flow is oscillated at a given frequency, all components of pressure drop may be additive, but as the frequency increases, lag occurs in the various components. Oscillatory instabilities can be present even in the absence of excursive instability and can occur between parallel tubes forming part of a flow circuit. System instability of the completed circuit can also occur.

Acoustic effects are often found either independently or coupled with other forms of instability. They result from the transmission of pressure disturbances and their reflection at turns, ends, reservoirs, etc. These disturbances can be caused by pumps, or by sudden changes in the system (opening or closing valves, for example). Acoustic effects are not usually dangerous except where they augment other modes of instability.

Flow instabilities result when a good portion of the total pressure drop occurs in the outlet head of the calandria. Flow instabilities can be reduced in such a system by throttling the liquid as it enters the calandria. This reduces the fraction of the total pressure drop occurring in the outlet head and stabilizes the system. Throttling the liquid inlet affects the performance in much the same way as lowering the liquid level in the body. The total fluid circulation is reduced. However, the whole problem can be eliminated with proper design of the inlet and outlet heads of the calandria.

Design of the top channel and nozzle (Figure 10-7) is one of the most important variables affecting the performance of natural circulation calandrias. An improperly designed top head can reduce the unit's capacity by 40%. The top channel should be constructed to provide a smooth transition from shell diameter to nozzle diameter. Re-entrant angles should be eliminated to reduce eddies and internal recirculation. Area of the vapor-outlet nozzle should be approximately equal to the flow area of the tubes. Diameter of the liquid-inlet nozzle should be one-half that of the vapor-outlet nozzle. Conventional bonnet-type heads with radial nozzles are usually acceptable for the bottom head (liquid inlet). Axial flow nozzles can be used if proper distribution is provided.

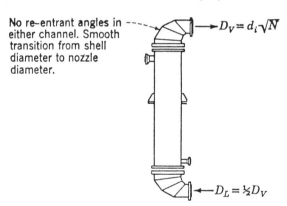

D_V = Vapor outlet nozzle I.D., in.
D_L = Liquid inlet nozzle I.D., in.
d_i = Tube I.D., in.
N = Number of tubes in bundle

Figure 10-7.

INTERNAL CALANDRIAS

When the shell diameter is large, the required vapor nozzle is large and is quite some distance above the top tubesheet. A large portion of the available driving force may be required to lift the two-phase mixture into the vapor body. Consequently, when shell diameters exceed 72 inches, the calandria is often designed to be integral with the evaporator body. This often provides a more economical design because circulation rates may be higher, and no inlet and outlet heads are required. Economies often result if one large unit is fabricated instead of several smaller ones.

Internal calandrias are not without problems, however. Hydrodynamics of internal calandrias are difficult to predict. Downcomers must be adequately provided; in most systems apparent liquid levels are actually controlled in the downcomers. A froth exists above the top tubesheet; this froth can build to appreciable heights. Liquid circulation must migrate across the tubesheet to the nearest downcomer. In doing so, it is agitated by the two-phase mixture leaving the tube ends. Froth can be reduced somewhat by extending the tube ends beyond the tubesheet to provide an undisturbed liquid flow path. However, the two-phase mixture must be separated before liquid can accumulate in this path; this is often not accomplished satisfactorily.

When liquid levels are measured, they are normally based on a calibration which assumes clear liquid and vapor at the two extremes of level measurement. Liquid level measurement is in reality a measurement of density differences. When a froth is present, the density is not known and liquid level measurements indicate only the apparent liquid level; the actual liquid level may be considerably higher. Consequently, operating liquid levels may be difficult to establish and maintain. Froths may alo affect the performance of internal calandrias. If deflectors are provided, they must be properly anchored and the pressure drop introduced included in the design of the system.

When froth exists, the froth density is a function of the heat transfer. As a result, apparent liquid levels often vary as steam flow varies. This can result in operational problems and flooding of systems, especially if trays are provided in the evaporator body. Excessive entrainment can also result.

In order to control internal calandrias, it is recommended that the bottom liquid level tap be a distance below the tubesheet equal to approximately 25% of the shell diameter. The top liquid level tap should be located a distance above the top tubesheet equal to approximately 250% of the shell diameter.

Downcomers must be sized to be self-venting and to minimize liquid holdup on the top tubesheet. In order to avoid entrainment of vapor in the downcomer, the superficial liquid velocity should not exceed 0.4 ft/sec, based on the expected liquid circulation rate. Several downcomers are usually preferred in order to reduce the flow path liquid must take in order to reach the downcomer.

If deflectors are used, the deflector diameter should be 6 to 12 inches greater than the diameter of the heating element. Deflector height above the tubesheet should be adequate to minimize pressure drop. Flow velocities around the deflector should be approximately half that leaving the tubes. Deflectors must be adequately supported to avoid mechanical problems.

Internal calandrias must be carefully evaluated. External calandrias are much more predictable; consequently less heat transfer surface may be required. This should be considered in any cost comparison.

FEED LOCATION

Cold feed should be introduced into the vapor body; introducing it at the calandria inlet may result in reduced heat transfer rates. Cold feed should be sprayed into the vapor body to improve direct contact heat transfer.

Flashing feed is often introduced at the inlet of the calandria. The effects of flashing must be evaluated when establishing natural circulation flow rates. It may be advantageous to introduce flashing feed in the vapor body.

SUMMARY

Liquid level in the vapor body is an important variable affecting operation of natural circulation calandrias. Normally units are operated with the evaporator liquid level at the top tubesheet of the calandria. For non-fouling fluids, the liquid level can be lowered to the optimum value in order to minimize heat transfer surface or maximize performance. The optimum value is approximately half the distance between the top and bottom tubesheets of the calandria, and will vary with each system. The liquid level should not be appreciably above the top tubesheet and certainly should not be maintained above the calandria outlet nozzle. Liquid levels above the vapor return will limit the performance of the calandria and may result in damage to the evaporator. Flow instabilities may also be experienced.

Internal calandrias present some unusual problems. Understanding the effects of the froth created above the top tubesheet will permit adequate design and operation. Downcomers must be properly sized. However, external calandrias generally are preferred.

11
Evaporator Types and Applications

Evaporators are often classified as follows:

(1) heating medium separated from evaporating liquid by tubular heating surfaces,

(2) heating medium confined by coils, jackets, double walls, flat plates, etc.,

(3) heating medium brought into direct contact with evaporating liquid, and

(4) heating with solar radiation.

Evaporators with tubular heating surfaces dominate the field. Circulation of the liquid past the surface may be induced by boiling (natural circulation) or by mechanical methods (forced circulation). In forced circulation, boiling may or may not occur on the heating surface.

Solar evaporators require tremendous land areas and a relatively cheap raw material, since pond leakage may be appreciable. Solar evaporation generally is feasible only for the evaporation of natural brines, and then only when the water vapor is evaporated into the atmosphere and is not recovered.

Evaporators may be operated batchwise or continuously. Most evaporator systems are designed for continuous operation. Batch operation is sometimes employed when small amounts must be evaporated. Batch operation generally requires more energy than continuous operation.

Batch evaporators, strictly speaking, are operated such that filling, evaporating, and emptying are consecutive steps. This method of evaporation requires that the body be large enough to hold the entire charge of the feed and the heating element be placed low enough not to be uncovered when the volume is reduced to that of the product. Batch operation may be used for small systems, for products that require large residence times, or for products that are difficult to handle.

A more frequent method of operation is semibatch in which feed is continuously added to maintain a constant liquid level until the entire charge reaches the final concentration. Continuous-batch evaporators usually have a continuous feed, and over at least part of the cycle, a continuous discharge. One method of operation is to circulate from a storage tank to the evaporator and back until the entire tank is at a specified concentration and then finish the evaporation in batches.

Continuous evaporators have continuous feed and discharge. Concentrations of both feed and discharge remain constant during operation.

Evaporators may be operated either as once-through units or the liquid may be recirculated through the heating element. In once-through operation all the evaporation is accomplished in a single pass. The ratio of evaporation to feed is limited in single-pass operation; single-pass evaporators are well adapted to multiple-effect operation permitting the total concentration of the liquid to be achieved over several effects. Agitated-film evaporators are also frequently operated once through. Once-through evaporators are also frequently required when handling heat-sensitive materials.

Recirculated systems require that a pool of liquid be held within the equipment. Feed mixes with the pooled liquid and the mixture circulates across the heating element. Only part of the liquid is vaporized in each pass across the heating element; unevaporated liquid is returned to the pool. All the liquor in the pool is therefore at the maximum concentration. Recirculated systems are therefore not well suited for evaporating heat sensitive materials. Recirculated evaporators, however, can operate over a wide range of concentration and are well adapted to single-effect evaporation.

There is no single type of evaporator which is satisfactory for all conditions. It is for this reason that there are many varied types and designs. Several factors determine the application of a particular type for a specific evaporation result. The following sections will describe the various types of evaporators in use today and will discuss applications for which each design is best adapted.

JACKETED VESSELS

When liquids are to be evaporated on a small scale, the operation is often accomplished in some form of jacketed kettle. This may be a batch or continuous operation. The rate of heat transfer is generally lower than for other types of evaporators and only a limited heat transfer surface is available. The kettles may or may not be agitated. Jackets may be of several types: conventional jackets (formed with another cylinder concentric to the vessel), dimpled jackets, patterned plate jackets, and half-pipe coil jackets. (See Figure 11-1.) This variety provides a great deal of flexibility in the choice of heat transfer medium.

Jacketed evaporators are used when the product is very viscous, the batches are small, good mixing is required, ease of cleaning is important, or glass-lined equipment is required.

72 Handbook of Evaporation Technology

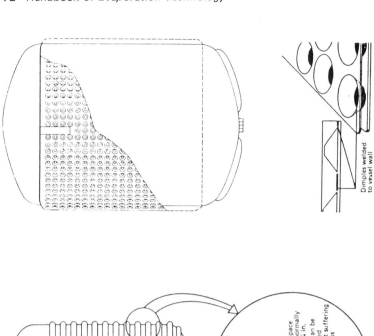

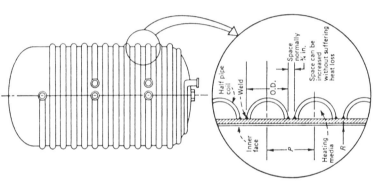

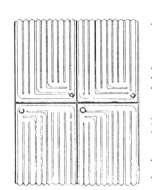

Figure 11-1: Jacketed vessels.

COILS

Coils for evaporator heating surfaces come in an almost unlimited variety of shapes and sizes. The most common application is to provide coils inside a tank with the evaporation process outside the coils. An appropriate heating medium flows inside the coils. Agitation can be provided to increase sensible heat transfer. An example of such a system is shown in Figure 11-2.

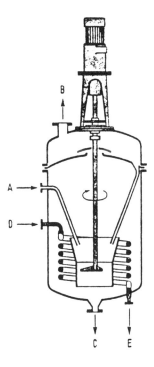

Figure 11-2: Stirrer evaporator with propeller type stirrer and built-in heating coil (for products with favorable flow behavior and average viscosity). A, product; B, vapors; C, concentrate; D, heating steam; E, condensate.

Evaporation can also occur inside coils with the heating medium outside the coil. Downflow or upflow of the process may be provided. Figure 11-3 shows a typical downflow unit with process inside helical coils.

Coils are used for the following reasons:

(1) small capacity
(2) product is difficult to handle
(3) high operating pressures for either process or heating fluid
(4) spiral flow is used to increase heat transfer or reduce fouling.

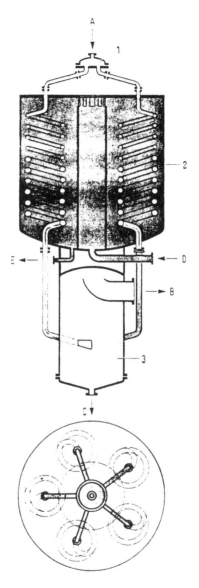

Figure 11-3: Typical downflow unit with process inside helical coils. A, product; B, vapors; C, concentrate; D, heating steam; E, condensate; 1, liquid distributor; 2, heating chamber; 3, separator.

HORIZONTAL TUBE EVAPORATORS

The first evaporator to receive general recognition was a design utilizing horizontal tubes. This type is seldom used except for a few special applications.

The simplest evaporator design is a shell and horizontal tube arrangement with heating medium in the submerged tubes and evaporation on the shell side. (See Figure 11-4.) Tubes are usually 7/8 inch to 1-1/2 inches in diameter and 4 to 16 feet long. The maximum area in a feasible design is about 5,000 sq. ft. The tube bundle is not removable. The tubes are sometimes spaced larger than normal and are prebent to facilitate cleaning.

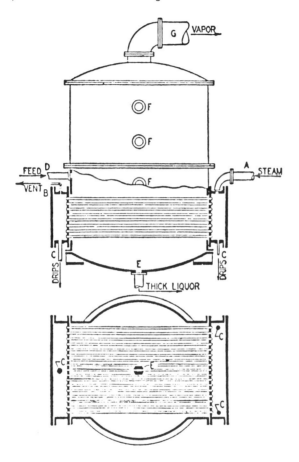

Figure 11-4: Horizontal-tube evaporator. A, steam inlet; B, vent for noncondensed gases; C, condensate outlets; D, liquor inlet; E, liquor outlet; F, sight glasses; G, vapor outlet. (Swenson Evaporator Co.)

Modifications include the use of U-bend tubes to facilitate bundle removal for inspection and cleaning.

Another modification is the kettle-type reboiler shown in Figure 11-5. This design permits the use of longer tubes; more heat transfer surface can be provided; the bundle can also be removed for inspection and cleaning. Bent-tubes can also be provided.

76 Handbook of Evaporation Technology

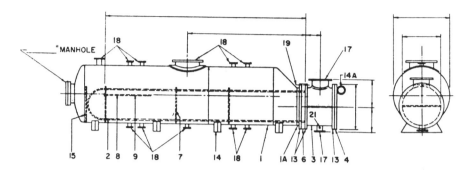

Figure 11-5.

Initial investment of horizontal tube evaporators is low. They are well adapted for nonscaling, low viscosity liquids. For severely scaling liquids the scale can sometimes be removed from bent-tube designs by cracking it off periodically by shock-cooling with cold water. Alternately, removable bundles can be used to confine the scale to that part of the heat transfer surface which is readily accessible.

Heat transfer rates may be low, the unit may be susceptible to vapor-binding, and foaming liquids cannot usually be treated. However, extended surfaces may be used to increase heat transfer and facilitate cleaning.

The short tube variety is seldom used today except for preparation of boiler feedwater. The kettle-type reboiler is frequently used in chemical plant applications for clean fluids.

The advantages of horizontal tube evaporators include relatively low cost in small-capacity applications, low headroom requirements, large vapor-liquid disengaging area, relatively good heat transfer with proper design, and the potential for easy semiautomatic descaling.

Disadvantages include the limitations for use in salting or scaling applications, generally. Bent-tube designs are relatively expensive.

Horizontal tube evaporators are best applied for small capacity evaporation, for clean or nonscaling liquids, when headroom is limited, or for severely scaling services in which the scale can be removed by thermal shocking bent-tube designs.

Horizontal Spray-Film Evaporators

A modification of the horizontal tube evaporator is the spray-film evaporator as shown in Figure 11-6. This is essentially a horizontal, falling-film evaporator in which the liquid is distributed by recirculation through a spray system. Sprayed liquid falls by gravity from tube to tube. Advantages include:

(1) noncondensables are more easily vented

(2) distribution is easily accomplished

(3) precise leveling is not required

(4) vapor separation is easily accomplished

(5) reliable operation under scaling conditions
(6) easily cleaned chemically.

Disadvantages include:

(1) limited operating viscosity range
(2) crystals may adhere to tubes
(3) cannot be used for sanitary construction
(4) more floor space is required
(5) more expensive for expensive alloy construction
(6) limited application for once-through evaporation.

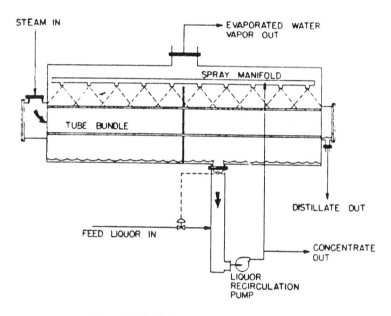

Figure 11-6: Horizontal spray-film evaporator.

Horizontal spray-film units are suitable for multiple effects and for vapor compression.

SHORT TUBE VERTICAL EVAPORATORS

Although the vertical tube evaporator was not the first to be built, it was the first type to receive wide popularity. The first was built by Robert and the vertical tube evaporator is often called the Robert type. It became so common that this evaporator is sometimes known as the standard evaporator. It is also called a calandria. Figure 11-7 illustrates this type of evaporator.

78 Handbook of Evaporation Technology

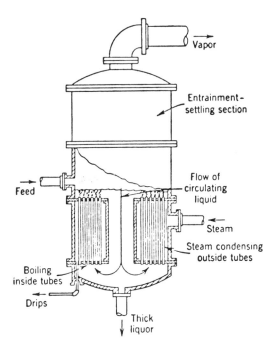

Figure 11-7: Cross-sectional diagram of standard vertical-tube evaporator with natural circulation. (Swenson Evaporator Co.)

Tubes 4 to 10 feet long, often 2 to 3 inches in diameter, are located vertically inside a steam chest enclosed by a cylindrical shell. The first vertical tube evaporators were built without a downcomer. These were never satisfactory, and the central downcomer appeared very early. There are many alternatives to the center downcomer; different cross sections, eccentrically located downcomers, a number of downcomers scattered over the tube layout, downcomers external to the evaporator body.

Circulation of liquid past the heating surface is induced by boiling (natural circulation). The circulation rate through the evaporator is many times the feed rate. The downcomers are therefore required to permit liquid flow from the top tubesheet to the bottom tubesheet. The downcomer flow area is generally approximately equal to the tubular flow area. Downcomers should be sized to minimize liquid holdup above the tubesheet in order to improve heat transfer, fluid dynamics and minimize foaming. For these reasons, several smaller downcomers scattered about the tube nest are often the better design.

Basket Type Evaporators

In the basket type evaporator (Figure 11-8), construction and operation is much the same as a standard evaporator except that the downcomer is annular. This construction often is more economical and permits the evaporator to be removed for cleaning and repair. An important feature is the easily installed deflector to reduce entrainment or "burping." A difficulty sometimes is associated with the steam inlet line and the condensate outlet line and differential thermal expansion associated with them.

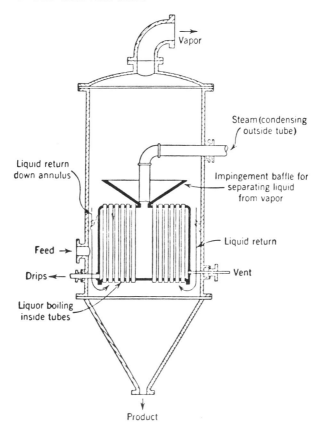

Figure 11-8: Cross-sectional diagram of basket-type evaporator. (Swenson Evaporator Co.)

Inclined Tube Evaporators

In an inclined tube evaporator the tubes are inclined, usually 30 to 45 degrees from horizontal. (See Figure 11-9.) In early designs the inclined calandria was mounted directly to the bottom head of a vapor body and the downcomer recirculating the product from the separator back to the bottom of the calandria was incorporated within the steam chest. Circulation in this configuration was sometimes impaired because of heat transfer across the downcomer. A first improvement was to insulate the downcomer. Circulation was further improved by providing a downcomer external to the evaporator.

Inclined tube evaporators sometimes perform well in foaming services because of the sharp change in flow direction at the vapor head. It sometimes offers advantages when treating heat sensitive materials. Inclined tube evaporators require low headroom and permit ready accessibility to the tubes.

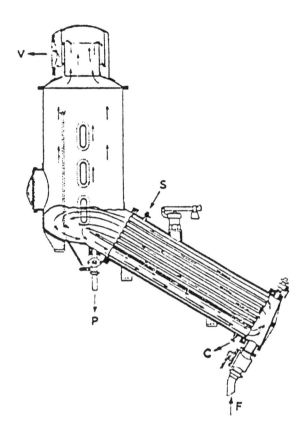

Figure 11-9: Natural-circulation inclined-tube evaporator with downcomer integral with steam chest. (Bennett, Sons & Shears Ltd.).

Propeller Calandrias

Circulation in the standard short tube evaporator depends upon boiling. Should boiling stop any solids present in the liquid will settle out. The earliest type of evaporator that perhaps could be called a forced-circulation system is the propeller calandria illustrated in Figure 11-10. Basically a standard evaporator with a propeller added in the downcomer, the propeller calandria often achieves higher heat transfer rates. The propeller is usually placed as low as possible to avoid cavitation and is placed in an extension of the downcomer. The propeller can be driven from above or below. Improvements in propeller design have permitted longer tubes to be incorporated in the evaporator. Propeller evaporators are sometimes used in Europe when forced circulation or long tube evaporators would be used in the United States.

Evaporator Types and Applications 81

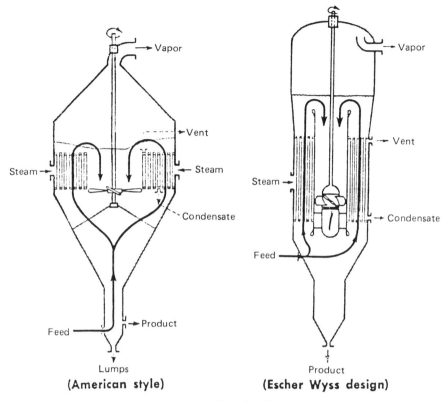

Figure 11-10: Propeller calandria evaporators.

Applications

The short tube evaporator offers several advantages: low headroom, high heat transfer rates at high temperature differences, ease of cleaning, and low initial investment. Disadvantages include high floor space and weight, relatively high liquid holdup, and poor heat transfer at low temperature differences or high viscosity. Natural circulation systems are not well suited for operation at high vacuum. Short tube vertical evaporators are best applied when evaporating clear liquids, mild scaling liquids requiring mechanical cleaning, crystalline product when propellers are used, and for some foaming products when inclined calandrias are used. Once considered "standard," short tube vertical evaporators have largely been replaced by long tube vertical units.

LONG TUBE VERTICAL EVAPORATORS

More evaporator systems employ this type than any other because it is versatile and often the cheapest per unit capacity. Long tube evaporators normally are designed with tubes 1 to 2 inches in diameter and from 12 to 30 feet in length. Long tube evaporators are illustrated in Figure 11-11.

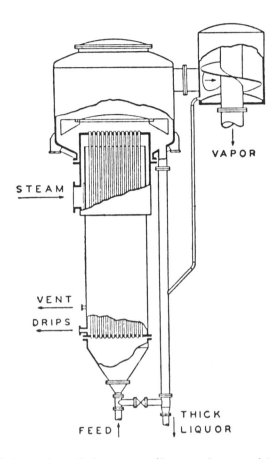

Figure 11-11: Long-tube vertical evaporator. (Swenson, Courtesy of American Institute of Chemical Engineers.)

Long tube units may be operated as once-through or may be recirculating systems. If once through, no liquid level is maintained in the vapor body, tubes are 16 to 30 feet long, and residence time is only a few seconds. With recirculation a level must be maintained, a deflector plate is often provided in the vapor body, and tubes are 12 to 20 feet long. Recirculated systems can be operated batchwise or continuously.

Circulation of fluid across the heat transfer surface depends upon boiling. The temperature of the liquid in the tubes is far from uniform and relatively difficult to predict. These evaporators are less sensitive to changes in operating conditions at high temperature differences than at lower temperature differences. The effects of hydrostatic head upon the boiling point are quite pronounced for long tube units.

Rising or Climbing Film Evaporators

The long tube evaporator described above is often called a rising or climbing film evaporator. The theory of the climbing film is that vapor traveling faster than the liquid flows in the core of the tube causing the liquid to rise up the tube in a film. This type of flow can occur only in a portion of the tube. When it occurs, the liquid film is highly turbulent and high heat transfer rates are realized. Residence time is also low permitting application for heat sensitive materials.

Falling Film Evaporators

The falling film version of the long tube evaporator (Figure 11-12) eliminates the problems associated with hydrostatic head. Liquid is fed at the top of long tubes and allowed to fall down the walls as a film. Evaporation occurs on the surface of the highly turbulent film and not on the tube surface. This requires that temperature differences be relatively low.

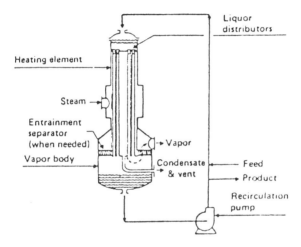

Figure 11-12: Vertical-tube, falling-film evaporator body.

Vapor and liquid are usually separated at the bottom of the tubes. Sometimes vapor is allowed to flow up the tube counter to the liquid. Pressure drop is low and boiling point rises are minimal. Heat transfer rates are high even at low temperature differences. The falling film evaporator is widely used for concentrating heat sensitive products because the residence time is low. Falling films are also used in fouling services because boiling occurs on the surface of the film and any salt resulting from vaporization is swept away and not deposited on the tube surface. They are also suited for handling viscous fluids. Falling film units are also easily staged.

The main problem associated with falling film units is the need to distribute the liquid evenly to all tubes. All tubes must be wetted uniformly and this may require recirculation of the liquid unless the ratio of feed to evaporation is relatively high. Recirculation can only be accomplished by pumping. Distribution can be achieved with distributors for individual tubes, with orifice plates above the tubes and tubesheet, or by spraying. Updraft operation complicates the liquid distribution.

Rising-Falling Film Evaporators

A rising and a falling film evaporator are sometimes combined into a single unit. When a high ratio of evaporation to feed is required and the concentrated liquid is viscous, a tube bundle can be divided into two sections with the first section functioning as a rising film evaporator and the second section serving as a falling film evaporator. The most concentrated liquid is formed on the downward passage. Figure 11-13 illustrates such a system. This system is also sometimes used when headroom is limited. Residence times are relatively low and heat transfer rates are relatively high.

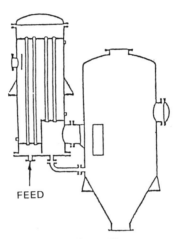

Figure 11-13: Rising-falling film concentrator (RFC).

Applications

The long tube vertical evaporator offers several advantages: low cost, large units, low holdup, small floor space, good heat transfer over a wide range of services. Disadvantages include: high headroom, recirculation is frequently required, and they are generally unsuited for salting or severely scaling fluids. They are best applied when handling clear fluids, foaming liquids, corrosive fluids, large evaporation loads. Falling film units are well suited for heat sensitive materials or for high vacuum application, for viscous materials, and for low temperature difference.

FORCED CIRCULATION EVAPORATORS

Evaporators in which circulation is maintained, regardless of evaporation rate or heat duty, by pumping the liquid through the heating element with relatively low evaporation per pass are suitable for a wide variety of applications. The forced circulation system is the easiest to analyze and permits the functions of heat transfer, vapor-liquid separation, and crystallization to be separated. Forced circulation systems are illustrated in Figures 11-14 and 11-15. Forced circulation systems are generally more expensive than natural circulation systems and are therefore used only when necessary.

Evaporator Types and Applications 85

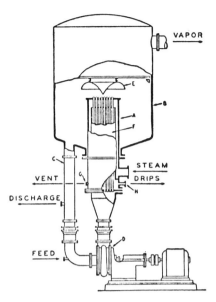

Figure 11-14: Forced-circulation evaporator with internal heating element: A, heating element; B, vapor head; C, liquor-return pipe; D, circulating pump; E, deflector; F, cylindrical baffle; G, noncondensed-gas vent; H, condensate outlet. (Swenson, Courtesy of American Institute of Chemical Engineers.)

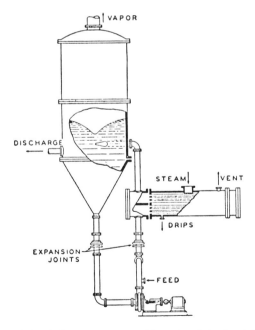

Figure 11-15: Forced-circulation evaporator with external horizontal heating surface. (Swenson, Courtesy of American Institute of Chemical Engineers.)

A choice of forced circulation can be made only after balancing the pumping energy cost, which is usually high, with the increase in heat transfer rates or decrease in maintenance costs. Tube velocity is limited only by pumping costs and by erosion at high velocities. Tube velocities are usually in the range of 5 to 15 feet per second.

Sometimes the pumped fluid is allowed to vaporize in the tubes. This often provides higher heat transfer rates but increases the possibility of fouling. Consequently this type of evaporator is seldom used except where headroom is limited or the liquids do not scale, salt, or foul the surface.

The majority of applications are designed such that vaporization does not occur in the tubes. Instead, the process liquid is recirculated by pumping, heated under pressure to prevent boiling, and subsequently flashed to obtain the required vaporization. These are therefore suited for vacuum operation. This type of evaporator is often called the submerged-tube type because the heating element is placed below the liquid level and use the resulting hydrostatic head to prevent boiling (often even in a plugged tube that is at the steam temperature). Often restrictions are provided in the return line to suppress boiling in order to reduce the headroom required.

The heating element may be installed vertically (Figure 11-14) usually single pass. Heating elements may also be installed horizontally (Figure 11-15), often two pass.

Circulation Pumps

Factors which must be considered when establishing the pumping rates include:

(1) maximum fluid temperature permitted

(2) vapor pressure of the fluid

(3) equipment layout

(4) tube geometry

(5) velocity in the tubes

(6) temperature difference between the pumped fluid and the utility fluid, and

(7) characteristics of pumps available for the service.

Circulation pumps should be selected so that the developed head is dissipated as pressure drops through the system. It is important that the pump and system match. The fluid being pumped is at or near its boiling point, and the required NPSH (net positive suction head) may be critical. The pump should operate at its design level. If it develops excessive head, it will handle more volume at a lower head. At the new operating point, the required NPSH may be more than is available, and cavitation will occur in the pump. If insufficient head is provided, the velocities may not be sufficiently high to prevent fouling; lower heat transfer rates may result; or the fluid may boil in the heating element with subsequent fouling or decomposition.

Applications

Forced circulation evaporators offer these advantages: high rate of heat transfer; positive circulation; relative freedom from salting, scaling, and fouling; ease of cleaning; and a wide range of application. Disadvantages include: high cost; relatively high residence time; and necessary pumps with associated maintenance and operating costs. Forced circulation evaporators are best applied when treating crystalline products, corrosive products, or viscous fluids. They are also well adapted for vacuum service and for services requiring a high degree of concentration and close control of product concentration.

PLATE EVAPORATORS

Plate evaporators may be constructed of flat plates or corrugated plates. Plates are sometimes used on the theory that scale will flake off such surfaces, which can flex more readily than curved surfaces. In some plate evaporators, flat surfaces are used, each side of which can serve alternately as the liquor side and the steam side. Scale deposited while in contact with the liquor side can then be dissolved while in contact with the steam condensate. There are still potential problems, however. Scale may form in the valves needed for reversing the fluids and the condensate frequently is not sufficient to dissolve the scale produced. Plates are often used as an alternative design to tubular equipment.

Spiral-Plate Evaporators

Spiral-plate evaporators may be used in place of tubular evaporators. They offer a number of advantages over conventional tubular equipment: centrifugal forces increase heat transfer; the compact configuration results in a shorter undisturbed flow length; relatively easy cleaning; resistance to fouling; differential thermal expansion is accepted by the spiral arrangement. These curved-flow units are particularly useful for handling viscous or solids-containing fluids.

A spiral-plate exchanger is fabricated from two relatively long strips of plate, which are spaced apart and wound around an open, split center to form a pair of concentric spiral passages. Spacing is maintained along the length of the spiral by spacer studs welded to the plates. In some applications both fluid-flow channels are closed by welding alternate channels at both sides of the spiral plate (Figure 11-16). In other applications, one of the channels is left completely open, the other closed at both sides of the plate (Figure 11-17). These two types of construction prevent the fluids from mixing.

The spiral assembly can be fitted with covers to provide three flow patterns:

(1) both fluids in spiral flow

(2) one fluid in spiral flow and the other in axial flow across the spiral

(3) one fluid in spiral flow and the other in a combination of axial and spiral flow.

Axial flow units perform well as natural circulation calandrias. Spiral-plate units are also effective for condensers and for heat recovery applications.

88 Handbook of Evaporation Technology

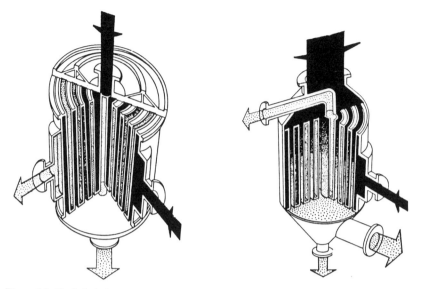

Figure 11-16: Spiral plate heat exchanger.   **Figure 11-17:** Spiral plate heat exchanger.

Gasketed-Plate Evaporators

The gasketed-plate evaporator is also called the plate-and-frame evaporator because the design is much like that of a plate-and-frame filter press. This evaporator is constructed by mounting a number of embossed plates with corner openings between a top carrying bar and a bottom guide bar. The plates are gasketed and arranged to form narrow flow passages when a series of plates are clamped together in the frame. Fluids are directed through the adjacent layers between the plates, either in series or parallel, depending upon the gasketing. Gaskets confine the fluids from the atmosphere. Figure 11-18 shows a typical flow in schematic form.

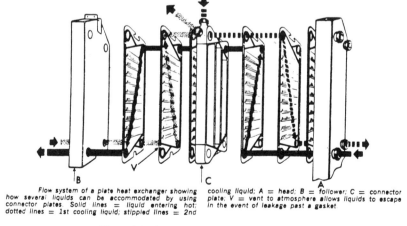

Flow system of a plate heat exchanger showing how several liquids can be accommodated by using connector plates. Solid lines = liquid entering hot; dotted lines = 1st cooling liquid; stippled lines = 2nd cooling liquid; A = head; B = follower; C = connector plate; V = vent to atmosphere allows liquids to escape in the event of leakage past a gasket.

Figure 11-18: Plate-and-frame heat exchanger.

Plate evaporators may be used as a heating element in a forced circulation system in which boiling is deliberately suppressed. Boiling may also occur within the gasketed-plate evaporator with the mixture of liquid and vapor discharged into a cyclone or other type of separator.

The design of the corrugations on the plate and the design of the flow patterns of the heating and process streams through the evaporator vary from one proprietary design to another. The units may be operated as a rising film, falling film, or rising-falling film. In some applications the rising and falling films are removed from the plate by the very high turbulence caused by extremely high vapor velocities. This action reduces the effective viscosity and prevents scaling.

The volume of product held in the evaporator is very small in relation to the large available heat transfer surface. Gasket-plate evaporators are well adapted to evaporating heat sensitive, viscous, and foaming materials. They afford fast startup and shutdown. They are compact with low headroom required. They are easily cleaned and readily modified.

Several different fluids can flow through different parts of the evaporator. A fluid may be removed for intermediate processing and returned for another heat transfer cycle. A fluid may also be subjected to a different heating and cooling media simultaneously or in series within the same evaporator.

A major disadvantage is the large gasketed area. However, interleakage of fluids cannot occur (without a plate rupture) in these units because all fluids are gasketed independently to the atmosphere. Leakage can be avoided by selecting adequate gasket materials and following proper assembly procedures.

Patterned-Plate Evaporators

Patterned plates are available in a variety of designs, configurations, and materials of construction. Evaporators may be fabricated from patterned plates to serve as alternatives to tubular elements. Patterned plates are often less costly than tubular elements. Patterned plates can also be used inside tanks instead of coils. Patterned-plate evaporators are shown in Figures 11-19 and 11-20.

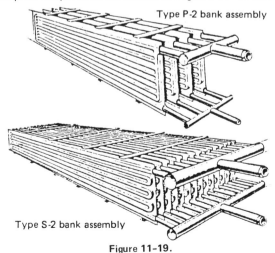

Figure 11-19.

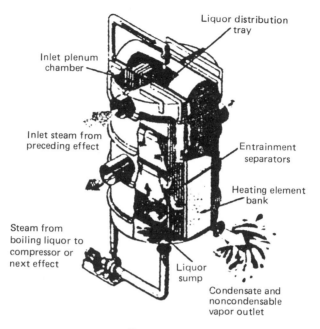

Figure 11-20.

MECHANICALLY-AIDED EVAPORATORS

Mechanically-aided evaporators can be very sophisticated or relatively simple. Mechanically-aided heat transfer is used for two reasons:

(1) to reduce the effects of fouling by scraping the fouling products from the heat transfer surface,

(2) to improve heat transfer by inducing turbulence.

Agitated Vessels

The simplest type of mechanically-aided evaporator is an agitated vessel with either jackets or coils as the heating element. Figure 11-21 illustrates three various types of agitated evaporators.

Agitated vessels are seldom used for evaporators except for the following applications:

(1) small systems

(2) products that are difficult to handle

(3) where mixing is important.

Scraped-Surface Evaporators

Provided with scraper elements that continuously sweep the heat-transfer

Evaporator Types and Applications 91

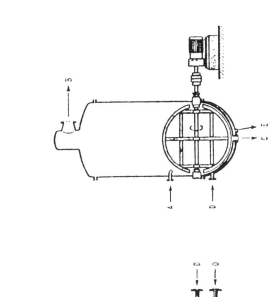
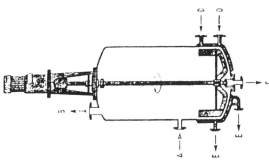
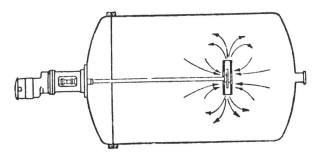

Stirrer Evaporator with Paddle Type Stirrers driven from the side and with heated semi-spherical bottom (for products with unfavourable flow pattern and very high viscosity)

Stirrer Evaporator with Anchor Stirrer and Heating Jacket (for high viscosity products)

Figure 11-21: Three types of agitated evaporators.

surface to reduce fouling and to increase heat transfer, these units (high and low speed) are widely used for viscous and rapidly fouling fluids. They are normally used as forced circulation systems in which boiling is suppressed. They can, however, be used in boiling applications by controlling a liquid level on the process side.

Low-Speed Units: Low-speed units usually consist of jacketed pipes connected in series (by special return bends), where the heat-transfer medium flows in the annulus between the two pipes. The scraper assembly consists of a series of blades attached to springs (which hold the blades against the inner pipe wall) that in turn are attached to a central shaft driven at 15 to 50 rpm. Figure 11-22 illustrates this type of evaporator.

High-Speed Units: High-speed units operate in the range of 200 to 2,000 rpm. Unlike low-speed units, the shaft occupies a major portion of the internal volume, so the process fluid flows in the comparatively small annnular space between the outside of the shaft and the heat-transfer surface. The rapidly rotating scraper blades create thin process-side films and violent agitation, resulting in high heat-transfer rates. These machines are applicable where low residence times are essential, to prevent fouling, for crystallization, and for heating viscous fluids. Figure 11-22 illustrates this type of evaporator.

Mechanically Agitated Thin-Film Evaporators

Thin-film evaporators are mechanically-aided, turbulent film devices. These evaporators rely on mechanical blades that spread the process fluid across the thermal surface of a single large tube. All thin-film evaporators have three major components: a vapor body assembly, a rotor, and a drive system (Figure 11-23).

Product enters the feed nozzle above the heated zone and is transported by gravity and mechanically by the rotor in a helical path down the inner heat transfer surface. The liquid forms a highly turbulent thin film or annular ring from the feed nozzle to the product outlet nozzle. Only a small quantity of the process fluid is contained in the evaporator at any instant. Residence times are low and gases or vapors are easily disengaged. The blades may also act as foam breakers. Typically about a half-pound of material per square foot of heat transfer surface is contained in the evaporator.

A variety of basic or standard thin-film evaporator designs is commercially available today. They are either vertical or horizontal, and can have cylindrical or tapered thermal bodies and rotors (Figures 11-24, 11-25).

The rotor may be one of several "zero-clearance" designs, a rigid "fixed-clearance" type or, (in the case of tapered rotors) an adjustable clearance construction may be used (Figure 11-26). One vertical design includes an optional residence-time control ring to manipulate the film thickness to some extent. The majority of thin-film evaporators are the vertical design with a cylindrical fixed-clearance rotor shown in Figure 11-23.

Application for Thin-Film Evaporators: Thermal separation in an evaporator may be conveniently characterized by the viscosity of the nonvolatile stream—the concentrate. Figure 11-27 illustrates various evaporator types and typical viscosity ranges for their useful applications. Unless other considerations are important (thermal stability, fouling tendencies), the terminal viscosity frequently dictates the type of evaporator selected. By far, most evaporation

Evaporator Types and Applications 93

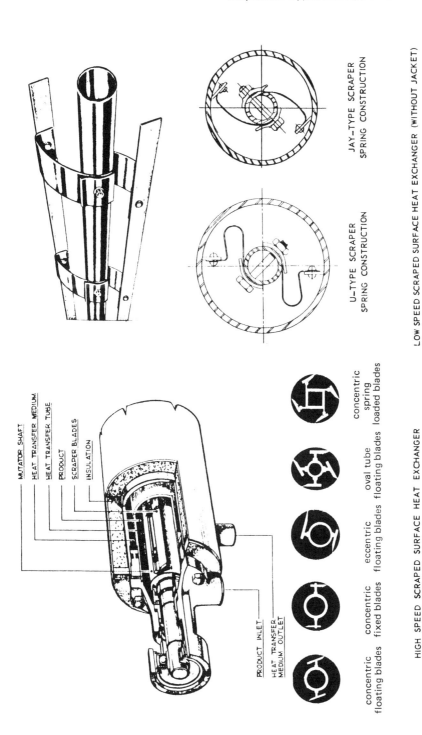

Figure 11-22: Scraped surface heat exchangers

94 Handbook of Evaporation Technology

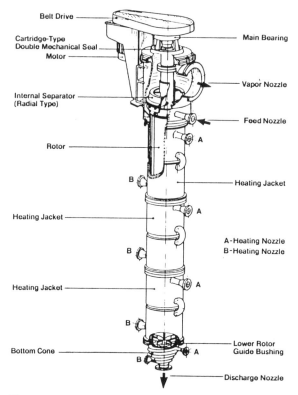

Figure 11-23: Vertical thin-film evaporator, cylindrical thermal zone.

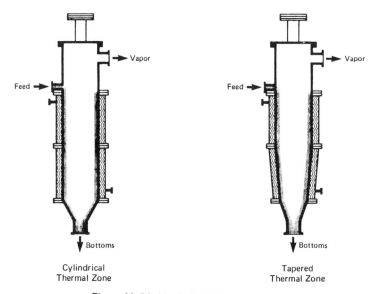

Figure 11-24: Vertical thin-film evaporators.

Evaporator Types and Applications 95

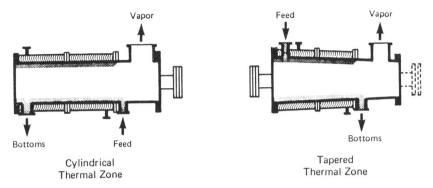

Figure 11-25: Horizontal thin-film evaporators.

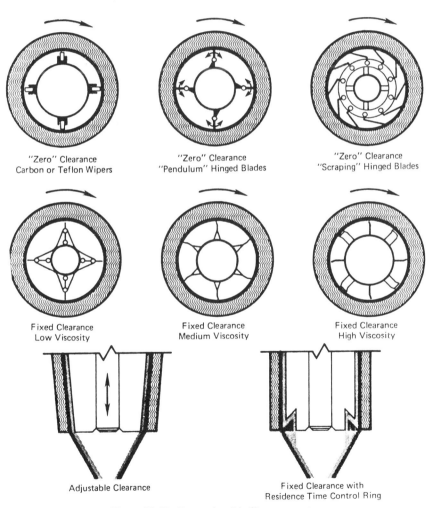

Figure 11-26: Rotors for thin-film evaporators.

96 Handbook of Evaporation Technology

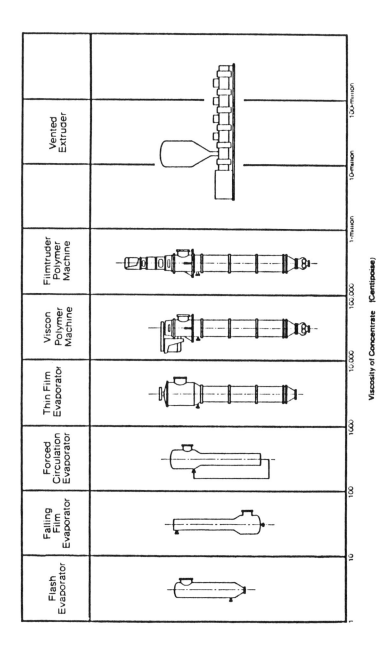

Figure 11-27: Evaporator design and cost are dependent upon the maximum viscosity of the concentrate.

applications involve nonviscous (less than 100 centipoise) fluids. Mechanically agitated evaporators are usually specified for terminal viscosities exceeding 1,000 centipoise and for heat-sensitive, foaming, or fouling products with lower viscosities. Chemical processing applications for thin-film evaporators may be classified into four broad categories:

(1) *Heat Sensitive Products.* Under high vacuum and moderate temperatures, the *mean* residence time of a vertical thin-film evaporator can be between 2 and 100 seconds; more importantly, the residence time *distribution* is very narrow (Figures 11-28, 11-29).

(2) *Materials that Tend to Foul.* Because of high turbulence imparted by the rotor and excellent localized mixing, fouling of the thermal surface is not generally a problem with thin-film evaporators. Approximately ten rotor blades pass each point on the heat transfer surface every second and the surface is "washed" each time.

(3) *Viscous Liquids.* Fluids with viscosities up to 50,000 centipoise can be processed in a standard thin-layer evaporator. "Zero-clearance" rotors do not exhibit the range of viscosity application that "fixed-clearance" rotors exhibit. Some firms manufacture "thin-film" vertical extruders" for the 50,000 to 20,000,000 centipoise range where fluids cease to flow under the influence of gravity alone. Thin-film evaporators, inherently low-pressure-drop devices, have mechanical turbulence and therefore relatively good heat transfer properties over a wide range of viscosities.

(4) *Miscellaneous Applications.* A variety of special evaporation problems have been solved in thin-film evaporators: evaporation and chemical reaction; two-phase flow (immiscible fluids, slurries, suspensions); cocurrent evaporation; high overhead splits; multiple-effect evaporation; and others.

Most industrial thin-film evaporators are used in complex chemical processes where several evaporators are employed to perform the required separation; a preconcentrator which may be a relatively inexpensive evaporator or perhaps some other separation device, and a thin-film evaporator as a "finisher" to reach the final concentration or degree of volatile recovery. In other applications a feed preheater is used and the feed material is flashed into the thin-film evaporator. In any case, all thin-film evaporator suppliers make discrete, standard sizes ranging up to about 450 square feet. It is important to properly utilize thin-film evaporator surface because of its relatively higher cost (20 to 30 times more expensive than tubular evaporators).

Process Considerations and Performance: The capacity, or performance, of a thin-film evaporator is often reported as a feed mass flow-rate scale factor (unit mass per unit surface area) for a specific process step. Performance is a function of a few process and equipment design variables:

(1) required material balance
(2) required energy balance
(3) heating temperature

98 Handbook of Evaporation Technology

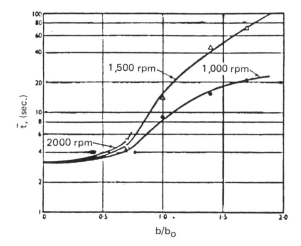

Figure 11-28: Mean residence time in a thin-film evaporator equipped with a residence time (RT) control ring at various rotor speeds. Feed rate is 157 lb/hr/ft^2. The ratio b/b_o is the RT ring width relative to the standard RT ring width.

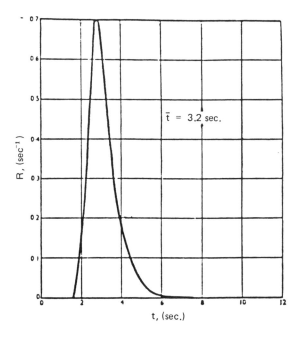

Figure 11-29: Residence time distribution in a thin-film evaporator. R is the relative frequency of fluid elements per unit time with a specific residence time t.

(4) operating pressure
(5) rotor speed
(6) materials of construction.

While mathematical models have been developed for predicting performance of thin-film evaporators, suppliers of thin-film machinery maintain extensive pilot plant testing laboratories for obtaining development and process data with small evaporators. Data obtained from one or two days of testing on several drums of feed material are sufficient for predicting and scaling performance to large evaporator applications.

Clearance between the rotor blades and the thermal wall is not a variable for most suppliers of thin-film evaporators. While the clearance for "fixed-clearance" designs should be close to zero, the rotor blades should not touch the wall. Thermal expansion and contraction of the rotor and the body and manufacturing considerations (especially of large rotors) dictate an optimum clearance: typically 1.25 millimeters for small rotors, up to 5 millimeters for large rotors.

The relationship between thermal performance and rotor clearance has been studied for a small vertical thin-film evaporator and is indicated in Figures 11-30, 11-31 and 11-32. Heat transfer rates are not strongly dependent on blade clearance or rotor speed.

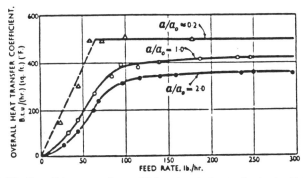

Figure 11-30: Overall heat transfer coefficient in relation to feed rate. Rotor speed 2,400 rpm. Δt 126°F. (a) blade tip clearance, (a_0) reference blade tip clearance.

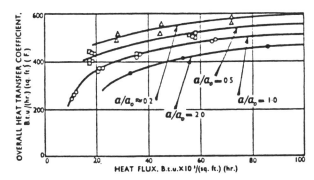

Figure 11-31: Overall heat transfer coefficient in relation to heat flux. Rotor speed 3,200 rpm.

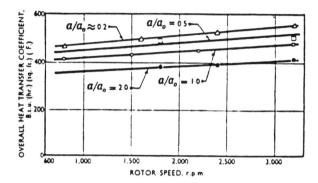

Figure 11-32: Overall heat transfer coefficient in relation to rotor speed. Heat flux constant at 55,000 Btu/(sq. ft.)(hr.).

Maintenance: Because thin-film evaporators are "rotating machinery," they are precision machines and preventative and corrective maintenance are important consideration. Bearings and mechanical seals must be properly applied and maintained. Properly maintained, thin-film evaporators provide long, trouble-free service. Premature failure can be caused by:

(1) loss of lubricant

(2) improper or dirty lubricants

(3) abnormal vibration

(4) process upsets which cause liquid to back up into the rotor area, vibration, or liquids or solids to enter the bearings.

SUBMERGED COMBUSTION EVAPORATORS

Submerged combustion is the process of burning a fuel in a specially designed burner under the surface of a liquid. The heat produced by combustion leaves the burner mostly as sensible heat in the hot combustion gases. The gas bubbles exit the burner directly into the surrounding liquid and heat is transferred from the gas to the liquid almost simultaneously. The gases are cooled and the liquid heated, so that by the time the combustion gases rise to the surface, they are practically at the same temperature as the liquid.

The submerged combustion evaporator construction is simple, requiring a tank, burner, combustion gas distributor, and a combustion control system. There are many different types of submerged combustion evaporators. Many of these evaporators use an external burner and combustor, where the combustion actually occurs. The hot gases thus produced are then bubbled through the solution being heated or evaporated. These evaporators tend to be less efficient than those actually burning the fuel under the liquid surface because of heat losses associated with moving combustion gases from the burner to the evaporator. Other types employ heat transfer surfaces (often coiled tubes) which in turn is heated directly by the combustion gases. (See Figure 11-33.) In some cases, coils immersed in water are heated by sparging steam into the water. However, this type of unit is not strictly a submerged combustion evaporator.

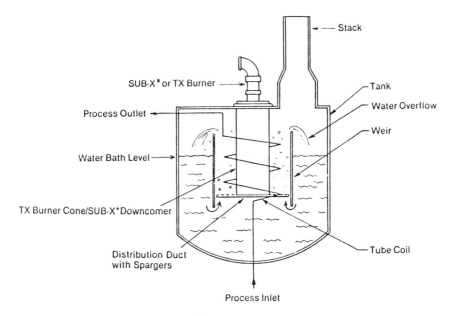

Figure 11-33: Simplified SUB-X®/TX submerged combustion vaporizer system.

Most submerged combustion units do not have heat transfer surfaces and are therefore used in applications where the lack of heat transfer surface can be put to advantage. Figure 11-34 illustrates such a system. Fuels can be either gaseous or liquid.

Submerged combustion evaporators frequently are smaller than other types. They can be economically constructed entirely in corrosion-resisting materials. For this reason they are often used when evaporating highly corrosive fluids.

Thermal efficiencies are high, ranging from 90 to 99% of the net heat input. However, since the vapor is mixed with large quantities of noncondensable gases, it is difficult to reuse the heat in this vapor. Sometimes the heat in the vapor can be used as a preheater or pre-evaporator before the submerged combustion unit. When the heat in the vapor cannot be recovered, low fuel costs greatly enhance the economics of the process.

Water vapor is produced in the combustion of fuel. To prevent its condensation into the liquid the gas and liquor temperatures must exceed 140°F. Above that temperature the water derived from combustion is not sufficient to saturate the dry combustion gases, and water will be evaporated from the liquor to achieve saturation. In many concentration operations, between 0.68 and 0.72 pound of water is evaporated per cubic foot of gas burned (1000 Btu/cu ft).

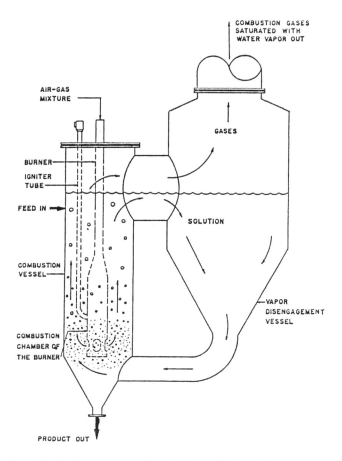

Figure 11-34: Evaporator vessels for submerged combustion evaporator.

In applications using heat transfer surfaces, when the submerged combustion unit is really only a heater, the water temperature must be below 140°F in order to avoid water makeup. These units normally operate such that the water from combustion exceeds the amount evaporated.

Bubbling combustion gases through the liquid results in a boiling point depression. Thus under submerged combustion conditions water boils at a temperature appreciably below its atmospheric boiling point of 100°C. The submerged combustion boiling point depends upon the amount of noncondensable gases bubbled through the liquid. The depression in boiling point is not limited to water only but is also noticed with aqueous salt and acid solutions. The lowered boiling point can be advantageous in some applications and detrimental in others. The reduced boiling point of most aqueous solutions can be approximated by multiplying the boiling point of the solution at atmospheric conditions by the ratio of the submerged combustion boiling point of water to its atmospheric boiling point. For submerged combustion conditions burning natural gas with 8% excess air, the boiling point of water has been found to be approximately 192°F. As the amount of excess air is increased, the boiling point decreases accordingly. Since submerged combustion can tolerate a maximum of 20% excess air before flame stability is impaired, the magnitude of boiling point reduction by combustion air alone is limited. In special cases, secondary air can be blown through the solution to further decrease the boiling point.

The problems resulting from submerged combustion evaporation have been related to entrainment and stack emissions.

Applications

Submerged combustion can develop the temperature required to boil any solution however high its boiling point. Submerged combustion evaporators are best suited for corrosive, scaling, salting, and highly viscous liquids which are not heat sensitive, and where water carbonation can be tolerated. They can also be used as crystallizing evaporators where control of crystal size is not important. Techniques have also been developed for concentrating heat sensitive materials. They are also widely used to concentrate waste streams. Although not an evaporator, submerged combustion heaters are widely used to heat corrosive, hard-to-handle, or cryogenic fluids.

FLASH EVAPORATORS

Originally applied for production of distilled water on board ships, flash and multistage flash evaporators have been extended to application on land for evaporating brackish and sea water as well as for process liquids. The principle of flash evaporation is simple in theory although highly developed and sophisticated in application. Water is heated and introduced into a chamber which is kept at a pressure lower than the corresponding saturation pressure of the heated water. Upon entering the chamber, a small portion of the heated water will immediately "flash" into vapor which is then passed through a moisture separator to remove any entrained liquid and condensed to form distilled product water. A series of these chambers can be held at progressively lower pressure with vapor flashing at each stage. Such a system is called a multistage flash evaporator. Figure 11-35 illustrates a basic flash evaporator cycle.

The flashing process can be broken down into three distinct operations: heat input; flashing and recovery; and heat rejection. The heat input section, commonly called a brine heater, normally consists of a tubular exchanger which transfers heat from steam, exhaust gas from a turbine, stack gases from a boiler, or almost any form of heat energy. The flashing and recovery sections consist of adequately sized chambers which allow a heated fluid to partially flash, thereby producing a mixture of vapor and liquid. The vapor produced in this process is passed through moisture separators and directed either to the heat recovery condensers (for multistage units) or to the third section, the reject condensers.

Since the evaporator does no work, the heat reject sections receive essentially all of the energy supplied in the heat input section of the evaporator.

In normal applications, the three sections are combined into one package. In single stage units there are no regenerative stages to recover the energy of the

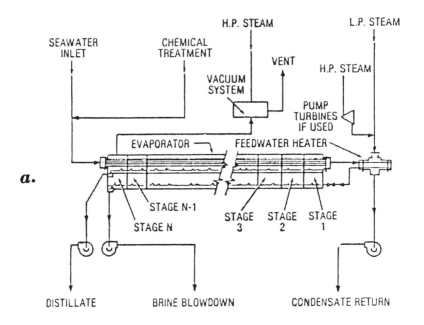
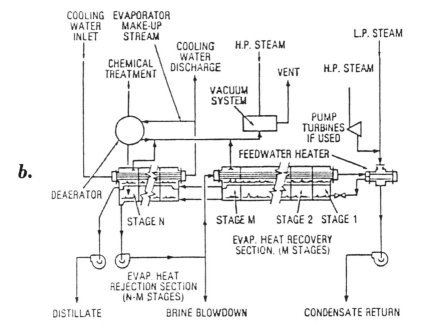

Figure 11-35: (a) Once through flash evaporator cycle. (b) Brine recirculation flash evaporation cycle.

flashed vapor. A multistage system extends the flashing and recovery zone by condensing the flashed vapor in each stage by heating the brine prior to the heat input zone. This reduces the amount of heat required for evaporation.

The number of stages or flashes is determined by the economics of each installation. Until recently, flash evaporators were limited to "waterpoor" areas where there was an abundance of low cost fuel. Now this type of evaporator can be applied to almost any water conversion unit. The flash evaporator is an extremely flexible system and can be made to operate with almost any form of heat energy. Proper instrumentation must be applied for multistage evaporators which incorporate a large number of stages. The inter-related variables of brine recirculation, makeup and blowdown flow rates, brine heater temperature, and final stage liquid level must be properly controlled. Once-through and brine recirculation systems are both used. Brine recirculation systems permit better chemical treatment to reduce scale.

Flash/Fluidized Bed Evaporators

A modification of the multistage flash evaporator is the multistage flash/fluidized bed evaporator as shown in Figure 11-36. The evaporator can be described as a countercurrent heat exchanger. Feed is heated in condensers, which are mounted on top of one another in the condensation column. After obtaining a final heat supply in the heat input section of brine heater, the brine flashes through all chambers mounted in the flash column and a gradual drop in saturation pressure and temperature in the chambers occurs, resulting in a partial evaporation of the brine in each flash chamber. The flash vapor flows from the flash column to the condensation column and finally condenses on the fluidized bed condenser surfaces that are cooled with incoming feed. The distillate is collected in the distillate trough and cascades down in the same way as the brine to the trough in the next chamber. Pumps are required for removal of brine and distillate and for feed supply.

The condensers consist of many vertical parallel condenser tubes filled with particles that are fluidized by upward liquid flow through the tubes. This principle permits it to function with extremely short condenser tubes per stage as a result of good heat transfer performance at low liquid velocities. Advantages include:

(1) chemicals are not required for scale control
(2) the system is easier to operate than conventional multistage flash evaporators.

Special Application

A flash evaporator system having no heating surfaces has been developed for separating salts with normal solubility from salts having inverse solubility. Steam is injected directly into the feed slurry to dissolve the normal-solubility salt by increasing the temperature and dilution of the slurry. The other salt remains in suspension and is separated. The hot dilute solution is then flashed to a lower temperature where the normal-solubility salt crystallizes and is separated. The brine stream is then mixed with more mixed salts and recycled through the system. This system can be operated as a multiple effect by flashing down to

106 Handbook of Evaporation Technology

the lower temperature in stages and using flash vapor from all but the last stage to heat the recycle stream by direct injection. In this process no net evaporation occurs from the total system and the process cannot be used to concentrate solutions unless heating surfaces are added.

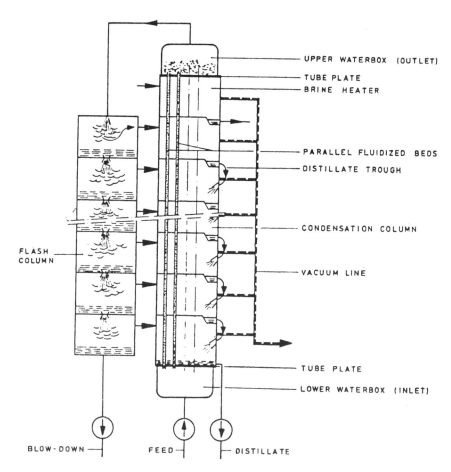

Figure 11-36: Multistage flash/fluidized bed evaporator.

SPECIAL EVAPORATOR TYPES

Special evaporator types are sometimes required when heat loads are small, special product characteristics are desired, or the product is especially difficult to handle.

Vertical Tube Foaming Evaporator

The vertical tube foam evaporator is used to evaporate cooling tower blow-down before disposal. It is essentially a conventional vertical tube, recirculating evaporator; the novel feature is the addition of a small amount of surfactant to the feed. The surfactant provides three advantages:

(1) rate of scale formation is reduced
(2) a stable two-phase fluid results which lowers hydrostatic losses enough to permit brine recirculation without a pump
(3) overall heat transfer is improved.

Operation can be either upflow or downflow.

Direct-Contact Multiple-Effect Evaporator

Submerged combustion (direct contact) can be combined with multiple-effect evaporators to combine the best features of each. The first effect is a direct contact evaporator in which combustion gases from burning a fuel directly contact liquor in a venturi scrubber to evaporate water and saturate flue gases. The second effect provides heat recovery by vacuum evaporation. Condensate heated by countercurrent scrubbing contacts the saturated flue gas to remove its latent heat. The heated condensate passes through a heat exchanger to heat liquid that is circulated and flash evaporated in a vacuum system. This procedure uses the latent heat for one or more effects of vacuum evaporation. The third effect of air evaporation heats liquor in the vacuum-evaporator's surface condenser. Additional heat is obtained by further cooling the flue gas with condensate and passing the heated condensate through a heat exchanger to heat the liquor. The heated liquor is circulated to an air evaporator, where it is contacted with air, thereby heating and saturating the air for added liquor evaporation and cooling.

The techniques can be applied to evaporation systems to use:

(1) any fuel economically
(2) heat generated in burning waste liquor
(3) flue heat from recovery or waste-heat boilers
(4) waste heat contained in a plume
(5) heat from a condensing system for vacuum or air evaporation.

Grainer

The grainer is a type of evaporator confined to use in the salt industry. The brine is contained in a shallow pan with temperature maintained somewhat below the atmospheric boiling point by circulation through external heaters or by steam tubes running the length of the granier below the liquid surface, or by both. Evaporation occurs at the quiescent air-water interface, forming crystals that develop a hopper shape as they grow on the surface. The peculiar crystal shape is the sole justification for the grainer. The predecessor to the grainer was the direct-fired flat-bottom pan still occasionally used.

Disk or Cascade Evaporator

This type of evaporator is used in the pulp and paper industry to recover heat and entrained chemicals from boiler stack gases and to effect a final concentration of the black liquor before it is burned in the boiler. These evaporators consist of a horizontal shaft on which disks are mounted perpendicular to the shaft. Sometimes bars parallel to the shaft are used. The assembly is partially immersed in the thick black liquor so that films of liquid are carried into the hot-gas stream as the shaft rotates.

Refrigerant Heated Evaporators

For highly heat-sensitive materials, especially enzymes, antibiotics, glandular extracts, fine chemicals, and certain foods, evaporating systems have been developed which use suitable refrigerants as the heat transfer medium. A heat pump (which will be discussed later) provides the heat required for evaporation.

The operating temperature may be varied over a wide range. Efficiency is high and no steam or cooling water is required.

Ball Mills

Ball mills are sometimes used for batch processes when the product is a solid. The feed enters the ball mill and is concentrated as evaporation occurs and the vapor is removed. The ball mill prevents the heat transfer surface from becoming fouled. The solid product is periodically removed. Residence times are relatively high.

Fired Heaters

Some high-temperature applications require evaporation to be accomplished in some type of fired heater. Most fired heaters are radiant heaters employing refractories. Other types employ finned fire tubes with combustion gases inside the fire tube. Refractories are not required because the fins permit rapid heat transfer. Forced-air-stoichiometric heaters employ a high-intensity combustion chamber from which combustion gases are discharged at high velocities into a heat transfer cabinet where convective heat transfer occurs. Various types of fired heaters are illustrated in Figures 11-37, 11-38, and 11-39.

The latter two units require no refractory; consequently they can be rapidly started up and shut down. Refractories must be heated or cooled slowly to avoid damage to the refractories.

Flame impingement upon the heat transfer surface can result in damage to the surface as well as degradation of the circulating fluid. It is preferred to have a single flow path when handling fluids that decompose at high temperatures, in order to avoid low velocities and stagnant areas and resultant poor heat transfer associated with poor distribution of the fluid to multiple flow circuits.

Electrical Heaters

Electrical energy as a source of heat for evaporation systems can effectively compete with conventional fuels only when the cost for electricity is low or some unusual process condition or configuration exists. Principal advantages of electrical heating methods are:

(1) ease of application and control

Evaporator Types and Applications 109

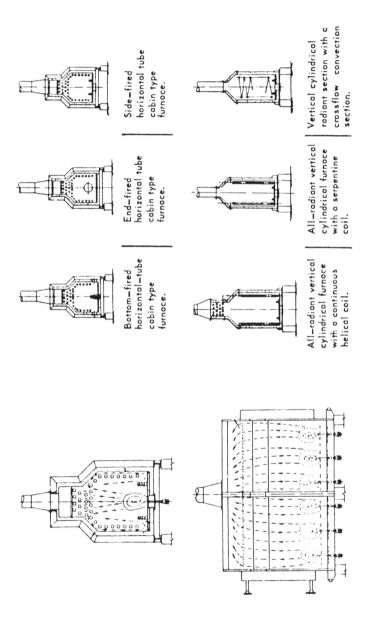

Figure 11-37: Radiant section fired heater.

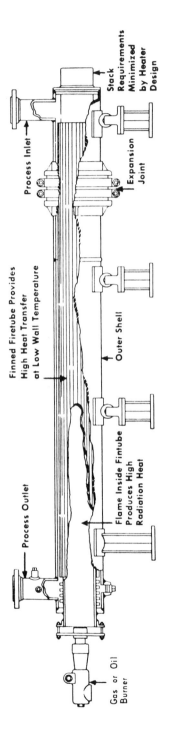

Figure 11-38: Fired heater—finned firetube.

Evaporator Types and Applications 111

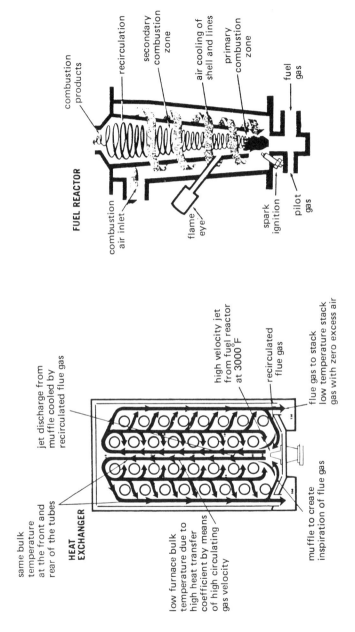

Figure 11-39: Forced air stoichiometric heater.

(2) uniformity of temperature and heating rate
(3) development of high temperature
(4) absence of oxidative or reductive environments.

The most widely used method of electrical heating is metal-sheathed resistance heaters. They are inexpensive, highly efficient, easy to control, sturdy, and easy to apply. Several forms are available:

(1) tubular heaters
(2) strip heaters
(3) cartridge heaters
(4) finned tubular heaters.

Rod, wire, and ribbon resistance heaters are also used. Silica carbide, quartz, and graphite heating elements are also available.

Other forms of electric heating include:

(1) induction heating
(2) infrared heaters
(3) dielectric heaters
(4) plasma torches.

12
Fouling

Fouling may be defined as the formation of deposits on heat transfer surfaces which impede the transfer of heat and increase the resistance to fluid flow. The growth of these deposits causes the thermal and hydrodynamic performance of heat transfer equipment to decline with time. Fouling affects the energy consumption of industrial processes and it can also decide the amount of material employed in the construction of heat transfer equipment. In addition, where the heat flux is high, fouling can lead to local hot spots and ultimately it may result in mechanical failure, and hence an unscheduled shutdown of the equipment.

The designer and operator of heat transfer equipment must be able to predict the variation of heat exchanger performance as the fouling proceeds. The designer needs this information to ensure that the users' requirements with regard to cleaning schedules can be met and maintained with a heat exchanger designed for a predetermined first cost. The users of heat exchangers subject to fouling must be able to formulate rational operating schedules, both for plant management and in order to obtain equipment that will meet the desired operating schedule.

Fouling is a phenomenon which occurs with or without a temperature gradient in a great many natural, domestic, and industrial processes. A temperature gradient complicates, but is frequently not essential to, the phenomenon. If fouling is defined as the accumulation of undesired solid material at phase interfaces for heat transfer surfaces, then five possible interfaces are encountered:

(1) gas-liquid
(2) liquid-liquid
(3) gas-solid
(4) liquid-solid
(5) gas-liquid-solid.

The first two involve direct-contact heat transfer between a fluid and a liquid

114 Handbook of Evaporation Technology

phase, where the crud which develops at the interface is mobile and relatively easy to remove. The last, commonly referred to as the triple interface, occurs in change-of-phase operations such as condensation and boiling. Current interest in fouling is focused on liquid-solid and gas-solid interfaces.

COST OF FOULING

Fouling is a topic of considerable economic and technical importance. The penalty for fouling can be attributed in roughly equal parts to:

(1) higher capital costs through oversized plants
(2) energy losses through increased thermal inefficiencies and pressure drops
(3) maintenance, including cleaning of heat exchangers and use of antifoulants
(4) loss of production during shutdown for cleaning or through reduced overall plant efficiency.

CLASSIFICATION OF FOULING

Because of the wider variety of fouling mechanisms from liquids than from gases, thermal fouling is classified with the liquid-solid interface as the prototype. The classification is based on the key physical/chemical process essential to the particular fouling phenomenon. Six primary categories have been identified:

(1) *Precipitation Fouling:* the crystallization from solution of dissolved substances onto the heat transfer surface, sometimes called scaling. Normal solubility salts precipitate on subcooled surfaces, while the more troublesome inverse solubility salts precipitate on superheated surfaces.

(2) *Particulate Fouling:* the accumulation of finely divided solids suspended in the process fluid onto the heat transfer surface. In a minority of instances settling by gravity prevails, and the process may then be referred to as sedimentation fouling.

(3) *Chemical Reaction Fouling:* deposit formation on the heat transfer surface by chemical reactions in which the surface material itself is not a reactant (e.g., in petroleum refining, polymer production, food processing).

(4) *Corrosion Fouling:* the accumulation of indigenous corrosion products on the heat transfer surface.

(5) *Biological Fouling:* the attachment of macro-organisms (macro-biofouling) and/or micro-organisms (micro-biofouling) to a heat transfer surface, along with the adherent slimes often generated by the latter.

(6) *Solidification Fouling:* the freezing of a pure liquid or the higher melting constituents of a multi-component solution onto a sub-cooled surface.

These categories do not denote the rate-governing process for the fouling. Almost any pair of the above fouling modes are mutually reinforcing. This is particularly true of corrosion fouling in conjunction with each of the other modes. One exception is scaling accompanied by particle deposition, which tends to weaken an otherwise tenacious scale.

NET RATE OF FOULING

The net rate of the formation of the deposits is the generally accepted starting point for all fouling models. Accordingly:

$$\text{Net Rate of Fouling} = \text{Rate of Formation of the Fouling Deposit} - \text{Rate of Removal of the Fouling Deposit}$$

To relate this expression to the fouling factor, it must be integrated over an appropriate time period to obtain the mass of the deposit.

This model views the net rate as the result of two competing processes. Formation involves the following processes acting in sequence:

(1) processes in the body of the fluid
(2) transport to the heat transfer surface
(3) attachment/formation of the deposit.

The removal term involves one or more of the following:

(1) dissolution; material leaves in molecular or ionic form
(2) erosion or reentrainment; material leaves in particulate form
(3) spalling or sloughing; material leaves as a large mass.

The first two processes for removal, where they occur, are probably continuous and affect all points on the surface of the deposit. Spalling or sloughing takes place at random times and at random locations.

The fouling characteristics of a fluid in contact with a heat transfer surface are a function of a variety of parameters:

(1) geometry of the heat transfer surface
(2) material of the heat transfer surface
(3) temperature of the interface between the fouling liquid and the heat transfer surface
(4) velocity past the heat transfer surface
(5) characteristic of the fouling fluid (a complicated parameter in itself).

116 Handbook of Evaporation Technology

Fouling is an extremely complex phenomenon. From a fundamental point of view, it may be characterized as a combined momentum, heat, and mass transfer problem. In many instances, chemical kinetics are involved as well as solubility characteristics and corrosion technology.

SEQUENTIAL EVENTS IN FOULING

For all categories of fouling, the successive events which commonly occur in most situations are up to five in number:

(1) *Initiation* (delay, nucleation, induction, incubation, surface conditioning)

(2) *Transport* (mass transfer)

(3) *Attachment* (surface integration, sticking, adhesion, bonding)

(4) *Removal* (release, reentrainment, detachment, scouring, erosion, spalling, sloughing)

(5) *Aging.*

Initiation

Initiation is associated with the delay period so often (but not always) observed before any appreciable fouling is recorded after starting up a process with a clean heat transfer surface. For precipitation fouling it is closely associated with the crystal nucleation process, and thus tends to increase as the degree of supersaturation is increased with respect to the heat transfer surface temperature, and as the general temperature level is increased for a given degree of supersaturation; the effect of operating velocity is not well known. For chemical reaction fouling, the delay period appears to decrease as the surface temperature is increased. For all fouling modes, the delay period decreases as the surface roughness increases. The roughness projections provide additional sites for nucleation, adsorption, and chemical surface-activity, while the grooves provide regions for deposition which are sheltered from the mainstream velocity. Surface roughness also increases eddy transport to the wall.

In the case of biofouling, the initial events involve the surface adsorption of polymeric glycoproteins and proteoglycons, traces of which act as surface conditioning films, to which micro-organisms subsequently adhere.

Transport

Transport is the best understood of the fouling stages. A key component must be transported from the bulk of the fluid to the heat transfer surface. The transport processes can operate in two ways. For particulate fouling, precipitation fouling, chemical reaction fouling, and fouling due to liquid solidification, the fouling material itself is transported to the surface. With corrosion fouling, the situation is different; fouling material is not transported but an ionic species which contributes to the corrosion reaction, which in turn results in the formation of reaction products which cause fouling. The transported ionic species can be viewed as a "nutrient" that "feeds" the growth of the deposit.

For microbial fouling, both the transport of the deposit material (microbes) and the transport of nutrients is involved in the formation of the deposits.

Attachment

Attachment of the fouling species to the wall follows transport of the key component to the wall region, where the solid which deposits is actually formed, except in the case of particulate fouling. The attachment/formation process is very complicated, involving both physical and chemical processes, and it is not as well understood as the mass transfer process. The attachment/formation processes depend on the presence or absence of a fouling deposit on the surface. The attachment/formation of the deposit at a clean surface occurs in a different way from attachment to a fouled surface.

Processes in the body of the fluid can affect formation in two ways:

(1) the bulk concentration of the deposited material will depend on conditions in the fluid at points far from the heat transfer surface.

(2) the form in which the fouling deposit is transported to the surface will be a function of processes in the bulk of the fluid.

Removal

Removal of the deposit may or may not begin right after deposition has started. The rate of removal is directly proportional to both the mass of deposit and the shear stress on the heat transfer surface, and inversely proportional to the deposit strength. The particular removal process will depend on the characteristics of the deposit and this may vary with time so that more than one removal process may be present during the period in which fouling is occurring.

Removal of the deposit by dissolution must be related to the solubility of the deposit material. Since the deposit at the time of its formation was presumably insoluble, then dissolution will only occur if there is a change in the character of the deposit, in the flowing stream, or both. Sloughing and spalling of materials may occur for a number of reasons: changes in dimensions of the deposit due to changes in temperature; death of a microbial deposit; diffusion of normal soluble species down the temperature gradient away from a heated surface; mechanical strength of the material. The essential aspect for sloughing is a change in the character of the deposit after its formation leading to an eventual weakening in its attachment.

Reentrainment of deposited material by the flowing fluid involves fluid mechanical forces and the mutual interaction between the elemental particles of the deposit.

The removal process is not nearly as well understood as the formation process. Understanding is important not only for fouling models, but also in the cleaning of fouled surfaces.

Aging

Aging of the deposit starts as soon as it has been laid down on the heat transfer surface. The aging process may include changes in crystal or chemical structure. Such changes, especially at constant heat flux and hence increasing deposit temperature, may strengthen the deposit with time. Changes in crystal structure or chemical degradation may result in gradually decreasing strength with time. Thus a deposit which is not hydrodynamically removable at the beginning of a run may suddenly become so after some times has passed.

PRECIPITATION FOULING

Precipitation fouling may be defined as the phenomenon of a solid layer deposition on a heat transfer surface, primarily as a result of the presence of dissolved inorganic salts in the flowing solution which exhibit supersaturation under the process conditions. Deposits formed under various conditions have different mechanical characteristics. The term "scaling" is generally used to describe a dense crystalline deposit, well bonded to the metal surface. It is often associated with the crystallization of salts of inverse solubilities under heat transfer conditions. When the deposited layer is porous and loosely adherent, it is described by terms such as "soft scale," "powdery deposit," or "sludge."

Precipitation fouling arises from the creation of supersaturation potential. Process conditions leading to supersaturation are:

(1) A solution is evaporated beyond the solubility limits of a dissolved salt.

(2) A solution containing a dissolved salt of normal solubility is cooled below its solubility temperature or a solution containing a dissolved salt of inverse solubility is heated above its solubility temperature. In both cases, there is a temperature gradient associated with the heat transfer operation. Thus, supersaturation conditions may exist with respect to the heat transfer surface without respect to the bulk temperature.

(3) Mixing of different streams may lead to the creation of supersaturated conditions.

The significant parameters determining the supersaturation potential responsible for the precipitation fouling in industrial processes and operations are as follows:

(1) composition of solution

(2) concentration effects occurring in the process

(3) temperature level of the process

(4) temperature gradient characterizing the heat transfer operation.

Industrial Systems

Precipitation fouling is encountered in industrial operations and processes with natural waters or aqueous solutions containing dissolved sc ling salts. With natural waters, some pretreatment is usually necessary to alleviate the problems of scaling as well as the problems of corrosion and microbiological fouling.

Cooling Water Systems: Water is used as a medium for heat rejection through heat transfer equipment. In once-through systems, there is a constant supply of potential scaling components, but deposit accumulation is relatively slower than in open recycle systems because no concentration effects are involved. Evaporation occurs in open recycle systems; the potential for scaling is magnified by the concentration effect and the necessity for pretreatment is a predominant consideration.

Steam Generating Systems: Precipitation fouling can be a major problem in the operation of steam generating equipment because of the combined influences of high temperature level, an appreciable concentration effect, a high temperature gradient, and the flow-boiling heat transfer mechanism. The water treatment system is an important part of the steam generation unit in order to alleviate the problems of deposit formation, reduce corrosion, and maintain steam purity. Scale formation may cause tube failure, particularly when corrosion also occurs.

Saline Water Distillation: The scaling problem is of particular significance in water desalination because the very low cost value of the fresh water product restricts treatment methods. Scale formation cannot be tolerated to any extent because of its effect on production rate and energy consumption. Current practice for all distillation methods relies on one of two approaches for controlling calcium carbonate:

(1) acid treatment for pH control. Principal disadvantage is that it requires close control because of the possibility of enhanced corrosion.

(2) polyelectrolyte treatment which acts to delay deposit formation and to deform crystal structure so that fouling is reduced. Additives are easier to handle than acid, but inhibitors lose their effectiveness as temperature increases.

No economical control method is available for calcium sulfate other than ensuring that the concentration-temperature process path is within the solubility confines of the various crystalline modification of calcium sulfate. The limiting top temperature at which seawater may be evaporated without calcium sulfate scale deposition is of major design significance.

Geothermal Brine: The search for new energy resource has focused in recent years on technological problems related to the development of geothermal energy supplies. Corrosion and scaling are among the difficult problems. Solutes in geothermal waters are of two general categories:

(1) soluble species that are minor contributors to scaling

(2) "equilibrium solubility" species, present in the brines at concentrations dictated by chemical-equilibria with the solid-rock phase. This is the primary cause of scaling.

The major contributor to scaling is silica. Supersaturation is created as a result of temperature drop and flashing of the brine. Calcium carbonate may also be a factor.

Potable Water Supply: The scaling problem in potable water supply lines is due mainly to calcium carbonate deposit accumulation on the pipe surface when the flowing water is supersaturated. Two situations are of practical interest:

(1) objectionable deposit formation that reduces the pipe carrying capacity

(2) desirable thin protective coatings which control corrosion in iron pipes.

Evaporation and Crystallization: Precipitation fouling is a problem in evaporation when the process fluid being concentrated contains dissolved components that acquire supersaturation conditions during the heat transfer operation. Two types of precipitating species may be distinguished: high solubility salts and sparingly soluble salts.

The former case is usually associated with the use of evaporators for crystallization objectives. Crystal growth on the heat transfer surface competes with the process of crystal growth on the greater deposition area of the suspended crystals. Supersaturation with respect to the heat transfer surface which has a higher temperature than the bulk solution is lower for normal solubility salts and higher for inverse solubility salts.

Where sparingly soluble salts such as calcium carbonate, calcium sulfate, and calcium oxalate are present, they are essentially contaminants that can form scale deposits. The supersaturation potential is caused by both concentration and temperature effects. The inverse solubility characteristic acts to favor deposition on the heat transfer surface. The contaminant may also be produced by a corrosion effect.

The most conducive situation for deposit accumulation is heat transfer under conditions where evaporation to dryness can occur locally. Forced circulation systems are the least conducive for deposit accumulation.

Examples of evaporation operations where precipitation fouling is significant include concentration of solutions of:

(1) phosphoric acid

(2) sodium sulfate

(3) sulfate waste liquor

(4) Kraft black liquor

(5) citrus waste liquors.

PARTICULATE FOULING

Particulate fouling may be defined as the accumulation of particles from liquid or gaseous suspensions onto heat transfer surfaces. Particulate fouling occurs in a wide range of situations. In liquid streams the best known example is probably that of corrosion products in boiler waters and reactor coolants. In gaseous stream a well known example is that of air-cooled heat exchangers as a result of dust particles deposited from the air to the cooler.

Particles in liquid and gaseous streams are common in both natural and process situations. When these are in turbulent flow in pipes and equipment they may deposit out as in particulate fouling of heat transfer surfaces. Particulate deposition fouling tends to increase asymptotically with time. Higher particle concentration and higher fluid/wall temperature increase foulant accumulation. The effect of fluid velocity is less clear, but generally the higher the velocity the lower the rate of fouling.

CHEMICAL REACTION FOULING

Deposits due to chemical reactions at a heat transfer surface will inhibit the transfer of heat. Two important examples are the polymerization of petroleum refinery feedstocks and the protein denaturalization that occurs in foodstuffs which come in contact with heated surfaces. Another situation of this type could be the chemical reactions that take place subsequent to the deposition of the foulant. This occurs, for example, in corrosion products particularly in the presence of boiling heat transfer.

Foodstuffs

Fouling of heat transfer surfaces by biological fluids is a complex phenomenon requiring an understanding of several fundamental processes:

(1) fouling: an adsorption phenomenon on surfaces

(2) fouling: a heterogeneous chemical reaction on heating surfaces

(3) fouling: a complex physiochemical process within the boundary layer of heating surfaces.

Adsorption onto surfaces represents the first determinant step occurring in fouling. Thermodynamically, fouling, as an adsorption phenomenon on surfaces, takes place spontaneously on every surface. The nature of a solid surface (surface free energy, strength, and distribution of active sites, roughness) can influence the history of fouling as well as the adhesion strength of the deposit. High-energy surfaces (or wettable surfaces), such as metal surfaces, are very susceptible to fouling and the adhesion strength increases with an increase in roughness. Low-energy surfaces (or unwettable surfaces), such as fluorocarbon surfaces, permit fouling also, but they exhibit larger induction periods and the adhesion strength is much smaller than for the majority of metal surfaces. Titanium apparently is an exception among the high-energy surfaces since it is considered wettable, but its behavior originates from the high resistance to corrosion (its surface presents less active sites). In flow systems, adhesion strength can be diminished by shear stresses favoring the "antifouling" properties of low-energy surfaces.

Mirror and polished surfaces do not always reduce fouling effects. In manufacturing processes, these surfaces receive an amorphous surface layer which is usually very active (electrolytically), with a large number of active sites. The adhesion strength of such surfaces is essentially the same effect within the range of finishes applied to such surfaces. The explanation is related to the comparable dimensions between the fouling components of biological fluids, especially proteins, and surface roughness. Consequently, the opinion that polished surfaces could prevent or reduce fouling must be reconsidered taking into account manufacturing costs and "antifouling efficiency."

The character of high-energy solid surfaces can be completely altered through depositions of very thin layers of low surface free-energy substances, surface coating. Industrial systems processing fluids with high fat content usually do not encounter fouling problems: this is a result of a surface coating with an "autophobic" layer contributed from the fluids themselves. Also fouling is a deposition process that alters the character of the metal surfaces; the resulting "fouled surfaces" maintain the general property of solid surfaces; they undergo adsorption from the adjacent solution.

Although fouled surfaces and surfaces of small particles in biological fluids have the character of solid surfaces, they have particularities associated with the superficial molecules that can extend into the solution. Liquid surfaces generated by boiling on heating surfaces and by dissolution of gases by heating biological fluids are potential adsorption surfaces which sometimes have severe effects in fouling. Fouling processes are determined by the nature of the adsorbed constituents of the biological fluids. Generally, all biomacromolecules tend to be strongly adsorbed by surfaces. Surface adsorption of macromolecules actually represents a first step in the reactions induced by heating surfaces. The macromolecules undergo profound unfolding and orientation changes, in comparison with their native state, and even reactions at surfaces, so generally the adsorption corresponds to a "surface denaturization."

Protein macromolecules build up the most stable "constructions" at surfaces due to their large molecular weight, flexibility, and large number of active sites. Adsorption of proteins on surfaces is therefore mostly an irreversible process. The conformational state of proteins on surfaces is determined by the surface and by the environmental conditions (pH and ionic strength).

Heterogeneous chemical reactions on heating surfaces contribute to the second determinant step occurring in fouling. In contrast with the bulk solution, the heating surface satisfies near ideal conditions for heat induced reactions: higher concentration in the adsorbed layer, higher reactivity of macromolecules through unfolding, and higher temperature. The principal types of reactions are those involving proteins and sugars/lipids. Inverse solubility salts also deposit on heating surfaces due to the higher surface temperature than in bulk solution.

Reactions at heating surfaces are very important from many aspects:

(1) they contribute to the transition of the deposition from the condition of "adsorbed layer" (revealing partly solid state, partly solution state, in the case of macromolecules) to the condition of "fouled surface."

(2) they are temperature-time dependent, thus producing a fouling rate.

(3) they adhere strongly onto surfaces, probably due to a low free sur-

face energy as well as the penetration into the interstices of the roughness.

The boundary layer of heating surfaces represent the place of surface adsorption as well as of heat-induced reactions. Its temperature, concentration, and velocity gradients introduce anisotropic conditions for the two complex processes taking place within it, and provide an explanation for the observed fouling dynamics (induction period, constant rate period, and falling-rate period).

CORROSION FOULING

Corrosion can produce deposits which will inhibit heat transfer. These deposits can originate from corrosion products which are transported from sites other than the fouled surface, alternatively the corrosion products can form on the heat transfer surface. The first case is particularly important in steam generators. Corrosion films can have serious consequences; impairment of heat transfer can lead to tube overheating and eventually to catastrophic failure. Sometime indigenous corrosion films are important; they can seriously impede heat transfer in superheater or reheater tubes, and cause excessive pressure drop of flow distribution problems in once-through boilers. Corrosion products carried into the boiler by the feedwater and deposited as particles contribute to the general fouling, particularly if they contain elements such as copper which destroy the porous structure of deposits and thereby interfere with boiling heat transfer.

In condensers, the corrosion films on the cooling water side of the tubes have important influences on corrosion mechanisms and fouling. In brass condensers, iron corrosion products may be introduced into the cooling water stream and induced to deposit on the tubes as a protective film. Also the growth of indigenous corrosion film interacts strongly with biological fouling. Thus, the copper oxide film on copper-bearing alloys inhibits fouling by marine organisms, presumably because of the toxicity of the solid oxide.

The three principal types of galvanic cells that take part in electrochemical corrosion reactions are:

(1) dissimilar-electrode cells, formed by immersion in a common electrolyte of dissimilar metals, or more frequently of similar metals with different degrees of surface irregularities or other imperfections.

(2) concentration cells, caused by bridging of industrial metals immersed in ionic solutions of different concentrations.

(3) differential-temperature cells, due to immersion in a common solution of identical metals at different temperatures.

All three types may be present in heat exchangers. Any object such as loose scale or other deposits resting on a metal surface can give rise to a concentration cell and hence to crevice corrosion, which is reported to be the most potentially damaging type of corrosion in a water cooling system. Corrosion products can nucleate crystallization from supersaturated solutions or subcooled melts, anchor suspended particles, serve as shelters for micro-organisms, and conceivably even catalyze certain reactions. Corrosion fouling is thus a potential promoter for all other types of fouling.

Integrated water treatment programs attempt to deal with both corrosion and fouling. One of the most effective corrosion inhibitors has been the chromate ion, but its extreme toxicity has necessitated the development of nonchromate treatments, for example the use of polyphosphates as both antiscalants and corrosion inhibitors. Polyphosphates, while they are among the least toxic of water treatment chemicals, have the disadvantage especially at higher temperatures of gradually decomposing to orthophosphates, which react chemically with certain cations to produce scale deposits. A large number of other additives (carboxylated polyelectrolytes) and combination of additives, as well as alternate treatment procedures, have been developed. It has been stated that the nonchromate additives, whose effectiveness depends on careful pH control, are better antifoulants than corrosion inhibitors.

BIOFOULING

Some 2,000 different species of organisms have been recorded as a marine fouling nuisance to some degree. Biofouling is usually distinguished between macro-organisms (such as clams, barnacles, cockles, and mussels) and micro-organisms (such as algae, fungi, and bacteria). The macro-organisms in their early life stages can pass through the water strainers in the intake to industrial plants and grow on warm heat exchanger surfaces. Some months later they can seriously impair the performance of these exchangers. The micro-organisms, especially bacteria, produce slime which adheres to heat transfer surfaces, thus adding to thermal resistance and pressure drop as well as entrapping silt or other suspended solids and giving rise to deposit differential corrosion. In addition, some anaerobic bacteria can produce hydrogen sulfide, which attacks steel. Biofouling may thus promote both particulate and corrosion fouling, the deposits from which may provide further velocity shelters for bio-organisms. Each class of bacteria has its own pH-range for optimum growth.

The original attachment of slimes to heat transfer surfaces is promoted by low fluid velocities, at least locally. Slime growth depends on a supply of oxygen, the transfer of which to the site of growth along with the transfer of other nutrients, may be the rate-controlling step. In that case, slime growth would be expected to increase with fluid velocity. If there is a removal or sloughing process, it too would be promoted by increased velocity which would counteract the deposition process to the point where fluid velocity increases would decrease the net fouling rate. Increased velocity may serve to compress the slime.

Slime growth is strongly temperature dependent. Temperatures below $0°$ to $20°C$ and above $40°$ to $70°C$ will kill most marine micro-organisms and the use of heat against both micro- and macro-organisms has in some cases proved to be an effective and economical fouling control method. The more common method is the use of either a biocide, which kills bio-organisms, or a biostat, which arrest their growth and reproduction. Chlorination, an example of biostat, remains the most frequently applied biofouling control method, although the fact that it is lethal to fish can limit its use in the case of once-through or open recirculation systems. Care must be taken that the chlorine residuals after chlorination do not exceed 0.5 ppm, above which waters become increasingly corrosive. Copper-bearing alloys (more than 60 to 70% copper) have built-in biocidal properties.

Microbial fouling of a surface is the net result of several physical, chemical, and microbial processes:

(1) transport and accumulation of material from the cooling water to the tube surface
(2) microbial growth within the film
(3) fluid shear stress at the surface of the film
(4) metal condenser material and roughness
(5) fouling control procedures.

Control of microbial results in control of macrobial fouling.

SOLIDIFICATION FOULING

Solid deposits can build up on heat transfer surfaces where the temperature is lower than the freezing point of the fluid or one of its constituents. The overall fouling problem occurs in three areas:

(1) cooling and subsequent fouling associated with pure substances such as water
(2) cooling of mixtures of related substances, such as paraffins
(3) cooling of mixtures such as humid air.

Several factors affect freezing fouling:

(1) flow rate
(2) crystallization and temperature
(3) nature of the surface
(4) concentration of the solid precursor.

There is no way of eliminating the fouling problem in respect to freezing pure substances unless the temperature conditions can be changed to avoid solidification temperatures on the surface. Additives can sometimes be used for mixtures. Scraped surface heat exchangers are also used to reduce the effects of freezing fouling.

FOULING IN EVAPORATION

Boiling normally gives rise to higher rates of scaling than sensible heating because of the higher temperature and heat fluxes involved, because of the concentration effect due to evaporation, and because of the higher mass fluxes of liquid to the heat transfer surface especially in the case of nucleate boiling. The principal scale-formers are the hardness scales in process evaporators and metal-oxide scale, especially magnetite, in steam generators.

Water-soluble scales and calcium carbonate account for many evaporator scaling problems. Calcium scales have a much greater impact on evaporator capacity and pose the more serious problem. The rate of calcium scaling is very dependent on temperature and is also directly proportional to solids content. Calcium scales can be controlled by minimizing calcium inputs and, to a great extent, through operation techniques. Special treatments are also sometimes effective.

Water-soluble scales can generally be easily removed by water washing. Higher fluid velocity helps in controlling scale formation.

Optimum Cleaning Cycles

Periodic shutdown and cleaning of a heat exchanger is often required due to its declining production rate with time. An evaporator undergoing scaling consequently delivers a steadily decreasing concentrated liquor rate of the required concentration; a sensible heat exchanger undergoing fouling delivers a steadily decreasing rate of product heated or cooled through the required temperature range. The problem of determining when to shut down for cleaning can be formulated in terms of either a maximum throughput or a minimum product cost cycle. In either case, the cleaning time is often constant and independent of the cycle throughput. Simple graphical procedures can be developed for both cases.

The quantity of scale deposited in an evaporator at any time is proportional to the quantity of liquid which has been evaporated and therefore to the heat which has been transferred up to that time. Evaporators normally operate with constant temperature difference.

In many cases it is more reasonable to assume that cleaning time is a weak linear function of production cycle. Thus, in general:

$$\theta_K = \theta_{cc} + aP \quad (12.1)$$

where θ_K = cleaning time
θ_{cc} = fixed unreducible portion of cleaning time
a = constant
P = cumulative product, at time θ

$$\theta_{cv} = aP \quad (12.2)$$

The criterion for the maximum throughput cycle is that R is a maximum.

$$R = P/(\theta + \theta_K) \quad (12.3)$$

The maximum throughput cycle is one in which the tangent from $\theta = -\theta_{cc}$ (at P = 0) to the curve of P versus θ touches the curve at (θ_{opt}, P_{opt}). The total cleaning time per cycle is $\theta_{cc} + aP_{opt}$. The graphical solution is shown in Figure 12-1.

The corresponding minimum cost problem can be formulated in a similar manner.

$$\theta_{cc}' = \theta_{cc}(H + K)/(H + F) \qquad (12.4)$$

where θ_{cc} = fixed unreducible portion of cleaning time
H = fixed overheat costs per unit time
K = variable (labor) costs per unit time during cleaning
F = variable (labor) costs per unit time during production

The term θ_{cc}' is a cost-modified value of the cleaning time constant, θ_{cc}, which becomes equal to θ_{cc} when K = F, but is commonly greater than θ_{cc}. The optimum throughput for a minimum cost cycle is such that the tangent from $\theta = -\theta_{cc}'$ (at P = 0) to the curve of P versus θ touches the curve at (θ_{opt}', P_{opt}'). Actual cleaning time is then $\theta_{cc} + aP_{opt}'$. The graphical solution is shown in Figure 12-2.

In both Figures 12-1 and 12-2 an alternative procedure is shown when the tangent is drawn to the curve of P versus ($\theta + \theta_{cv}$) instead of P versus θ. The virtue of these simple graphical methods is that they do not require a mathematical equation relating P and θ, the curve of which can be based directly on operating data. The minimum cost procedure can be modified for a variable rather than constant energy cost per unit mass of product.

DESIGN CONSIDERATIONS

If the heat transfer coefficients for an exchanger are high, the design may be "fouling limited." Designs in such cases must avoid areas of low flow velocity on the fouling side(s) of the exchanger. Parallel exchangers require well balanced flows. When a mechanical cleaning method is adopted, access to the fouled surface is important.

For augmented heat transfer surfaces, it is useful to distinguish between enhancement of a plain surface in order to increase the heat transfer coefficient and extension or enlargement of a plain surface in order to increase the heat transfer area. In some cases, both effects occur simultaneously, but often one or the other predominates. Turbulators, for example, enhance, while conventional fins extend. For a given degree of uniform fouling, an enhanced surface suffers a larger heat transfer penalty than a plain surface, whereas an extended surface experiences a smaller penalty than a plain surface. In many cases, extended surfaces facilitate cleaning procedures.

FOULING: PHILOSOPHY OF DESIGN

A fouling factor is incorporated in the thermal design of most heat transfer equipment. Selection of fouling factors must be given proper emphasis because the fouling factor used can greatly increase the cost of a given heat exchanger. In all too many cases, the design determines the fouling factor; the fouling factor does not determine the design! Hence, "you get what you expect" is in many cases true. Design for a high fouling factor and it will be achieved; design

128 Handbook of Evaporation Technology

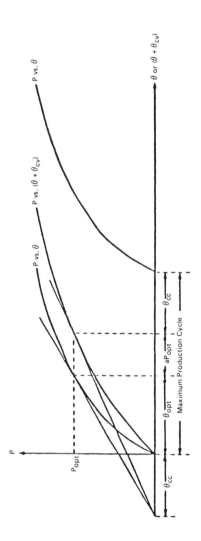

Figure 12-1: Maximum throughput cycle for cleaning time θ_K which is a linear function of P. $\theta_K = \theta_{cc} + \theta_{cv}$, $\theta_{cv} = aP$.

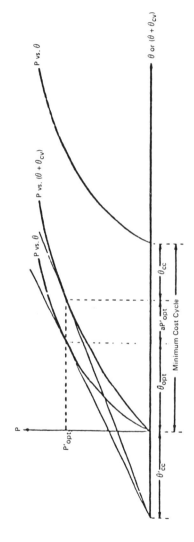

Figure 12-2: Minimum cost cycle for cleaning time θ_K which is a linear function of P. $\theta_K = \theta_{cc} - \theta_{cv}$, $\theta_{cv} = aP$.

for a low fouling factor and it will also be achieved. Reasons for this will be discussed later.

For services in which fouling is a real problem, tubular equipment may not be the proper choice. Fouling can often be reduced or eliminated by using other types of equipment: spiral plate, helical coils, gasketed plate, scraped surface, falling films, and others. In some calandria services, fouling can be reduced by using forced circulation or falling film units rather than natural circulation. Special coatings may also be applied to reduce fouling. Antifoulants may be effective in some services.

Fouling in shell-and-tube exchangers can be reduced by attention to many design details. Some of these are discussed below.

Tubeside Velocities

Tubeside velocities should be sufficient to provide reasonable distribution and to reduce fouling. When a minimum velocity has not been established, tubeside velocities should be as high as pressure drop will permit but should not be so high as to result in tube erosion. In general, increased velocities decrease the rate of fouling.

Recommended velocity ranges for various services are given below:

Service	Velocity Range
Cooling water	4-8 feet per second
Forced circulation calandrias	10-15 feet per second
Sensible heat, liquids	3-10 feet per second

When cooling with seawater, the dimensions of inlet and outlet heads are important to avoid fouling and erosion. Recommended dimensions are given in Figure 12-3. Seawater should be placed on the tubeside.

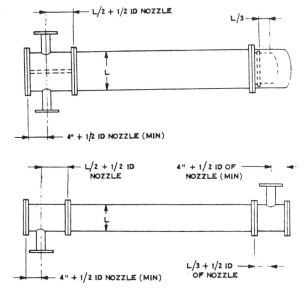

Figure 12-3: Recommended dimensions of inlet and outlet heads.

130 Handbook of Evaporation Technology

Shellside Design

The most important aspect of shell-and-tube exchanger design is to provide a proper shellside flow pattern. Clearances necessarily have to be provided between the baffles and the shell and between the baffles and the tubes. These clearances should be minimized in order to reduce fluid bypassing. The tubes do not completely fill the shell, and sealing devices must be employed to reduce fluid bypassing through the area represented by this "clearance." Multipass tubeside exchangers are often oriented such that spacing for the tubeside pass partitions creates a fluid bypass lane parallel to the direction of flow. Such lanes must be sealed or blocked to reduce the fluid bypassing. Typical bypass areas and sealing devices are illustrated in Figure 12-4.

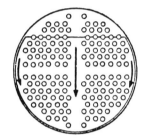

Tube layout with large bypass lanes

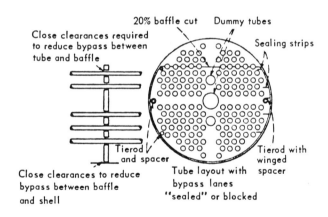

Figure 12-4: Tie rods and spacers placed to reduce bypassing.

If proper attention is not given to preventing fluid bypassing through these various flow areas, several things may happen. First, the heat transfer coefficient is reduced because only part of the fluid passes across the tubes. In the case of crossflow bundles, the available temperature difference is also reduced because only part of the shellside fluid is being cooled (or heated) and the uncooled (or unheated) bypass stream is not adequately mixed with the "working" fluid until it reaches the outlet nozzle. In such cases, the exchanger may appear to be "fouled" when in reality it is performing adequately when bypassing is properly taken into account.

Second, the velocities are also reduced because of bypassing. The reduced velocities may result in excessive fouling from solids deposition or from high tube-wall temperatures. Third, venting of noncondensables may not be properly achieved with resultant reduction of condensation heat transfer rates.

Another important aspect of the shellside design is the baffle cut. With crossflow baffles, the baffle cut should be 20% of the shell diameter. Larger baffle cuts result in areas of the bundle having little or no flow across them. These relatively stagnant areas permit solids deposition (silting). Because heat transfer is reduced at the low flows, tube-wall temperatures are high, which may result in scale formation, localized boiling, and excessive corrosion from the scale and deposits in these stagnant areas.

Shellside Velocities

If minimum velocities have not been established, velocities should be as high as pressure drop will permit. In general, increased velocities decrease the rate of fouling. When cooling water is on the shellside, the maximum velocity at the centerline of the bundle should be in the range of 2 to 5 feet per second. Baffles should have horizontal cuts; vertically cut baffles result in silting and improper venting. In many cases, omitting tubes from the baffle windows will eliminate many fouling and corrosion problems resulting from areas of stagnant water. Shellside velocities can frequently be increased for the same pressure drop by increasing the spacing between tubes. More baffles will be required, but higher velocities will result.

Rod baffles have been reported to reduce fouling of shellside fluids in some applications. The apparent reason is the vortices created downstream of the rods and the elimination of stagnant flow areas.

Tube Wall Temperatures

Scale formation may be aggravated with excessive tube-wall temperatures. Most cooling waters also scale at tube-wall temperatures above $55°C$.

Conclusions

Now, back to a statement made earlier: the fouling factor used for design is the fouling factor achieved. If a large fouling factor is incorporated in the design and shellside design criteria are properly followed, the exchanger will be much larger than required when started up. To achieve control, the shellside fluid may be throttled or bypassed. This results in lower velocities which may be too low for proper operation. This is especially true when one of the fluids is cooling water and the amount of cooling must be controlled. The reduced water velocity may result in fouling until the design fouling factor is realized. After taking some test data, the designer prides himself for selecting the correct fouling factor and uses that value from then on. He would have been much wiser to design for a lower fouling factor which would have required proper velocities and would not have resulted in conditions which permitted his "correct" fouling factor to be developed.

It should be apparent that you cannot tell someone what fouling factor to use for design without telling him how to design the unit. Clearances, velocities, tube-wall temperatures, and other factors influence the selection of fouling factor. They must, therefore, be as carefully considered as allowable pressure drop and temperature approaches. Proper attention to these design details will not only result in lower equipment cost, but will also reduce maintenance costs by reducing fouling.

It may be preferable to specify the amount of overdesign required for a specific service, rather than specify fouling factors. All too often fouling factors really serve as "fudge" or "ignorance" factors. A well-conceived set of criteria for equipment design will avoid many problems associated with fouling.

13
Evaporator Performance

The primary measures of the performance of an evaporator are the capacity and the economy. Capacity is defined as the amount evaporated per hour. Economy is the ratio of the mass evaporated to the mass of heat transfer fluid (often steam) used. When evaporating water with steam, the economy is nearly always less than 1.0 for single-effect evaporators; but in multiple-effect evaporators and vapor compression systems it is considerably greater.

Other performance variables of interest include product losses and decrease in performance as scaling, salting, or fouling proceed.

Designers of evaporator systems strive to achieve high heat transfer rates. This can be justified only if the cost for achieving high rates is less than the benefits enjoyed. High rates of heat transfer in theory must often be proved in practice. Evaporators designed for high rates of heat transfer are generally more affected by traces of scale or noncondensable gases.

Heat transfer rates in evaporators are often a function of the available temperature difference. Often it is difficult to determine the temperature of the liquid at all locations along the heating surface. The vapor pressure of most aqueous solutions is less than that of water at the same temperature. Consequently for a determined pressure the boiling point of the solution is higher than that for water. This increase in boiling point over that for water is known as the boiling-point elevation of the solution. Boiling-point elevations are slight for dilute solutions but may be large for concentrated solutions. Data on boiling-point elevations for many commonly encountered solutions is available or can be easily estimated. The effect of any hydrostatic head or pressure drop must also be considered when determining temperature differences.

VENTING

The condensation of vapors is frequently impeded by the presence of noncondensable gases. This is especially true in condensers operating under vacuum. Evaporators, in particular, are susceptible to noncondensable problems because

134 Handbook of Evaporation Technology

they frequently operate under vacuum and because the solution being evaporated usually contains dissolved gases, may contain entrained gases, and may liberate other gases on being concentrated. Although air is the usual contaminant, other gases, as long as they do not dissolve easily in the condensate, affect heat transfer in almost the same manner as air.

Two problems are involved in the design of condensers for vapors containing noncondensable gases:

(1) effect of noncondensables on heat transfer rates
(2) physical arrangement of the heating surface to properly move the vapors past the heating surface and the condensables out the vent.

Effect of Noncondensables on Heat Transfer

Noncondensable gases reduce heat transfer rates in two methods:

(1) the heat transfer coefficient is reduced because heat must be transferred across a gas film
(2) if the noncondensable concentration is significant, the available temperature difference is reduced because the condensation is no longer isothermal.

The reduction of heat transfer coefficient can be treated as an empirical "fouling factor." Data published indicates that the apparent fouling factor can be correlated for condensing steam in the presence of noncondensables as below:

$$h_f = 4350/C \qquad (13.1)$$

where h_f = apparent fouling coefficient for noncondensables Btu/(hr)(sq ft)(°F)
C = concentration of air, weight percent

The fouling factor would be $1/h_f$.

Design Precautions

Several precautions must be exercised when designing condensers in which noncondensable gases (of any concentration) are present:

(1) Each condenser should be vented separately to an acceptable vent system. If a vent header is provided, each vent line must be separately controlled.
(2) For multiple-effect evaporators, the vent should be routed to the condenser. It is not generally good practice to vent an earlier effect to a later effect.
(3) There must be a properly defined flow path for the vapor between the inlet and the vent outlet.
(4) The vapor flow path preferably should result in vapor velocities as constant as possible.

(5) The vents must be at the end of the vapor flow path regardless of where this is located.

(6) Possible variations in heat transfer conditions on the cooling side should be considered.

Venting is especially important for a vapor compression evaporator because reduced evaporator capacity can result in compressor surge.

Proper venting arrangements serve not only to improve heat transfer but also to reduce corrosion. High concentrations of noncondensables result in some redissolving in the condensate which accelerates corrosion either in the evaporator or of the condensate return system.

Enhanced heat transfer surfaces must be provided with adequate venting in order to fully achieve the benefits of the enhancement. Higher rates of heat transfer suffer greater loss of efficiency for the same inert load.

Tubeside Condensation

Equipment with tubeside condensation are more easily vented, provided condensate is not permitted to flood the heat transfer surface. Horizontal units may experience maldistribution or higher rates of condensing at the top of the unit than at the bottom of the unit. Multipass tube arrangements help to avoid this problem. Condensate loading in the tubes must not be so high as to prevent proper venting. This is discussed in Chapter 22.

Shellside Condensation

Venting of shellside condensers is more difficult. Several methods are available. The best method depends upon the size of the condenser and the method of condensate removal.

Small Units: Small shell sizes (less than 20 inches in diameter) can usually be adequately vented by proper baffling. The baffling forces the noncondensables to the bottom condensate outlet where they are removed with the condensate. If steam traps are used, they must be adequate for the expected noncondensable load. Large noncondensable loads may not be adequately handled by traps; a condensate pot with a liquid level control and separate vent may be required in such cases. It's good practice to decrease the baffle spacing as the vapor flow decreases in order to maintain adequate vapor velocity to sweep noncondensables to the outlet. Figure 13-1 indicates proper baffling for venting of small units. A tubesheet vent is also required as shown in Figure 13-2.

The method described above is not adequate for situations in which condensate levels are expected on the shellside. Operation in this manner is not recommended as discussed in Chapter 31. If a level must be maintained in the shell, venting can be accomplished as described below.

Intermediate Units: Shell sizes between 20 and 50 inches in diameter can be vented as shown in Figure 13-3. The vent header must be designed as a manifold in order to vent uniformly along the shell length. Total cross-sectional area of the holes should be no more than half that of the half-pipe header. Half-pipe header area must be at least equal to that of the vent nozzle. Perforations should be equally spaced no greater than 3 inches apart. One perforation should be

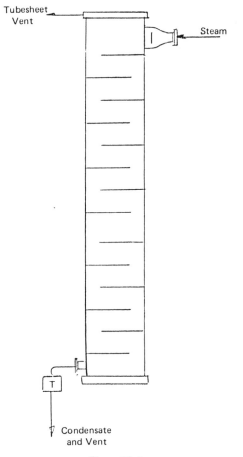

Figure 13-1.

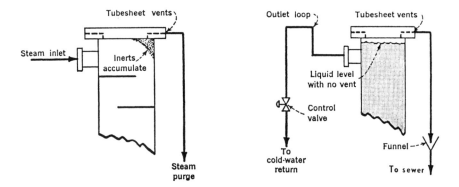

Figure 13-2: Tubesheet vents.

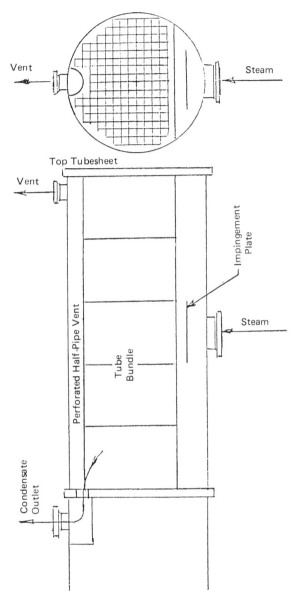

Figure 13-3.

located at the bottom of each tube support. Pressure drop can be conservatively estimated as 2 velocity heads, based on velocity through the holes.

Vapor velocity is not constant in this design. Consequently the design recommended for large units (described below) can often be justified for intermediate units.

Large Units: Shell sizes greater than 50 inches in diameter should be vented as shown in Figure 13-4. The tube bundle is often not concentric with the shell. This provides better vapor distribution. Venting is accomplished through a perforated pipe extending the length of the shell. The perforated pipe should also be designed as a manifold in order to uniformly vent along the shell length.

Condensate connections for large units can easily be provided through the bottom tubesheet. This permits maximum use of heat transfer surface by maintaining liquid level necessary for condensate removal.

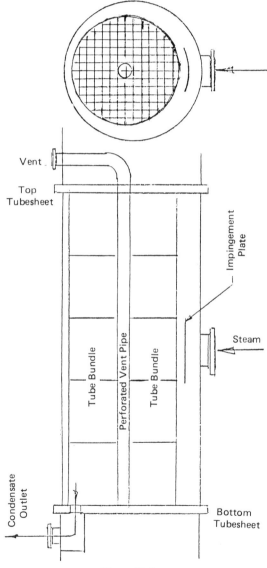

Figure 13-4.

Water-Cooled Equipment

For vertical exchangers with cooling water on the shellside, it is essential to provide means for venting gases that are released as the water is heated. If these gases are not continuously removed, they collect in pockets; the shellside heat transfer is reduced; and corrosive attack may occur in the gaseous region. In vertical units the gases tend to collect just beneath the top tubesheet and the tubes in this area corrode rapidly when not surrounded by liquid. Corrosion and stress cracking may occur in this area and solids may also build up on the tubes. Vent connections should be located at positions which enable these gases to be vented and to insure that the tubesheet is swept with water. (See Figure 13-2.) Sometimes tubes in this area are "safe-ended:" a section of tubing not corroded or stress cracked is welded to the process tube with the resulting safe-end placed in the unvented section of the exchanger. Safe-ending is relatively expensive.

Provisions must be provided for checking the vent flow. Sight cones or funnels are the most reliable method. However, these require an adequate receptable vehicle for the vent flow and may represent a load on wastewater treatment units. Magnetic flowmeters can be used but are not entirely reliable. If the water side operates under a vacuum or siphon condition, vent lines may need to be provided with some means of priming if vent flow is stopped for any reason.

TIME/TEMPERATURE RELATION

To prevent degradation of heat-sensitive materials during processing it has long been recognized that a low operating temperature and a short residence time are essential. This would normally exclude the recirculation type evaporators from such services except where high concentrations must be achieved. An average residence time expressed as holding volume divided by discharge rate was frequently used in the past for both single-pass and recirculation evaporators. However, statistical analysis of several types of evaporators has revealed that the actual time of replacement of 97% of the feed in a recirculating evaporator is about 3.2 times the average residence time as defined above. It takes longer to replace a larger fraction of feed.

The actual residence time achieved in any evaporator can be calculated from the equation below:

$$x = 1 - e^{-\theta/r} \qquad (13.2)$$

where x = fraction of feed removed
 θ = time
 r = ratio of holding volume to discharge rate (time)

Experience has shown that the time factor is more important than the temperature level in the time/temperature relation, and so the earlier emphasis on maintaining low operating temperatures in evaporators has shifted towards accurate control of low residence times and relatively higher operating temperatures. This has resulted not only in better products but also lower investment and operating costs.

PRESSURE VERSUS VACUUM OPERATION

The driving force in evaporation is the temperature difference between the heating medium and the evaporating liquid. Thus there is no advantage in maintaining a vacuum on the liquid if temperature difference can be provided by increasing the temperature of the heating medium. An advantage of pressure operation is that boiling heat transfer rates are higher. Cooling water may not be required if the vapor from the last effect can be used for heat energy. Also the problems and costs associated with vacuum producing equipment can be eliminated. Mechanical compression systems generally are more economical if operating pressures are above atmospheric. Operation at higher pressures (higher temperatures) in many cases is acceptable if adequate attention is given to residence time.

ENERGY ECONOMY

The single largest cost of evaporation is the cost of energy. Single-effect evaporators have limited application: batch systems; small size units; those that must operate at low temperatures to avoid product degradation; those that require expensive evaporator construction. Energy efficiency can be improved by using multiple effects, by using heat pumps, by heat exchange, and by using energy efficient condensers.

Energy can never be used—the first law of thermodynamics guarantees its conservation. When we normally speak of "energy use" what we really mean is to lower the level at which energy is available. Energy has a value that falls sharply with level. Accounting systems need to recognize this fact in order to properly allocate the use of energy level.

The best way to conserve energy is not to "use" it in the first place. Of course, this is the goal of every process engineer when he evaluates a process. But once the best system, from an energy point of view, has been selected, then the necessary energy should be used to the best advantage. The most efficient use of heat is by the transfer of the heat through a heat exchanger with process-oriented heat utilization or by generation of steam at sufficient levels to permit it to be used in the process plant directly as heat. When heat is available only at levels too low to permit recovery in the process directly, thermal engine cycles may be used for energy recovery. Heat pumps may also be used to "pump" energy from a lower to a higher level, enabling "waste" heat to be recovered through process utilization.

Thermal efficiencies of heat exchangers are high: 90 to 95%. Thermal efficiencies of thermal engine cycles are low: 10 to 20%. Heat pumps permit external energy to be reduced by a factor of 4 to 25; however, the energy required in a heat pump is in the work form, the most expensive energy form.

Utility consumption, of course, is one of the major factors which determine operating cost and, hence, the cost of producing the product for which a plant has been designed. In order to select the proper equipment for a specific application it is important to be able to evaluate different alternatives which may result in a reduction of utility usage or enable the use of a less costly utility. For example, the choice of an air-cooled condenser versus a water-cooled condenser can be made only after evaluating both equipment costs and the costs of cooling water and horsepower.

Steam

When heating with steam, a selection of the proper steam pressure level must be made when designing the unit. No definite rules for such selection can be established because of changing plant steam balances and availability. However, it is generally more economical to select the lowest available steam pressure level which offers a saturation temperature above the process temperature required. Some evaporator types require relatively low temperature differences. Some products may require low temperatures in order to reduce fouling or product degradation.

Superheated steam in general presents no design problems as calandrias are well baffled to serve as a self-desuperheater and are vented to prevent gas blanketing from any inert gases present in the steam. Any superheat is quickly removed; the mechanism is primarily one of condensing and re-evaporation, maintaining tubewall temperatures essentially the same as that experienced when condensing saturated steam. Desuperheating is accomplished at the expense of the temperature difference available. The mechanical design of the equipment must be adequate for the temperature conditions that can be imposed by the superheat in the steam.

Cooling Water

Maximum outlet temperatures for cooling water are usually dictated by the chemistry of the cooling water. Most cooling waters contain chlorides and carbonates; consequently heat transfer surfaces (tube wall) temperatures must not exceed certain values in order to minimize formation of calcareous deposits, or scaling, which reduces heat transfer and leads to excessive corrosion. In addition, velocity restriction must be imposed and observed to prevent corrosion and fouling as a result of sedimentation and poor venting. Stagnant conditions on the water side must *always* be avoided. In some plants water consumption is dictated by thermal pollution restrictions.

Do not impose unnecessary restraints on the pressure drops permitted across the water side of condensers. All too often, specified design values for pressure drop are too low and much higher values are realized when the unit has been installed and is operating. Not only does this result in more expensive equipment, but frequently no monitoring of the water flow is undertaken and cooling water consumption is excessive, increasing operating costs.

Tempered Water Systems

Because cooling water consumption is governed by factors other than energy conservation and because cooling water velocities must be maintained above certain values, tempered water systems can be effectively used at locations where cooling water temperatures vary with the season of the year. At some locations a 30°C difference between summer and winter water temperatures is experienced. At such locations a tempered water system may be used in order to reduce both pumping costs and maintenance costs. A tempered water system requires a pump to recycle part of the heated cooling water in order to maintain a constant inlet water temperature. Figure 13-5 presents a schematic diagram of such a tempered water system. Annual pumping costs may be reduced by as much as 25% with such a system.

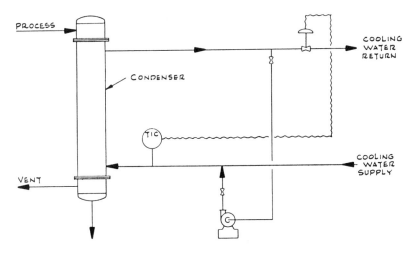

Figure 13-5: Tempered water system.

Pumping costs for such a system are reduced because the pump has to develop only the pressure loss across the heat exchanger, which is usually much less than the pressure difference across the supply and return water headers.

Tempered water systems have the disadvantage that a pump is required for each condenser in order to realize all the potential energy savings. However, the consequences of losing a pump are not great because the heat transfer can be met at the expense of increased pumping costs. Also, there are several months of the year when the tempered water pumps are not needed. This period can be used for pump maintenance. Tempering all the water to a unit in order to reduce the number of pumps can be done, but the power savings is greatly reduced because the pumps must usually develop a greater head. However, maintenance costs will be reduced as a result of proper velocity control.

Heat Exchange

Energy economy can be improved by properly exchanging heat between entering cold streams and leaving warm streams. Ideally, all streams should enter and leave at the same temperature; obviously this cannot be achieved. However, heat exchange will permit this goal to be more closely approached.

Usable heat may be present in the product and/or condensate leaving the system. This can frequently be used to preheat feed. Several heat exchangers are available for such application:

(1) shell-and-tube

(2) spiral tube

(3) spiral plate

(4) gasketed plate.

The gasketed plate exchanger is highly efficient and can be used in a cost-effective manner. Gasketed plates should be considered when heat exchange opportunities exist.

Evaporative-Cooled Condensers

Evaporative-cooled condensers in many applications give greater heat transfer than air-cooled or water-cooled condensers. The evaporative equipment can do this by offering a lower temperature sink.

Evaporative-cooled condensers are frequently called wet-surface air-coolers. Perhaps the best description for this type of equipment is a combination shell-and-tube exchanger and cooling tower built into a single package. The tube surfaces are cooled by evaporation of water into air. Figure 13-6 shows a typical evaporative-cooled unit.

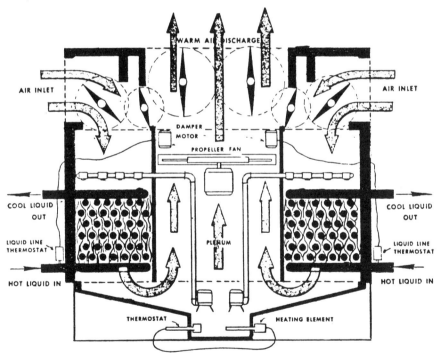

Figure 13-6: Evaporative-cooled exchanger.

A recirculating pump draws water from the basin under the unit and pumps it through a system of sprays from which the water is directed over the tube surfaces. Air is induced or forced over the wetted tube surfaces and through the rain of water droplets. The unsaturated air intimately contacts the wetted tube surfaces and water droplets and evaporation of part of the water occurs, thus cooling both the tube surfaces and the water simultaneously. Evaporation is used to increase the rate of heat transfer from the tubes. Extended surfaces are not required.

The evaporative-cooled condenser is basically a water-saving device. From a heat transfer standpoint, the performance of the evaporative-cooler is influenced only by the wet-bulb temperature and the density of the air entering the cooler.

General acceptance of the wet air-cooled heat-transfer surface has been deterred mainly by concern over the rapid deterioration of wet metallic surfaces exposed to air, particularly in a chemical plant environment. Two other problems must also be considered: control and protection from freezing.

Extensive experience has proved that a bundle of smooth tubes continuously flooded by massive quantities of water will not develop encrustations provided the water is properly treated and the surface temperature does not exceed 65°C. A continuous water blanket over the tubular surfaces prevents precipitation of dissolved minerals. Cocurrent flow of water and air provides better operation.

Treatment of the spray water is comparable to that of cooling tower water. Chromate treatment is not required; polysulfate treatments are adequate. Only a small amount of water is required to maintain the evaporative cooling cycle. The makeup water rate is determined by the amount of water that is lost due to evaporation, blowdown, and entrainment. Higher cycles of concentration can be achieved with evaporative-cooled equipment than with cooling towers.

Control or capacity modulation is provided by varying the amount of air pumped or by a partial return of warm, moist, exhaust air. In the latter process the humidity of the system is in effect regulated to permit process control. Varying the amount of air pumped is generally preferred. This generally requires speed control or stepped operation of multiple fans because controllable-pitch fans are generally not recommended for the wet air service.

Protection from freezing is obtained by varying the air flows in the inlet and exhaust openings in relation to the air recycle. If these openings are completely closed and the fan is stopped, the spray water is heated. Thus, a blanket of heated water continuously covers the tubes. The water sump can also be heated when shutdown.

Perhaps the greatest advantage of the evaporative cooler is a lower temperature heat sink (the wet-bulb temperature of the air). With some overlap in operating temperatures, the evaporative cooler has application where the air-cooled equipment must drop out of the picture. The evaporative cooler has limited use for process temperatures above 75°C because high tube-wall temperatures result in rapid buildup of scale on the tubes. In most cases, the evaporative-cooled condenser has application when process temperatures are between 75°C and 5°C above the design wet-bulb temperature of the ambient air.

Power consumption and space requirements are generally less than for an air-cooled or a cooling tower, cycle-water system. When compared to an open cycle, the water requirements are less than 5% of a sensible heat system; this results in smaller pumps and lower piping costs. The water side of evaporative-cooled equipment can be easily inspected and cleaned (if necessary) while the unit is in service.

Air-Cooled Condensers

Air-cooled condensers are especially attactive at locations where water is scarce or expensive to treat. Even when water is plentiful, air coolers are frequently the most economical selection. Elimination of the problems associated with the water side of water-cooled equipment, such as fouling, stress-corrosion cracking, and water leaks into the process, is an outstanding advantage of air-cooled equipment. In many cases, carbon-steel tubes can be employed in air-cooled condensers when more expensive alloy tubes would otherwise have been necessary. The use of air cooling may eliminate the need for additional investment in plant cooling-water facilities.

Maintenance costs for air-cooled equipment are usually much less than that for water-cooled equipment. Power requirements for air coolers vary throughout the day and the year if the amount of air pumped is controlled. Water rates can be varied to a lesser degree because daily water temperatures are more constant and because water velocities must be kept high to reduce maintenance.

The initial investment for an air-cooled condenser is generally higher than that for a water-cooled unit. However, operating costs (and maintenance costs) are usually considerably less. These factors must be considered when selecting water or air as the cooling medium.

Air-cooled condensers employ axial-flow fans to force or induce a flow of ambient air across a bank of externally finned tubes. Finned tubes are used because air is a poor heat transfer fluid; the extended surface enables air to be used economically. Several types of finned-tube construction are available; the most common types are extruded bimetallic fintubes and footed tension-wound fintubes. The most common fin material is aluminum.

A typical air-cooled condenser has a bank of finned tubes, a steel supporting structure with plenum chambers and fan rings, axial-flow fans, drive assemblies (usually electric motors with cog or V-belt drive), and miscellaneous accessories, such as vibration switches, fan guards, fencing, hail screens, and louvers. Figure 13-7 shows a typical air cooler. These components may be arranged in any of several ways to best suit a particular application and equipment layout.

Air-cooled equipment generally requires more space than other types. However, they can be located in areas that otherwise would not be used (e.g., atop pipe racks).

A forced-draft unit has a fan below the tube bundle which pushes air across the finned tubes. An induced-draft unit has a fan above the tube bundle which pulls air across the finned tubes. Theoretically, a forced-draft unit requires less horsepower because air is pumped at the highest available density (lowest available temperature). However, in practice induced-draft units require no more power than forced-draft units and in many cases actually require less power.

An induced-draft unit provides more uniform air distribution across the tube bundle since the air velocity approaching the tube bundle is relatively low. The better distribution also results in better heat transfer and lower pressure drop across the first row of tubes. Induced-draft units are less likely to recirculate the relatively hot exhaust air because the exhaust velocity is several times that of a forced-draft unit. For these reasons, less heat transfer area (and less pressure drop) is required in an induced-draft unit. This more than offsets any possible power savings (or structural cost savings) that may result from forced-

146 Handbook of Evaporation Technology

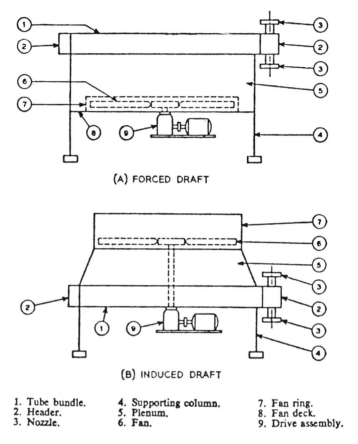

Figure 13-7: Typical components of an air-cooled heat exchanger.

1. Tube bundle.
2. Header.
3. Nozzle.
4. Supporting column.
5. Plenum.
6. Fan.
7. Fan ring.
8. Fan deck.
9. Drive assembly.

draft construction. Induced-draft units provide better protection from the elements and more precise control when sudden temperature changes occur. All these reasons contribute to make induced-draft units the best selection for most applications.

Air-cooled condensers are normally controlled by using controllable-pitch fans. Good air distribution is achieved if at least 40% of the face area of the bundle is covered with fans. Controllable-pitch fans permit only the air required for heat transfer to be pumped. An important added advantage is the reduction of the power required for operation when ambient air temperature is lower than that used for design. Controllable-pitch fans can result in a 50% reduction in the annual power consumption over fixed-blade fans. However, power savings can be effected only upon proper use during operation. Energy savings using controllable-pitch fans is discussed in Chapter 31.

Air-cooled condensers may be designed without fans. Natural-draft units employ a stack or chimney to create the head necessary to move the air. Natural-draft units require more heat transfer area because achievable air velocities are low relative to those achieved with fan driven units. Natural-draft units should be considered at high temperature levels, especially when the exhaust air temperature exceeds 100°C. Natural-draft units save power and increase reliability, since there is no mechanical equipment.

Air-cooled condensers for cold climates may require protection from freezing. Louvers, fans, motors, and drive systems should be top quality with adequate safety factors. Several methods are used to prevent freezing in cold climates:

(1) Induced draft with internal recirculation
(2) Forced draft with internal recirculation
(3) Forced draft with top internal recirculation
(4) Forced draft with external recirculation.

Pumping Systems

There are many ways to squander energy in pumping systems. As energy costs have continued to climb, it has often been found that a complete pumping unit's initial investment can be less than the equivalent investment value of one electrical horsepower. A discussion of energy savings for pumping systems is presented in Chapter 25.

Calandria circulating pumps require a certain available NPSH. This is usually obtained by elevating the evaporator, often with a skirt. Quite often the designer establishes the skirt height before he selects the calandria circulating pump. In the interest of economy he provides a skirt as short as possible, often without realizing that he will be forever paying an energy penalty for a smaller initial capital savings. More efficient pumps often require greater NPSH. Therefore, it's prudent to check the NPSH requirements of pumping applications before establishing skirt heights of evaporator systems.

High Temperature Heat Transfer Media

Steam is the most commonly used heat transfer medium. However, to obtain high temperature with condensing steam requires high pressures. Other heat transfer fluids are available which permit high temperature operation at low or moderate pressures. Both liquid- and vapor-phase systems are used. Liquid systems are almost always more economical: liquids are easier to confine; smaller transfer lines are required reducing heat losses; control is more easily achieved when process temperature levels vary widely. Vapor-phase systems are often used when small heat loads are required or when the heat transfer system is dedicated to a particular service.

Several types of fluids are available. Some of the more common include: eutectic mixtures of organic compounds; mineral and petroleum oils; liquid metals; molten salts; gases.

Caution should be exercised in selecting the proper media for a particular service. Each heat transfer fluid has specific properties which must be evaluated for any application. Some of the things that should be considered include: stability at expected operating temperatures; cost of the heat transfer fluid; heat transfer properties; pumping characteristics; toxicity; flammability; ease of handling; reclaimability.

Heat Pumps

Heat pumps (or refrigeration cycles) involve the use of external power to "pump" heat from a lower temperature to a higher temperature. The working fluid may be a refrigerant or a process fluid.

Heat pumps use energy that often would otherwise be thrown away. The external energy input is reduced by a factor of 4 to 10 depending on the temperature difference and temperature level of the heat pump system. However, the energy now used is the work form as opposed to the heat form previously. Work is the most expensive energy form. The expense of installing the heat pump is not justified unless the external energy reduction factor is greater than the cost ratio of power to heat at the temperature level required. In the past a factor of 4 to 10 reduction in external energy input would be marginal at best, and the heat pump has usually been used when the condensing temperature was low enough to require refrigeration anyway. Changing energy economics require that we now examine the alternative system much more frequently. Heat pumps will be further discussed in Chapter 16.

Multiple-Effect Evaporators

The use of multiple effects in evaporation is widely practiced. The basic principle is to use the heat given up by condensation in one effect to provide the reboiler heat for another effect. These systems will be further discussed in Chapter 15. In some cases, it is not practical to condense the vapors directly in the calandria, for example, when the condensing vapor is under high vacuum and the piping and pressure losses would be prohibitive. In such cases, a circulating fluid may be used as shown in Figure 13-8. This has the advantage of better heat transfer rates but has the disadvantage of reduced temperature driving forces.

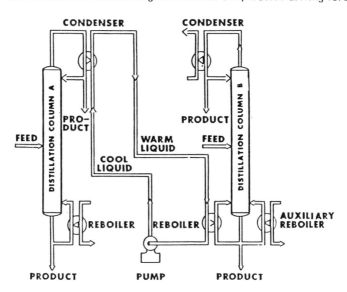

Figure 13-8: Heat recovery with tandem distillation columns when it is not practical to condense overhead vapors directly in reboiler of tandem column.

Thermal Engine Cycles

As already mentioned, energy recovery through a heat exchanger is the most efficient. Therefore, recovery of heat should be through process oriented heat utilization or generation of steam to be used in process plants. When this is not possible because of process characteristics or temperature levels available, a thermal engine cycle may be considered for energy recovery. Thermal engine cycle systems may be used to produce useful electrical or shaft power without added fuel expense. Thermal energy at low temperature can be converted into motive power with a thermal engine cycle utilizing fluorocarbons (or other refrigerant) as the working fluid.

Such a thermal engine cycle is shown in Figure 13-9. Evaporation, expansion, condensation, and pressure rise are repeated in a simple Rankine cycle. In the simplest form, "waste heat" is applied to a boiler which provides saturated or superheated vapor to the expander, and the fluid passes on to a condenser, which provides liquid to the pump. The pump raises the pressure and resupplies fluid to the boiler, thereby completing the cycle. The working fluid condenser heat is rejected to a cooling fluid in the condenser, either cooling water or air. The expander shaft work is ultimately used as shaft power to drive compressors or pumps, or to drive a generator to produce electrical power.

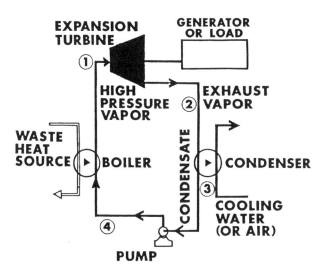

Figure 13-9: Schematic diagram of thermal engine cycle.

The thermal efficiency of a power cycle is a function of the source and sink temperatures, as well as the enthalpy characteristics of the working fluid. The total efficiency of waste heat recovery is a function of the thermal efficiency of the cycle and the mechanical efficiencies of the turbine, pump, and generator. In typical systems only 10 to 20% of the available heat is recovered.

The economic justification for this method of heat recovery depends on the source and sink temperatures, the heat load, and the site location. Other factors must also be considered: the nature of the heat source and the process characteristics. The recovery efficiency increases with an increase in the source temperature or a decrease in the sink temperature. Recovery of heat is more economical with increasing available heat. Generally, thermal engine cycles are economically attractive if the waste heat available exceeds 25 million Btu/hr. Even though the recovery efficiency is low, a significant amount of mechanical energy can be developed. A direct conversion of thermal energy into power would be far more efficient, but this is presently not feasible at low temperature.

A high recovery efficiency can be achieved by installing thermal engine cycles in series (usually with different working fluids) between the waste heat source and a cold heat sink. Whenever a waste heat source and a cold heat sink are available near each other in a plant, this cascading concept should be considered for energy recovery. Mixed refrigerants can also be used to extend the economic range of thermal engine cycles.

STEAM CONDENSATE RECOVERY

The process of recovering steam condensate for reuse as boiler feedwater or in processes requiring high quality water has been practiced in varying degrees for many years. The percentage of steam returned as condensate has ranged from a high of over 95% to a low of virtually none, except for that used in feedwater heating. The plants which have achieved a high level of condensate return have usually been the ones with a shortage of fresh water available. Rising energy costs and treatment costs makes it necessary to use this valuable commodity.

The benefits derived from maximizing steam condensate recovery in a specific plant depend on the quality of water used for boiler feedwater and process water makeup. Plants using lower quality makeup (such as sodium zeolite softened water) will benefit in more areas than plants using high quality makeup (such as produced by demineralization or evaporation), although the economic savings may be greater for high quality makeup systems. The elements of potential savings are listed below:

(1) Incentives in high quality makeup systems:
 (a) direct heat recovery
 (b) direct reduction in raw water requirements, water treating plant operating costs, and water treating plant capital costs
 (c) reduction in sewer flows with resultant savings in wastewater treatment costs.

(2) Incentives in low quality makeup systems:
 (a) direct heat recovery
 (b) direct reduction in raw water requirements, water treating plant operating costs, and water treating plant capital costs
 (c) reduction in sewer flows with resultant saving in wastewater treatment costs

(d) heat savings as a result of reduced boiler blowdown

(e) reduction in quantity of internal boiler feedwater chemical additives needed

(f) increase in boiler steam generating capacity.

Problem Areas

Several problems are present when condensate return is practiced. Most are associated with condensate contamination.

Contamination: Contamination of condensate occurs from two sources:

(1) *Iron:* In systems which use cold lime treated makeup water, iron is frequently encountered in sufficient concentrations to create problems with iron deposition on the boiler tubes. In some cases, iron pickup in the condensate return system can be minimized by feeding neutralizing or filming amines to the steam, but in some instances the use of amines is prohibited because of the possibility of product or process contamination. Ion exchange beds regenerated with sodium chloride brine can be used to remove iron from returned condensate fairly economically. Low levels of iron in condensate used as makeup to the boiler feedwater system can sometimes be tolerated by the addition of sequestering agents or dispersants to the boiler.

(2) *Process Leaks:* Process leaks can occur with varying frequency. Instrumentation must be provided to monitor condensate quality to prevent damage to the boiler. Normal practice is to sewer condensate upon evidence of contamination until it is determined that the contamination is not severe enough to cause problems in the steam generation system. Procedures must be provided to identify and isolate causes of contamination rapidly. Provisions for segregating the system into small segments should be provided.

Return Line Corrosion: Condensate (especially when cold line feedwater makeup treatment is used) is normally quite corrosive to steel piping systems. Stainless steel and some of the aluminum alloys are a better choice for piping materials.

Steam Traps: Steam traps have historically been high maintenance items and if not given proper attention, result in wasted steam and piping damage as a result of water hammer. Indicators are now available at nominal cost which are designed to identify leaking traps upon visual inspection. A good inspection and maintenance program is still required, however.

Improper Steam and Heat Balance: Many plants have an excess of low pressure steam and recovering heat from condensate results in increased venting of low pressure steam. It is advantageous to recover condensate even with low pressure steam venting, but the economic savings are not as substantial.

Instrumentation Used for Monitoring Condensate Quality

Condensate quality must be continuously monitored to avoid the problems discussed above. Many instruments are used for this purpose.

Conductivity: Conductivity measurement has historically been the most important method of condensate quality monitoring. Conductivity meters are relatively inexpensive, are quite reliable, and require a minimum of maintenance. They will detect most common inorganics except silica.

pH: pH meters are sometimes used for monitoring condensate, often as a backup for a conductivity instrument, or where there is a relatively high normal background of inorganic materials in the condenser. pH meters are not extremely reliable and require frequent maintenance.

Carbon and Hydrogen Analyzers: Carbon and hydrocarbon analyzers are used when the possibility of organic contamination of the condensate is likely to occur without simultaneous contamination by inorganics which will activate a conductivity alarm. These analyzers are quite expensive and require frequent maintenance for reliability.

Summary

Higher costs for energy, water treating chemicals, and environmental protection make it well worthwhile to recover as much condensate as practical. Capital expenditures will probably be required to upgrade return systems, segregate clean and contaminated streams, provide monitoring instrumentation, and provide a treatment system, but the economic justifications have risen tremendously.

14
Vapor-Liquid Separation

Product loss requirements may vary over a wide range. Provisions to reduce product losses have far less effect on the cost of an evaporator system than does heat transfer surface. Product losses in evaporator vapor occur as a result of entrainment, splashing, or foaming. Foaming properties of the liquid may at times dictate the selection of evaporator type.

ENTRAINMENT

Losses from entrainment result from the presence of droplets in the vapor that cannot separate because of the vapor velocity. Entrainment is thought to be due mainly to the collapse of the liquid film around vapor bubbles. This collapse projects small droplets of liquid into the vapor. The extent of entrainment is a function of the size distribution of the droplets and the vapor velocity. Little information is available regarding size distribution of the droplets that are formed or the influence of evaporator type and liquid properties on size.

Entrainment may be generated from atomization at a liquid surface by a high gas velocity, from a spray contacting operation, or from the disengaging of gas and liquid in a contacting operation. This latter mechanism is the most common.

Bubbles breaking the surface cause drops of different sizes to be propelled upward. The smaller ones are caught in the upward flowing gas and carried overhead as entrainment, while the larger ones return to the surface. The largest drop carried up is dependent on the gas velocity and density, and on other physical properties. At very high gas velocities, large drops produced at the surface are shattered into smaller droplets and all the generated entrainment is carried overhead, flooding the device. This breakup phenomenon occurs when the inertial forces, which cause a pressure or force imbalance, exceed the surface tension forces which tend to restore a drop to its natural spherical shape. The gas

154 Handbook of Evaporation Technology

velocity at which this flow crisis develops is the flooding velocity and is given by the following equation:

$$v_{gf} = 0.7[(\rho_l - \rho_v)\sigma/\rho_v^2]^{1/4} \qquad (14.1)$$

where v_{gf} = flooding velocity, ft/sec
ρ_l = liquid density, lb/cu ft
ρ_v = vapor density, lb/cu ft
σ = surface tension, dynes/cm

Equipment containing both gases and liquids in which the gas flows vertically upward will be flooded at velocities exceeding that predicted from the equation above. In practice, most equipment is designed to operate well below the flooding limit. Factors such as disengaging height, convergence effects, and nonuniform gas velocities prevent operation at velocities exceeding roughly half the flooding velocity.

The amount of entrainment from an upward flowing gas can be estimated and is a function of gas velocity, gas and liquid densities, and surface tension. Actual entrainment rates can be from 0.1 to 2.5 times what is estimated by various correlations.

Entrainment can be separated from a gas stream with a variety of mechanisms, including gravity, inertial impaction, interception, centrifugal force, and Brownian motion. Separators can be classified according to mechanism, but it is more useful to categorize them by construction type. Separators in common use include:

(1) flash tanks
(2) vane impingement separators
(3) wire mesh separators
(4) Karbate strut separators
(5) centrifugal separators
(6) cyclones
(7) special separator types.

Flash tanks are generally used when the liquid entrainment exceeds 20% of the gas flow on a weight basis. Flash tanks may be either vertical or horizontal. Proper sizing of a flash tank should result in a residual entrainment under 3% of the gas rate.

Vane impingement, wire mesh, centrifugal, and Karbate strut separators are commercial proprietary designs and all compete for similar applications. Performance and cost, however, can vary widely from one type to another. The designer should understand the advantages and disadvantages of each type and the level of separation that each can effect.

Except for flash tanks and some special separators, the efficiency of all these separators tends to increase with increasing velocity up to a maximum allowable limit. In this region the efficiency seems to depend primarily on gas velocity and particle size, and to be somewhat insensitive to gas and liquid physical properties. Except for the cyclone and some special separators, there is a recognized and predictable maximum allowable velocity. The following equation is commonly used:

$$v_m = F[(\rho_l - \rho_v)/\rho_v]^{1/2} \qquad (14.2)$$

where v_m = maximum gas velocity, ft/sec
F = experimentally derived constant, ft/sec
ρ_l = liquid density, lb/cu ft
ρ_v = vapor or gas density, lb/cu ft

This equation, and Equation 14.1, indicate that the term $v\rho_v^{1/2}$ is of primary importance in separator sizing. The flow area must be carefully specified to avoid any ambiguity in application of Equation 14.2.

FLASH TANKS

The simplest approach to disentraining liquid is the flash tank. The velocity is reduced to a value which permits the droplets to settle rather than to be carried by the vapor. Evaporator bodies are generally sized such that particles larger than a certain size will be disentrained. Flash tanks may be either vertical or horizontal. Vertical separators are usually more economical when the diameter is less than 8 feet; for larger sizes, horizontal separators may be less expensive.

Vertical vapor bodies may be sized using a value of 0.2 for F in Equation 14.2. Horizontal vapor bodies should be sized with a target value of F = 0.15 based on the surface area at the gas/liquid interface. Recommended vertical flash tank dimensions are shown in Figure 14-1. Liquid holdup requirements may control the size of flash tanks. In such cases, the tank should be oversized with respect to the gas flow in order to acheve practical vessel dimensions. A height-to-diameter ratio of 2 to 5 is usually economical.

Operation of horizontal flash tanks is improved when a distribution pipe is provided. Distributors can be installed normal to the tank axis with perforations directing the flow horizontally. Alternatively, the distributor pipe can be installed parallel to the tank axis with perforations directing the flow perpendicular to the tank axis. The inlet nozzle should always be located well above the maximum liquid level.

Both vertical and horizontal flash tanks are susceptible to low frequency motion induced by sloshing. Consequently they should be rigidly supported.

Often it is not practical to provide a vapor body large enough to accomplish the desired disentrainment. *Entrainment separators* are frequently used to reduce product loss. A number of proprietary designs are available.

WIRE MESH SEPARATORS

Knitted mesh, either metal or plastic, is often used inside vapor heads for entrainment separation. Knitted mesh can be used effectively in applications

156 *Handbook of Evaporation Technology*

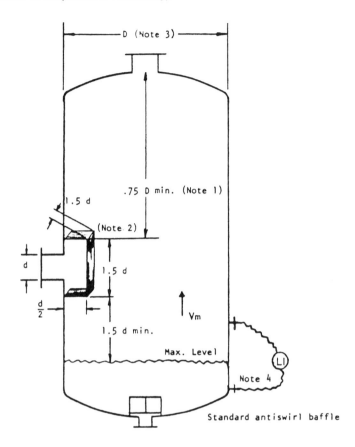

Note 1 — If wire mesh is installed, change dimension from .75 D to D and install top of mesh pad D/2 below outlet nozzle.

Note 2 — If erosion is suspected, omit baffle or replace with a perforated distributor pipe with horizontal holes. Avoid tangential inlets.

Note 3 — Size from $V_m = .2\sqrt{\Delta\rho/\rho_G}$ ft/s

Note 4 — Provide volume for minimum of one minute holdup between level taps.

Figure 14-1: Standard flash tank dimensions.

where it will not be fouled or plugged from solids in the liquid. Wire mesh is sometimes washed in service to remove solids. Any significant corrosion will cause rapid disintegration of the mesh. The performance of a knitted wire mesh depends on wire surface and free volume. Vapor flows easily through the mesh while droplets of liquid impinge upon the wire surfaces, coalesce, flow back, and drop away from the bottom of the demister pad. Vapor velocity is of major importance in obtaining best performance. Highest efficiency is obtained within certain limits. Knitted mesh has very good separation characteristics when the gas density is low and poor characteristics when it is high. Separator efficiency (percent of inlet entrainment removed) can be expected to increase from 80% at 2 ft/sec to 95% at 3 ft/sec to 99+% at 5 ft/sec. The maximum velocity is determined by flooding of the mesh. Knitted mesh separators are generally sized by using a value of F = 0.35 in Equation 14.2.

Pressure drop is typically 0.1 to 1.0 inch of water. It can be conservatively estimated at 100 velocity heads per foot of mesh thickness, based on superficial velocity through the pad. High pressure drops for properly sized pads indicate that the pad is probably either plugged or flooded, or both.

A mesh pad can be of one-piece construction (for diameter less than 3 feet) or it can be fabricated in strips for installation through a manway or large nozzle. This latter construction is usually preferred in order to facilitate pad replacement. Mesh pads should be supplied with both top and bottom support grids to increase mechanical integrity. Pads larger than 6 feet in diameter will require support beams. The pad should be placed on a support ring, welded around the vessel circumference. The pad should be anchored to this ring. The bottom grid is anchored to the ring with tie wire supplied with the pad. For severe services, the mesh should be anchored with J bolts rather than tie wire. The pad, grids, and tie wire (or J bolts) will be supplied by the mesh vendor. Support beams and rings should be part of the vessel fabrication.

VANE IMPINGEMENT SEPARATORS

Vane impingement separator elements consist of parallel zig-zag plates or vanes with included collection pockets, similar to those shown in Figure 14-2. As an entrainment laden gas moves through the separator, the liquid particles are forced onto the vane surfaces by the inertial impaction mechanism. The resultant liquid film flows to the collection pockets, where it drains by gravity.

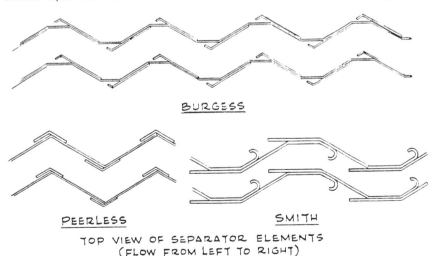

Figure 14-2: Vane impingement separator elements.

158 Handbook of Evaporation Technology

Each change of direction is a stage of separation, and by employing a number of such flow reversals, manufacturers are able to supply vane separators with very good efficiency characteristics. These separators are available from a number of manufacturers.

Vane impingement separators are built in a variety of configurations as shown in Figure 14-3. The most common is the line separator. It is often more economical to provide a line separator with necessary piping rather than provide an internal separator. Vertical gas separators are used when very high separation efficiencies are required or when the liquid entrainment to the separator is in the range of 5 to 20 weight percent of the gas flow. The vertical line separator contains two separation devices, an impingement baffle with flow direction change followed by a boxed vane arrangement.

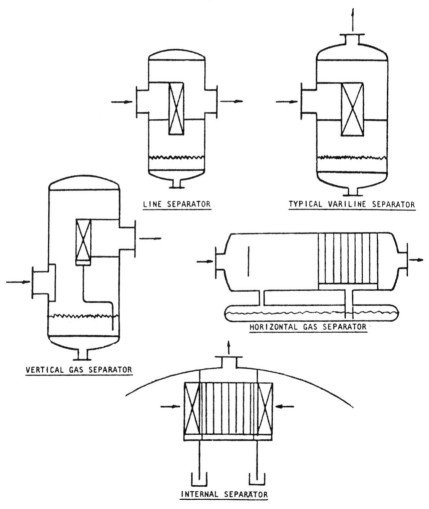

Figure 14-3: Types of vane impingement separators.

Vane impingement separators are generally purchased from the vendor on a performance basis. Sizes will vary somewhat from one manufacturer to another; consequently, the exact size cannot be determined before selection of the manufacturer. The pressure drop across vane impingement separators is approximately 0.25 psi plus 1.5 velocity heads based on the nozzle velocity. The total pressure drop will usually be less than 0.5 psi.

Liquid loading is limited to roughly 5 to 10% of gas mass flow. Pockets can plug with solids and are difficult to clean unless the vanes are removable.

CENTRIFUGAL SEPARATORS

Centrifugal separators impart a spin or rotation to the entrainment laden gas. Blades or a special passage are used to impart the rotation and the resultant centrifugal force. Entrained liquid collects on the wall and flows through a drain connection. Centrifugal separators are distinguished from cyclones because they have a somewhat different configuration and must usually be purchased from the manufacturer on a performance basis. Centrifugal separators are usually line-type separators, but other configurations are available. Centrifugal separators can be installed inside a vapor body by attaching it to the outlet nozzle as shown in Figure 14-4.

Centrifugal separators cannot be sized with the F factor approach used with other types of separators because the maximum allowable velocity is a weaker function of gas density than the square root dependence in the F factor correlation. Maximum velocities for centrifugal separators depend upon the design and are different for each of the several manufacturer[1] units. Pressure drop varies from 1 to 10 velocity heads depending upon the manufacturer. Velocity head should be based on inlet (outlet) nozzle velocity.

Centrifugal separators have lower separation efficiency than vane separators. They must be provided with a trap device for liquid collection, and have a limited liquid handling capacity. Line-type units can be operated with velocities as low as 30 ft/sec without significant loss in separation efficiency.

CYCLONES

Cyclone separators admit entrainment laden gas through a tangential inlet. The gas then spirals around the central tube, called a vortex finder, and finally exists through it. Cyclone separation efficiency is not as good as with other separators, but it is reasonably good. There is no well-defined upper limit on the cyclone inlet velocity; but it appears prudent to limit the velocity to a maximum of approximately 300 ft/sec. The pressure drop across a cyclone can be estimated as 8 velocity heads, based on the inlet velocity. Good separation in a cyclone usually results in high velocities with resultant high pressure drops. In evaporator design, pressure drop results in a loss of available temperature difference decreasing the efficiency of the system or requiring additional heat transfer surface. Recommended cyclone configuration is shown in Figure 14-5.

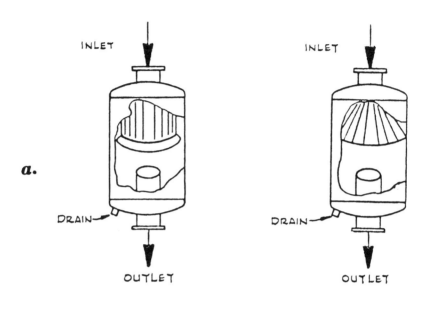

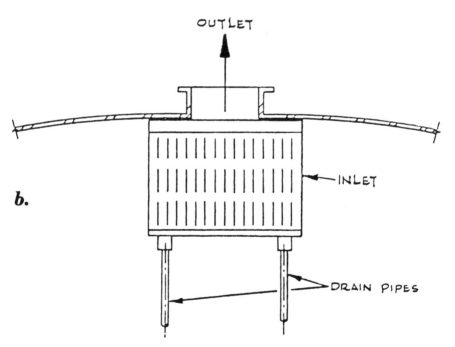

Figure 14.4: (a) Typical line centrifugal separators. (b) Typical internal centrifugal separator.

Vapor-Liquid Separation   161

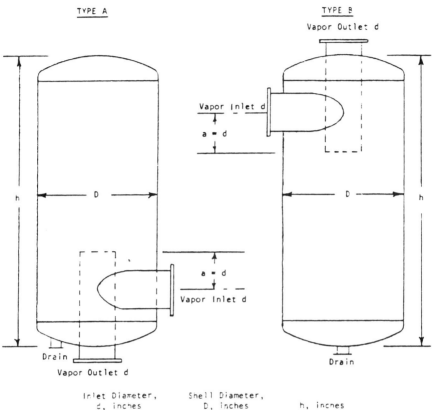

Inlet Diameter, d, inches	Shell Diameter, D, inches	h, inches
1	4	8
1-1/2	6	12
2	8	16
3	10	21
4	14	28
6	20	42
8	30	60
10	36	72
12	42	84
14	48	98
16	54	112
18	60	126
20	66	140
24	78	168

NOTES:

1. Antiswirl baffle should be provided for the drain nozzle.

2. Type A cyclone will have a somewhat higher efficiency than Type B cyclone.

Figure 14-5: Standard cyclone configuration.

OTHER SEPARATORS

Other separator types have been used and include:

(1) streamline struts
(2) banks of tubes
(3) zig-zag baffles (no liquid collection pockets)
(4) packed beds
(5) fibrous filters
(6) Brink mist eliminators.

Brink mist eliminators are filtering separators consisting of a bed of fine glass or synthetic fibers sandwiched between two screen supports. High velocity (H-V) and high efficiency (H-E) models are available. The H-V model can separate particles more efficiently than the conventional mesh because the fibers are much smaller in diameter. The liquid handling capacity of H-V separators is somewhat less than conventional mesh separators. The H-V separator uses inertial impaction but the H-E unit is based on capture by Brownian motion and is used for particles in the submicron range. Velocities are quite low (often 0.5 ft/sec); consequently large capacity requires large surface areas.

COMPARISON

Table 14-1 offers a comparison of various entrainment separators.

SOLIDS DEPOSITION

Industrial experience with separator fouling and plugging has yielded some qualitative observations. Vertical collection surfaces stay cleaner longer than horizontal surfaces because of better liquid drainage. Intermittent washing with sprays is beneficial, but the detail design of the washing unit and the operating procedure vary with each specific case. The wash represents a reduction in evaporation efficiency also.

Guidelines include:

(1) deposition rate decreases as the slurry flux on the surface increases
(2) deposition rate decreases as the liquid film thickness becomes greater
(3) small droplets are more susceptible to being caught in eddies which carry them to relatively inactive separator surfaces
(4) small droplets have a higher deposition rate than large drops.

Table 14-1: Comparison of Separators

Separator Type	Advantages	Disadvantages
Vane (Line Separator)	1. High Efficiency 2. Good turndown characteristics 3. Relatively inexpensive 4. Low pressure drop-typically less than .5 psi	1. Liquid loading is limited to roughly 5-10% of gas mass flow rate 2. Pockets can plug with solids 3. Difficult to clean unless vanes are removable
Vane (Vertical Gas Separator)	1. Better efficiency than line separator 2. Good turndown characteristics 3. Will separate liquid slugs 4. Will operate efficiently with high liquid loading	1. Pockets and drain piping can plug with solids 2. Difficult to clean unless vanes are removable 3. Relatively expensive
Mesh Separator	1. Generally good efficiency at low pressure 2. Inexpensive if installed in existing vessel 3. Available in a variety of materials 4. Very low pressure drop-less than one inch water	1. Very susceptible to plugging 2. Expensive if new tank is needed to house mesh pad 3. Highly susceptible to corrosion
Karbate Separator	1. High efficiency 2. Good turndown characteristics 3. Very low pressure drop-less than one inch water	1. Only available in Karbate 2. Liquid loading limited 3. Somewhat difficult to clean 4. A trap device is needed for liquid collection
Centrifugal (Line)	1. Inexpensive	1. Lower efficiency than vane line separators 2. A trap device is needed for liquid collection 3. Limited liquid handling capacity
Cyclone	1. Can be designed in-house 2. Less susceptible to plugging than other types 3. Can be operated at high inlet velocity (with high pressure drop)	1. Lower efficiency than vane line separators 2. A trap device is usually needed for liquid collection
Fiber Mat (Brink)	1. Very high efficiency 2. Relatively low pressure drop (approximately 6-8 inches water)	1. Relatively expensive 2. Highly susceptible to plugging

FALLING FILM EVAPORATORS

For falling film evaporators, disentrainment apparently occurs in the tubes. Droplets ejected from the film on one side of the tube are absorbed by the film on the opposite side. Consequently, falling film units are well suited for mechanical vapor compression systems.

FLASHING

Entrainment losses from flashing may occur in evaporators. If the feed is above the boiling point and is introduced below or near the liquid level, entrain-losses may be excessive. Excessive entrainment may also be experienced during startup if adequate control of liquid level and pressure is not provided.

SPLASHING

Splashing losses can be prevented by providing adequate distance between the liquid level and the top of the vapor body. The required distance depends upon the nature of the boiling. A distance of 8 to 12 feet is generally adequate for short-tube vertical evaporators in which the two-phase mixture is traveling upward. Deflectors may be used to reduce the height, but they should be used with caution and must be properly designed. Less distance is usually needed in other types where the mixture enters tangentially or is otherwise directed away from the vapor outlet.

Splashing losses often result from sudden liberation of large volumes of vapor in the two-phase mixture or below the liquid level. Splashing can frequently occur during startup or when liquid levels are permitted to rise for some reason.

FOAMING

Foam often results from the presence of colloids, surface tension depressants, or finely divided solids. These often concentrate at the liquid level and can sometimes be removed by a surface blowdown. Antifoam agents may be needed. Other methods of combatting foam include impingement or jets upon the foam surface, impingement at high velocities against baffles, operating at low liquid levels, ultrasonic jetting.

Frequent causes of foaming are an air leak below the liquid level or the presence of dissolved gases in the liquid. When foam begins to build in an evaporator, the foam layer can often continue to grow until it fills the evaporator and is carried out the top. This can sometimes be stopped by momentarily interrupting evaporation.

The designer should distinguish between liquid carryover resulting from entrainment phenomena and that resulting from foam. Foam cannot be broken with conventional entrainment separators unless the foam is highly unstable. The following generalizations on foam formation can be helpful in identifying potentially troublesome situations:

(1) Pure liquids do not foam.
(2) Substances that concentrate at the gas/liquid interface and lower surface tension tend to enhance foaming.
(3) In a binary solution, foaming tendency is increased if the more volatile component has the lower surface tension.
(4) High liquid viscosity stabilizes a foam.
(5) Finely divided particles increase foam stability.
(6) Small bubbles form more stable foams than large bubbles.

15
Multiple-Effect Evaporators

Multiple-effect evaporators are widely used and the principles well known. A multiple-effect system may be considered as a number of resistances, in series, to the flow of heat. The driving force causing heat to flow is the difference in temperatures of the steam condensing in the first effect and the temperature of the heat sink, often cooling water. Some of the available driving force is lost when the liquid exhibits a boiling-point elevation, the total loss equal to the sum of the boiling-point elevations in all the effect.

Other resistances result from pressure loss in vapor and liquid lines, from the cooling water temperature rise, and from the approach between cooling water and condensing temperature in the condenser. The largest resistances are those to heat transfer across the calandria in each evaporator effect and across the condenser if a surface condenser is used. The resistances of the heating surface are equivalent to the reciprocal of the product of the area and the overall heat transfer coefficient (1/UA).

Neglecting all resistances except those due to heat transfer and assuming they are equal, it is apparent that if the number of resistances (effects) is doubled, the flow of heat (steam consumption) will be reduced by half. With half as much heat, each effect will evaporate about half as much water but, since there are now twice as many effects, total evaporation will be the same. Using these simplifying assumptions, the same evaporation would be obtained regardless of the number of effects (of equal resistance), and the steam usage would be inversely proportional to the number of effects or the total heat transfer surface provided.

The resistance analogy can be used to explain design and operation of evaporator systems. The designer provides as many resistances (effects) in series as he can justify in order to reduce the flow of heat (steam consumption). If the overall heat transfer coefficient in one effect is lower than the others, the designer does not increase only the area in that effect to maintain equal resistance. He can more economically (less area) achieve the desired evaporation by adding only part of the increased area to the high resistance effect and the rest to each of the other effects. This reduces individual resistances but the total resistance is the same.

Usually, heat transfer rates decrease as temperature decreases so that the last effects have the lowest rates of heat transfer. By leaving the resistances of these effects higher, the designer can increase the temperature difference across them, increasing temperatures and heat transfer rates in all the earlier effects. It has been shown that the lowest total area is required when the ratio of temperature difference to area is the same for all effects. When the materials of construction or evaporator type vary among effects, lowest total cost is achieved when the ratio of temperature difference to cost is the same for each effect. However, in most cases where evaporator type and materials of construction are the same for all effects, equal heat transfer surfaces are supplied for all effects.

In addition to the reduction in steam usage, there is also a reduction in cooling water required to operate the last effect condenser. Approximately 30 pounds of cooling water must be provided for each pound of steam supplied to the first effect.

The increased energy economy of a multiple-effect evaporator is gained only as a result of increased capital investment. The investment increases almost at the same rate as the required area increases. A five-effect evaporator will usually require more than five times the area of a single effect. The only accurate method to predict changes in energy economy and heat transfer surface requirements as a function of the number of effects is to use detailed heat and material balances together with an analysis of the effect of changes in operating conditions on rates of heat transfer.

The distribution in each effect of the available temperature differences between condensing steam and process liquid can be allocated by the designer. Once the evaporator is put into operation, the system establishes its own equilibrium. This operating point depends upon the amount of fouling and the actual rates of heat transfer. Usually it is best not to interfere with this operation by attempting to control temperatures of different effects of an evaporator. Such attempts result in a loss of capacity since control usually can be accomplished only by throttling a vapor imposing an additional resistance. The pressure loss results in a loss of driving force and reduction in capacity.

The designer has a number of options to achieve the greatest energy economy with a given number of effects. These are usually associated with the location of the feed in respect to the introduction of the steam. Figure 15-1 illustrates several methods of operation.

FORWARD FEED

The feed to a multiple-effect evaporator is usually transferred from one effect to another in series so that the final product concentration is achieved in only one effect. In forward feed operation, raw feed is introduced in the first effect and passed from effect to effect parallel to the steam flow. The concentration of the liquid increases from the first effect to the last and product is withdrawn from the last effect. Forward feed operation is advantageous when

168 Handbook of Evaporation Technology

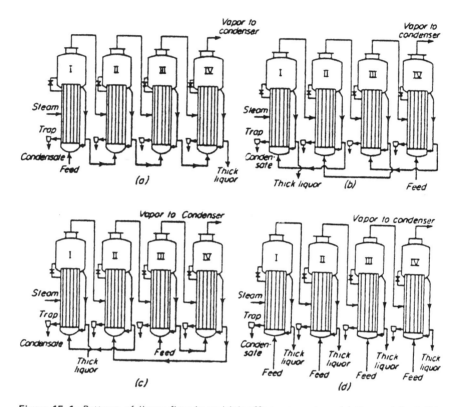

Figure 15-1: Patterns of liquor flow in multiple-effect evaporators: (a) Forward feed. (b) Backward feed. (c) Mixed feed. (d) Parallel feed.

the feed is hot or when the concentrated product would be damaged or foul at high temperature. Forward feed is the simplest arrangement when liquid can be transferred by pressure difference alone, removing the need for pumping between effects. When a feed is cold, forward feed has a lower energy economy because some steam must be used to provide sensible heat to heat the cold feed to the liquid temperature of the first effect. If forward feed is required and the feed is cold, energy economy can be improved by preheating the feed in stages with vapor from intermediate evaporator effects. This usually requires little increase in total heat transfer surface since the feed must be heated anyway. Further, heat exchangers are usually less expensive than evaporators, especially if gasketed plate exchangers can be used.

BACKWARD FEED

In backward feed operation the raw feed enters the system at the last (coldest) effect and product is withdrawn from the first effect. Pumps must be used to transfer liquid from lower pressure effects to higher pressure effects. This method of operation can be used to advantage when feed is cold since the entering liquid needs to be heated to a lower temperature than in forward feed operation. Backward feed is also used when products are viscous and higher temperature increases the rate of heat transfer. When the product is viscous but a hot product is not needed, the liquid from the first effect is sometimes flashed to a lower temperature in one or several stages and the flashed vapor added to the vapor from one or more later evaporator effects.

MIXED FEED

In mixed feed operation the feed enters as intermediate effect, flows in forward feed through the later effects, and is then pumped back to the earlier effects for further concentration. Operation in the earlier effects can be either backward feed or forward feed. This eliminates some of the pumps needed in backward feed and permits final evaporation at the highest temperature. Mixed feed operation is used only for special applications. Sometimes liquid at an intermediate concentration and certain temperature is required for additional processing. The feed temperature may be close to that of an intermediate stage and mixed feed may result in greater energy economy.

PARALLEL FEED

Parallel feed operation requires the introduction of raw feed and the withdrawal of product from each effect. In parallel feed there is no transfer of liquid from effect to effect. It is used primarily when the feed is saturated and the product is a solid. All effects may operate at the desired product concentration or they may operate at different concentrations and final product obtained by blending.

STAGING

Often in multiple-effect evaporators the concentration of the liquid being evaporated changes drastically from effect to effect—especially in the latter effects. In such cases, this factor can be used to advantage by "staging" one or more of the latter effects. Staging is the operation of an effect by maintaining two or more sections in which liquids at different concentrations are all being evaporated at the same pressure. The liquid from one stage is fed to the next stage. The heating medium is the same for all stages in a single effect, usually the vapor from the previous effect. Staging can substantially reduce the cost of an evaporator system. The cost is reduced because the wide steps in concentration from effect to effect permit the stages to operate at intermediate concentrations which result in both better heat transfer rates and higher temperature differences. Staging can reduce the total heat transfer surface by as much as 25% when high boiling point rises occur.

HEAT RECOVERY SYSTEMS

Other methods to improve energy economy are frequently employed in evaporators. For greatest efficiency, product and condensate should leave the system at lowest possible temperatures. Heat should be removed from these streams by heat exchange with feed or evaporating liquid at the highest possible temperatures. This may require separate heat exchangers adding to the system complexity. They can be justified only for large systems. Such heat exchangers are generally required for mechanical vapor compression to recover the heat of compression. Gasketed plate heat exchangers are cost-effective for such applications.

Energy economy can also be improved by flashing the condensate from one effect to the next. Flashing of first effect condensate generally results in a reduction of overall plant efficiency; normally this is returned to the boiler unflashed and heat is recovered only when it would otherwise be lost. Condensate from other effects should not be flashed if it can be used hot as boiler feedwater or for some other purpose. The loss in thermal efficiency by flashing can often be justified since the vapor can be used directly in later effects. However, it is more efficient to preheat feed with the hot condensate.

CALCULATIONS

In evaporator calculations, three relations must be satisfied: material balance; required rates of heat transfer; and heat balance. These relations must be applied to each effect in a multiple-effect system as well as to the total system. The equations describing evaporator systems can be solved algebraically but the process is tedious and time consuming. Trial-and-error procedures are generally used as follows:

(1) Assume values for boiling temperatures in all effects.

(2) From enthalpy balances determine rates of steam and liquid flows from effect to effect.

(3) Calculate rates of heat transfer and required heat transfer surface using the previously established temperature distribution.

(4) If the heating surface distribution is not as desired, establish a new temperature distribution and repeat steps 2 and 3 until desired results are achieved.

In practice these calculations are best accomplished with a computer.

OPTIMIZATION

The purpose of an evaporator system is to produce the required amount of specified product at the lowest total cost. The designer of an evaporator system must consider a wide range of variables in selecting the system which provides the best balance between operating cost and capital investment.

The evaporator type, the number of effects, and the type of compression system are the most important variables. Others include:

(1) initial steam pressure and its cost or availability
(2) final pressure and its relation to cooling water temperature and wet- and dry-bulb air temperatures
(3) final pressure and its relation to product quality
(4) cost and availability of electrical power
(5) materials of construction
(6) maintenance and reliability
(7) should more than one type of evaporator be used in the system
(8) method of feeding and heat recovery
(9) entrainment and separator type
(10) evaporator accessory equipment
(11) control required.

Few of these variables are completely independent.

Some guidelines can be offered:

(1) the optimum number of effects increases as the steam costs increase or plants become larger
(2) mechanical vapor compression looks attractive if electrical power is available at reasonable costs and boiling point rises are low
(3) thermal compression looks attractive when electrical power is costly and boiling point rises are low
(4) when boiling point rises are large, combinations of vapor compression preconcentrators followed by multiple-effect systems look attractive
(5) larger plants enable heat recovery systems to be more easily justified
(6) mechanical vacuum pumps are generally more energy efficient than steam jets.

Frequently it is necessary to consider other factors besides the evaporator design. Steam generation, electrical power, and cooling water distribution and supply costs may be sufficiently reduced to justify particular system designs. Sometimes evaporator efficiency can be sacrificed in order to utilize all of the heat available in a waste heat stream. Final operating pressure may be determined by the ability to recover the heat of condensation in some other part of the plant.

16
Heat Pumps

One method to increase energy efficiency is to use the principles of the heat pump.

CONVENTIONAL HEAT PUMP

Conventional heat pump operation is indicated in Figure 16-1. A refrigerant, upon boiling, absorbs the heat that would otherwise be rejected in a condenser. The refrigerant vapor is compressed to a pressure adequate to permit the vapor to be condensed in the calandria, thereby providing the heat needed for evaporation. The condensate from the calandria is flashed into the condenser thereby completing the cycle.

OVERHEAD VAPOR COMPRESSION

Overhead vapor compression can be utilized when the vapor is suitable as a refrigerant. Such a system is illustrated in Figure 16-2. The condenser is eliminated, along with its associated temperature driving force. The overhead vapor enters the compressor in a form equivalent to a refrigerant which has just absorbed the heat in the condenser, condenses in the calandria, and is flashed to the overhead pressure before being withdrawn. This alternative, when it can be used, is a lower capital alternative than the conventional heat pump, and offers potentially lower energy requirements. The advantage can be lost, however, if the overhead vapor properties are not close to those of the optimum refrigerant.

CALANDRIA LIQUID FLASHING

Calandria liquid flashing can also be used when the liquid product has suit-

Heat Pumps 173

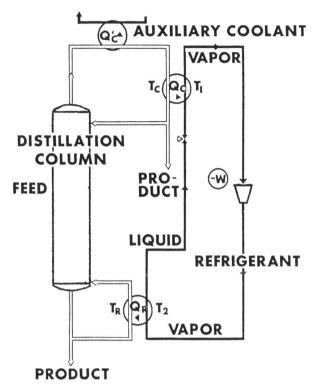

Figure 16-1: Distillation column with conventional heat pump.

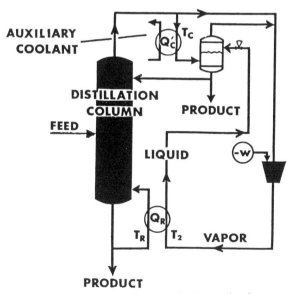

Figure 16-2: Distillation column with heat pumping by overhead vapor compression.

able working fluid properties. Figure 16-3 illustrates such a system. In this case, calandria liquid is flashed to a lower pressure and is vaporized as it absorbs heat by condensing the overhead vapor. The vaporized stream is then compressed to the evaporator pressure and returned as vapor to the bottom of the column. In this case, the calandria liquid is equivalent to that of a refrigerant which has just condensed, providing the calandria heat requirement. The calandria and its associated temperature difference is eliminated. The energy requirement is slightly higher for this case than for overhead vapor compression but the equipment arrangement is similar. In practice, the choice between the two is largely determined by the operating pressure. High pressures favor flashing of the calandria liquid, while low pressures favor compression of the overhead vapor.

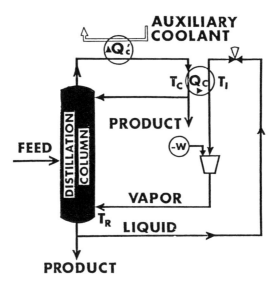

Figure 16-3: Distillation column with heat pumping by reboiler liquid flash.

17
Compression Evaporation

Compression evaporation could be defined as an evaporation process in which part, or all, of the evaporated vapor is compressed by means of a suitable compressor to a higher pressure level and then condensed; the compressed vapor provides part of all of the heat required for evaporation. Compression evaporation is frequently called recompression evaporation. All compression methods use the vapors from the evaporator and recycle them to the heating side of the evaporator. Compression can be achieved with mechanical compressors or with thermal compressors. Thermal compression uses a steam jet to compress a fraction of the overhead vapors with high pressure steam. Mechanical compression uses a compressor driven by a mechanical drive (electric motor or steam turbine) to compress all the overhead vapors.

Energy economy obtained by multiple-effect evaporation can often be equalled in a single-effect compression evaporation system. To achieve reasonable compressor costs and energy requirements, compression evaporator systems must operate with fairly low temperature differences, usually from 5 to $10°C$. As a result, large heat transfer surfaces are needed partially offsetting the potential energy economy.

When a compression evaporator of any type is designed, the designer must provide adequate heat transfer surface; in fact, he may want to provide more surface than required to compensate for reduced heat transfer as fouling occurs. If there is inadequate surface to transfer heat available after compression, the design compression ratio will be exceeded. Under these conditions a thermocompressor may become unstable and "backfire" and the mechanical compressor may require more energy than provided. Generally, it is wise to be on the safe side and provide extra heat transfer surface for compression evaporators.

18

Thermal Compression

Thermal compression is the term used to describe the application of a steam jet ejector or thermocompressor to an evaporator in order to increase the steam economy. A typical system is illustrated in Figure 18-1. Steam jet thermocompressors can be used with either single or multiple-effect evaporators. As a rule-of-thumb, the addition of a thermocompressor will provide an improved

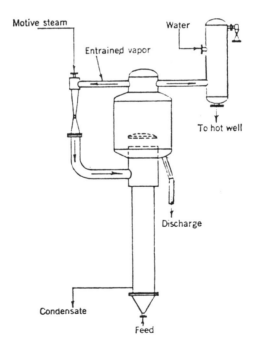

Figure 18-1: Evaporation with a steam jet thermocompressor.

steam economy equivalent to the addition of one other effect but at a considerably lower cost. Thermocompressors have low thermal efficiencies which further diminish when the jet is not operated at its design point. In typical operation, thermocompressors can entrain one unit of vapor per unit of motive steam. Thermocompressors are available in a wide range of materials of construction and exhibit a wide range of design operating conditions. They should be considered only when high pressure steam is available and the evaporator can be operated with low pressure steam. Motive-steam pressures greater than 60 psig usually are required to justify the use of thermocompressors. Space limitations favor thermal compression. Steam condensate from thermal compression systems often is contaminated with product traces and may have to be treated before being returned to the steam generator.

THERMOCOMPRESSOR OPERATION

The thermocompressor design is very similar to that of steam jet ejectors used for producing vacuum. The standard types of ejectors used for thermal compression are shown in Figure 18-2. Each ejector consists of three basic

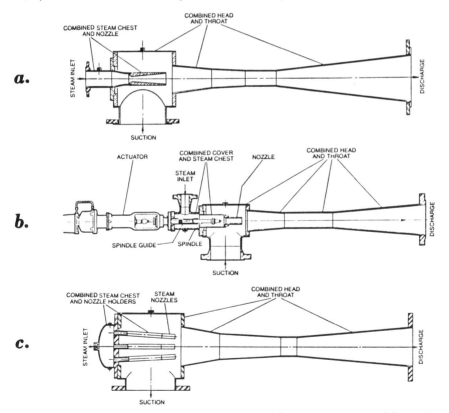

Figure 18-2: Ejectors used for thermal compression. (a) Single nozzle type. (b) Spindle operated type. (c) Multiple nozzle type.

parts: nozzle, mixing chamber, and diffuser. For a discussion of the operating principle, see Chapter 23. The higher pressure motive-steam enters the nozzle section and expands through the nozzle into the mixing chamber. As the vapor expands, it produces a suction that entrains the vapors from the suction side. The suction and motive steam are compressed through the diffuser and are discharged against the back pressure, which in this case is the operating pressure in the steam chest for the evaporator. If the motive-steam is dry and saturated, the vapors will be slightly superheated.

THERMOCOMPRESSOR CHARACTERISTICS

Thermocompressors are basically single point designed; the design and operation are best at one specified design condition of suction pressure and load, motive-steam pressure, and vapor rates. The ejector design may be either critical or noncritical. Critical designs occur at compression ratios above 1.8 and result in sonic vapor velocity in the diffuser. Since ejectors designed for higher compression ratios in which the performance is critical cannot be controlled by varying the motive-steam flow, any change in suction pressure or discharge pressure reaching the critical compression ratio will cause the thermocompressor to "break." A discussion of this phenomenon is given in Chapter 23. When operating in the "broken" condition, the variation in the suction and discharge pressures will result in loss of capacity, pump cavitation, and liquor flashing in the body from unstable pressures within the evaporator. For these reasons, thermocompressors are designed for noncritical operation.

For noncritical designs, the changes in suction pressure and vapor rates from the evaporator can be adjusted by varying the motive-steam pressure or flow. The design of the thermocompressor for the evaporator is therefore based on the performance curve, such as the general one shown in Figure 18-3. The design point is chosen on the basis of the most typical evaporator operating conditions. If a constant suction load is to be maintained and the discharge pressure increases, the suction pressure (pressure in the evaporator body) will increase to maintain the design compression ratio. A reduction in the suction pressure (when the vapor rate decreases) will reduce the discharge pressure. An increase in the discharge pressure decreases entrainment in the thermocompressor and thus reduces the suction load and heat recovery. If the discharge pressure is further reduced, unstable operation will occur and the unit breaks. If the operating pressure and therefore motive-steam flow are increased, the higher discharge pressure can be maintained at the same suction load. However, the amount of steam flow entering the steam chest is increased, and the evaporator must be balanced to maintain constant capacity. Control by increasing the motive-steam pressure or inlet pressure has a limited range of application, depending upon the hardware design. In addition, under a critical flow any reduction in motive-steam pressure below the design pressure causes the unit to break.

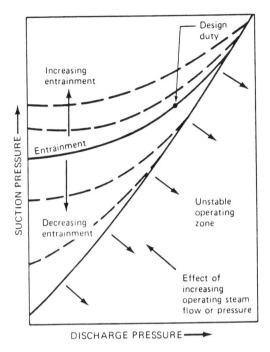

Figure 18-3: Typical thermocompressor performance curve.

THERMOCOMPRESSOR TYPES

Thermocompressors can be of two basic types: fixed nozzle types or units with a regulating spindle. Fixed-nozzle thermocompressors are more efficient and are used when the thermocompressor can be operated under steady-load conditions. Thermocompressors with regulating spindles are designed for use when performance approaches critical conditions and changes in suction load and suction and discharge pressures are expected.

Single Fixed Nozzle

The design and construction of a single nozzle fixed orifice thermocompressor is similar to that of a standard steam jet ejector and is shown in Figure 18-2(a). Single nozzle units are used for either critical or noncritical flow, but usually for one set of design conditions only. A modest degree of variation can be achieved by throttling the motive steam but this decreases the efficiency of the thermocompressor. Fixed nozzle units are used when the thermocompressor will be operated at a steady load.

The prime advantage of a single fixed-nozzle thermocompressor is its compactness. Single nozzle units are preferred when large compression ratios are required. Single nozzle units are used for small capacity application and when several parallel units are needed to provide some degree of control.

Multiple Nozzle

The construction of a multiple-nozzle fixed-orifice thermocompressor is the same as that of the single nozzle unit, except that it employs several steam nozzles as shown in Figure 18-2(c). The usual configuration has one nozzle on center with the remaining nozzles equally spaced peripherally around it. This type of unit can be used for both critical and noncritical applications. Its characteristic performance curve is similar to that of a single nozzle fixed orifice unit.

The multiple nozzle unit offers a substantially higher efficiency than a single nozzle unit operating under the same conditons. In most cases steam use can be reduced 10 to 15% with greater savings possible in some applications. Multiple nozzle units are more compact than single nozzle units when large capacity is required.

Single Nozzle, Spindle Operated

For spindle operated thermocompressors, shown in Figure 18-2(b), the nozzle design is quite different from that of fixed nozzle units. The nozzle and spindle assembly combines a rounded entrance orifice with a straight section into which a tapered spindle is guided. Operation is much like that of a needle valve. The length of the spindle travel is based upon the design requirements.

Unlike a control valve where energy is lost as flow is throttled, a spindle reduces the flow rate without reducing the energy available per unit flow. This permits the thermocompressor to use the maximum energy available at all ranges of flow and extends the capacity variation of a single unit.

Spindle operated units can be provided with either manually or automatically operated spindles. Thermocompressors with manually-operated spindles are designed for applications where loads will remain steady but where changes in operating conditions may occur and when operating conditions are not sufficiently established and some flexibility in nozzle orifice size is preferred. Thermocompressors with automatically-controlled spindles are used where pressure, suction, or discharge conditions vary and it is necessary to control discharge pressure or flow.

A characteristic performance curve for a spindle-operated thermocompressor is shown in Figure 18-4.

ESTIMATING DATA

The ratio of motive steam to load gas can be estimated using the correlation below. This relation should be used with caution! It is quite empirical and limited to ideal gases and to saturated steam at pressures below 300 psig. It should be used for values of R_m between 0.5 and 6.

$$R_m = (M_m/M_L)^{1/2} 0.40 e^{4.6 \ln (P_D/P_L)/\ln(P_M/P_L)} \tag{18.1}$$

where R_m = mass motive gas per unit mass load gas
M_m = molecular weight of motive gas
M_L = molecular weight of load gas
P_m = absolute pressure of motive gas

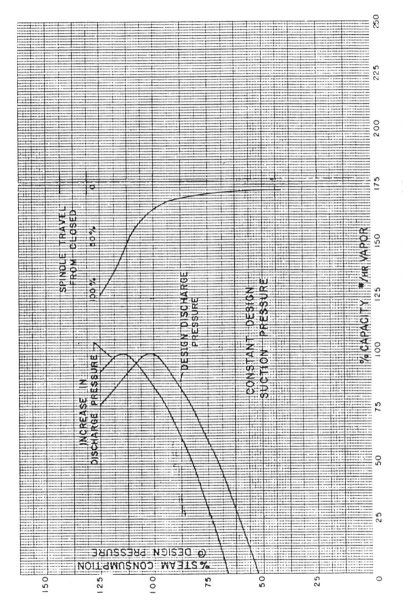

Figure 18-4: Characteristic performance curve of spindle operated thermocompressor.

182 Handbook of Evaporation Technology

P_L = absolute pressure of load gas
P_D = absolute pressure of discharge gas

Figure 18-5 and Figure 18-6 show typical capacity rates and sizing curves for fixed multiple nozzle and regulating spindle thermocompressors, with steam as both motive and suction fluids. These curves should be used to approximate thermocompressor size and performance. The actual efficiency can vary with the size of the thermocompressor, the control variability, the system, and the manufacturer.

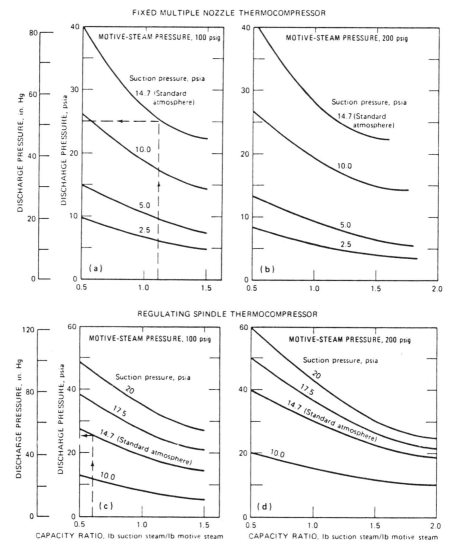

Figure 18-5: Capacity ratio curves of steam ejectors for thermal recompression.

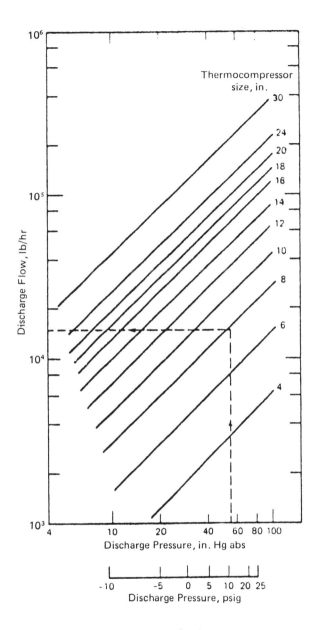

Figure 18-6: Sizing chart for thermocompressors.

Figure 18-7 can be used to estimate the purchase cost for fixed-multiple-nozzle and automatically-controlled-spindle thermocompressors.

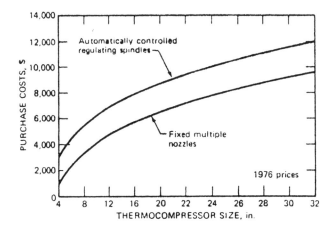

Figure 18-7: Purchase costs for thermocompressors. Basis: Material of construction, cast-iron–carbon steel bodies; nozzles, 304 stainless steel. No auxiliaries.

CONTROL

It is relatively easy to design a thermal compression evaporator for an operating point. However, once the thermocompressor has been designed and fabricated, its performance characteristics are fixed. The designer of an evaporator should carefully consider the effects of changing operating points. The problem is simplified with a single-effect evaporator where an absolute pressure controller can be used to maintain constant suction pressure. Even in this simplified case, the discharge pressure will vary as the motive-steam rate is varied to change capacity, or as the rate of heat transfer decreases due to fouling. As the discharge pressure varies, the capacity of the thermocompressor may or may not decrease depending on the thermocompressor's critical back pressure for the operating motive steam. For control, it could be desirable to throttle the motive steam flow, but this can introduce serious problems. The characteristics of the thermocompressor make it difficult to impossible to predict performance at conditions removed from the design point. Under these conditions, accurate prediction of evaporator performance at other than design conditions becomes impossible.

Because of the unpredictable performance of thermocompressors, control of evaporators using them is more difficult than for a conventional system where it is necessary to set only steam and feed rates to maintain a constant evaporation rate. One way to provide flexibility with better operating stability is to use two or more thermocompressors in parallel. This permits capacity control without loss in energy economy.

APPLICATION

Thermal compression, if applied correctly, offers, for a much lower installation cost, energy saving equivalent to the addition of another effect. Other advantages of thermocompressors are their simplicity of construction, reliability of mechanical performance, low maintenance costs, and ease of startup. Therefore under the right circumstances thermocompressors should be considered.

Thermal compression evaporators are well suited for single or double effect systems where low operating temperatures and improved economy are desired. They also find application when space is limited. For practical application the temperature difference across the thermocompressor should be less than 15°C. The application of thermocompressors is limited when the boiling point rise is high, high pressure steam (pressures above 60 psig) is not available, or the steam condensate must be returned to the steam generator. Thermal compression systems are not as flexible as some other systems and this sometimes limits their application. Thermal compression is a good tool for upgrading existing evaporator systems and this application is discussed in Chapter 40.

Perhaps the highest thermal vapor compression technology in any field can be found in the water desalination industry. Technology in that industry has advanced to designs with more than eight successive effects with steam economies over 12. Thermal compression evaporators beyond seven effects in process industries does not appear to be economical because mechanical vapor compression systems offer a better alternative.

19
Mechanical Vapor Compression

Mechanical vapor compression evaporator technology has been used for over 100 years and is a proven means to significantly reduce energy required for evaporation. It has seen rapid growth in the past two decades, but many are not familiar with its application. A vapor compression evaporator system, as shown in Figure 19-1, is similar to a conventional single-effect evaporator, except the vapor released from the boiling solution is compressed in a mechanical compressor. Compression raises the pressure and saturation temperature of the vapor so that it may be returned to the evaporator system chest to be used as heating steam. The heat present in the vapor is used to evaporate more water instead of being rejected to a cooling medium in a condenser.

The compressor adds energy to the vapor to raise the saturation temperature of the vapor above the boiling temperature of the solution by whatever net temperature difference is desired. The compressor is not completely efficient, having small losses due to mechanical friction, and larger losses due to nonisentropic compression. The additional energy required because of nonisentropic compression is not lost from the system, however, and serves to superheat the compressed vapor. The compression energy added to the vapor is of the same magnitude as energy required to raise feed to the boiling point and make up for heat losses and venting losses. It is usually possible to operate with little or no makeup heat in addition to the compression energy by proper heat exchange between the condensate and the product with the feed.

Compressor power is proportional to the increase in saturation temperature (ΔT) produced by the compressor. An evaporator designer must optimize power consumption and heat transfer surface. Mechanical vapor compression achieves the highest possible energy savings; these energy savings are reached at the expense of much higher investments and, in most cases, increased electrical power consumption.

Selecting the most economical method of evaporation requires knowledge of the costs of energy, capital, operating labor, spare parts, and maintenance over the life of the system. Energy and capital costs dominate and show the

Mechanical Vapor Compression 187

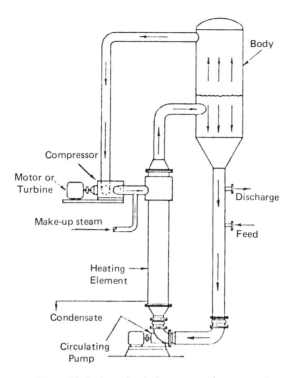

Figure 19-1: A mechanical recompression evaporator.

greatest variance among evaporator types. It is often difficult to assign true costs to energy sources. Each case must be considered individually. The economics can be calculated only if accurate costs are known; approximations and educated guesses will not provide the correct answer.

THERMODYNAMICS

The temperature-entropy diagram of a vapor compression system is shown in Figure 19-2. The liquid enters at **A** on the left and is heated to its boiling point **B** in a feed/condensate heat exchanger at pressure P_1. It is then evaporated at constant pressure and constant temperature to point **C**. The cross-hatched area between the line **B-C** and absolute zero represents the amount of heat that has to be added during the evaporation. The vapor at point **C** is now compressed in the compressor to the higher condenser pressure, P_2, at point **D**. The vapor is actually often slightly superheated before compression. The vapor at higher pressure is now condensed in the evaporator steam chest along the line **D-E-F**. As it condenses, it releases its heat of vaporization and provides the heat requirement of the cross-hatched area. The area bounded by **BCDEF** represents the compressor work. The ratio of this area to the cross-hatched area is the thermodynamic advantage of the vapor compression process. Finally, the condensate is cooled in the feed/condensate exchanger to point G. Two conclusions can be drawn:

(1) The lower the ΔT, the higher the thermodynamic advantage and the lower the energy consumption (less compressor power is required).

(2) The lower the ΔT, the more evaporator surface will be required and the higher will be the capital cost.

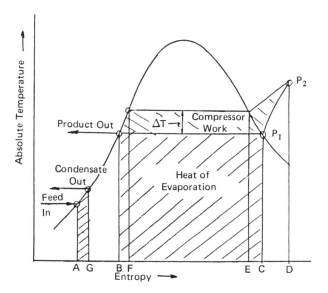

Figure 19-2: Temperature-entropy diagram of a vapor compression system.

FACTORS AFFECTING COSTS

Capital costs for vapor compression systems are usually higher than for multiple-effect systems, because they are usually designed for a lower temperature difference across the evaporator. As a result, greater heat transfer surface is required. In addition, the compressor and drive are relatively expensive. Equipment costs are affected by variables such as feed properties, materials of construction, available energy supply, and calandria type. Expected maintenance and necessary spare parts must be also evaluated. Heat exchangers must be provided to exchange heat between feed and condensate and between feed and product. Other costs that must be considered, but are less easily defined, are the penalty for interrupted operation during repairs and the part load response.

Careful analysis of life cycle costs and equipment size is important. Little can be done to correct an error in compressor sizing without incurring significant expense.

COMPRESSOR SELECTION

The compressor is perhaps the least understood component of a vapor compression system. It is a very simple piece of equipment, once a few simple facts are understood. However, the flow phenomena and mechanical dynamics within the machine are not simple, but they are generally of more concern to the designer than to the operator of the equipment.

The major factors to consider in selecting a compressor are the flow rate, compression ratios, variability of evaporator flow rates and operating conditions, vapor temperature, expected maintenance, and installed cost. Generally, the costs for all mechanical compressors limit their application to situations in which the compression ratio is less than 2, since multistage or a series of single-stage compressors would be required at higher compression ratios.

Compressors applied to evaporator systems include positive-displacement, and radial-flow and axial-flow centrifugals. Each has advantages for certain applications.

Positive-Displacement Compressors

Positive-displacement compressors, such as rotary lobe, screw type, or sliding-vane types, pump a relatively constant volume of vapor against whatever back pressure (resistance to heat transfer) is developed by the evaporator. The mass flow rate depends upon the inlet vapor density. Rotary lobe compressors are used for compression ratios up to about 2.0, the compression ratio being limited by mechanical and thermal distortion of small clearances inside the machine. The efficiency of positive-displacement compressors is between 60 and 75%.

The maximum inlet capacity of a single-stage rotary lobe compressor is about 30,000 cubic feet per minute which is equivalent to 60,000 lb/hr of water vapor at atmospheric pressure. Positive-displacement compressors pump against whatever resistance is developed by the evaporator; discharge pressures will increase and evaporator rate will remain relatively constant as the evaporator becomes fouled. Evaporator capacity may be varied when using positive-displacement compressors by:

(1) using a variable-speed driver

(2) using multiple compressors in parallel

(3) varying evaporator condensing pressure to change vapor volume.

Positive-displacement compressors are most often used for low-flow applications and for slightly higher compression ratios than are centrifugal types. Single-stage positive displacement compressors would appear to be better suited to compression evaporation than centrifugal machines because of lower cost and their characteristic fixed capacity. However, they are not available in a wide range of materials of construction and are limited in capacity and efficiency. Consequently, their application is limited.

Axial-Flow Centrifugal Compressors

Axial-flow centrigugal compressors direct the flow of vapor parallel to the compressor shaft. Their capacity range is much higher than that of radial-flow compressors and approaches a maximum at 400,000 inlet cubic feet per minute or approximately 800,000 lb/hr of water vapor at atmospheric conditions. The total efficiency is 85%, the highest for any compressor.

At compression ratios above 1.5, a multistage unit must be used. Axial-flow compressors are more expensive than radial-flow compressors, and, except for situations in which the capacity is needed or the efficiency justifies the added cost, a radial-flow compressor is normally used.

Radial-Flow Centrifugal Compressors

Radial-flow centrifugal compressors are commonly used to compress larger volumes of vapor at lower compression ratios than can be handled by positive-displacement machines. Radial-flow compressors are the most common compressors for large evaporation applications where the evaporation rate exceeds 20,000 lb/hr. In these compressors, the vapor changes direction radially at the inlet in the compressor. For most services, their design simplicity and commercial flow ranges bracket those of the axial-flow and positive-displacement types. Their capacity ranges are 1,000 inlet cubic feet per minute (2,000 lb/hr of water vapor) to 225,000 cubic feet per minute (450,000 lb/hr of water vapor) for atmospheric suction conditions. Efficiency for radial-flow compressors is 75 to 80%. The single-stage compression ratio is limited to about 1.7 at inlet volumes of 225,000 cubic feet per minute. The compression ratio is limited by rotor dynamics, gearing, and impeller strength. The basic advantage for radial-flow units when compared to axial-flow units is a greater range of performance to match standard methods of pressure and temperature control, as well as their capability for turndowns of 50% at constant speed and a lower sensitivity to changes in suction and discharge pressures.

One manufacturer offers compressors which operate successfully at higher speeds by using better impeller materials, integral gearing, and operation above first critical speed. Compression ratios up to 2.0 with a maximum inlet capacity of 55,000 cubic feet per minute (110,000 lb/hr of water vapor at atmospheric conditions). Higher compression ratios may be obtained by operating two or more single-stage machines in series or by using a multistage compressor. Using more than one single-stage compressor in series is often less expensive than using a multistage compressor.

The best compressor type for most current applications is the single-stage radial-flow centrifugal compressor. It is a simple and dependable machine that has proven itself for heavy duty use over a broad range of industrial applications.

FACTORS INFLUENCING DESIGN

Selection of compression ratio is a major design decision and requires optimization of capital cost and energy consumption. The proper selection requires knowledge of energy costs and required return on investment. The decision will also be influenced by feed characteristics, material of construction, type of evaporator, hours of operation per year, turndown required, and type of drive.

Evaporator condensing pressure also has a strong influence on capital and operating costs. The most desirable suction pressure is slightly above atmospheric. Vacuum operation may be required to achieve low process temperatures. However, vacuum operation has the following disadvantages:

(1) The vapor volume is greater; thus larger compressors and vapor piping are required.
(2) Air leaks into the system and increases noncondensables and the effects they produce.
(3) A vacuum producing system is required.
(4) Mechanical design of equipment must be adequate for vacuum operation.
(5) Lower temperatures and pressures may reduce rates of heat transfer.

Many of these disadvantages may also be present for any evaporator system. Mechanical vapor compression systems can be operated under vacuum, and many times are the best economical choice.

Boiling point rise is a major factor in design. High boiling point rises make vapor compression uneconomical because of the high compression ratio required with its resultant high energy use and, in some cases, high capital cost because of the need for multistage compressors or compressors in series. Boiling point rises greater than $6°C$ generally preclude the economical application of vapor compression.

Compressors designed to operate at very low ΔT ($3°C$ or less) have efficiencies that are not economically attractive. Compressor efficiency tends to peak at ΔT's of $10°$ to $15°C$. Often multiple-effect vapor compression systems are economical because the vapor flow volume is reduced approximately as the number of effects increases and the compressor ΔT is the sum of the ΔT's for all effects. Such applications may result in reduced investment as well as energy consumption.

Mechanical vapor compression can often be used in preconcentrators where the major portion of the evaporation is accomplished. The remaining concentration can then be done in some other evaporator system.

These factors must be considered:

(1) A rigorous heat balance for the plant is necessary to determine the best compressor drive. Any motive power can be used; the choice requires individual analysis of all factors present at a given location.
(2) The availability of cooling water should be determined. Mechanical vapor compression has no cooling-water requirements. This may be important where expansion of existing facilities is required or when water is expensive.
(3) The amount of automation required must be established.
(4) Space availability should be evaluated. Vapor compression units can be made extremely compact.
(5) The effects of any required maintenance must be known.
(6) The proper control scheme must be implemented.
(7) The calandria type must be properly determined.

(8) The proper safety factor for heat transfer surface must be studied.

DRIVE SYSTEMS

Any motive power is a logical candidate for driving the compressor. A few of the possibilities include:

(1) *Electric Motor:* When power costs are favorable (when compared to steam) and power is readily available, electric motors are indicated. In small units, this is the most practical and economical choice. Most larger units also are driven with electric motors.

(2) *Steam Turbine:* Compressors driven by steam turbines may offer many advantages. They are more easily adapted to speed control and can be used as a reducing station if backpressure turbines are indicated. The turbine exhaust steam can be condensed in the same evaporator or in auxiliary equipment such as product heaters, final concentrators, etc.

(3) *Diesel Drives:* Diesel or gas engine drives offer considerable additional savings in primary energy costs when compared to electric motor drives. Heat recovery of the exhaust heat in the exhaust gas and cooling water can largely be utilized. The useful energy yield can be therefore improved. An added advantage of such a system is that it is self-starting and requires no external source of heat.

The choice of drive is determined by the type of energy available in a plant and its cost. Turbine steam rate depends on inlet steam pressure and temperature, exhaust conditions, number of turbine stages, and turbine efficiency. With a turbine drive, evaporation rate may be controlled by varying compressor speed, making sure that the turbine and compressor critical speeds do not fall in the range of interest for the system.

Electric motor and gear drives are less expensive than steam turbines but until recently have not been suited for compressor speed control.

CENTRIFUGAL COMPRESSOR CHARACTERISTICS

Centrifugal compressors are dynamic compressors and are essentially constant pressure, variable volume machines and must be regulated in almost all applications. Before control can be discussed, the operating characteristics must be understood. The discussion to follow will be concerned with radial-flow compressors. Axial-flow compressors have similar characteristics curves, although the change in pressure rise is much greater with equivalent changes in compressor rotational speed.

Constant Speed

Figure 19-3 depicts a typical centrifugal compressor performance map operating at some fixed speed. The map is usually based on constant inlet conditions. The shape of the head curve is shown for a particular impeller design. Centrifugal compressor impellers can be designed to some extent to achieve more nearly the characteristics required for the system. For practical reasons, manufacturers usually offer standardized impeller designs with standardized casing size.

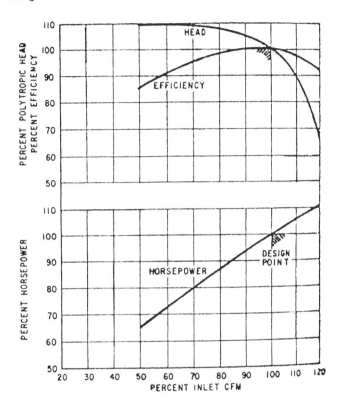

Figure 19-3: Typical constant speed performance curves for a centrifugal compressor.

Figure 19-4 shows the characteristic curve for a given compressor operating at constant speed but under varying inlet conditions. Performance is affected by the pressure, temperature, molecular weight, and specific heat ratio for the fluid being compressed.

Capacity Limitations

There are definite limitations of the stability range of a centrifugal compressor. Surge limits the minimum capacity and normally occurs at 50% of the design inlet capacity at design speed. There is a maximum discharge pressure and minimum vapor rate at which operation becomes unstable for an evaporator. At surge conditions the inlet and discharge pressures and the flow rates rapidly

194 Handbook of Evaporation Technology

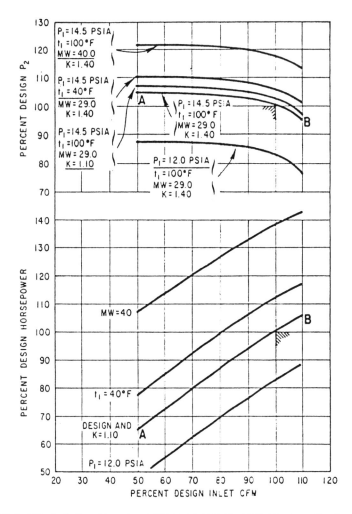

Figure 19-4: Effect of changing gas conditions on centrifugal compressor operating at constant speed.

oscillate as the compressor tries to find a condition to satisfy its stable-performance characteristics. The surge occurs because the process requires a higher compression ratio than the compressor can achieve. The instability caused by surge will cause flashing in the evaporator body, loss of capacity, variable product composition, and excessive entrainment. If surging becomes violent or prolonged, the compressor will be damaged. To operate at flows below the surge flow requires controls.

While the stability range of a centrifugal compressor is commonly indicated from the rated point to the surge point, the unit can operate stably to the right of the rated point. The greater the load demand on a centrifugal compressor, the greater the "fall-off" in delivered pressure. The upper limit of capacity is determined by the phenomenon of "stonewall." Stonewall occurs when the velocity of the fluid approaches its sonic velocity somewhere in the compressor, usually at the impeller inlet. Shock waves result which restrict the flow, causing a "choking" effect—rapid fall off in discharge pressure for a slight increase in volume throughput. The maximum volume is limited to about 120 to 130% of design. Stonewall is usually not a problem when compressing water vapor or air; however, the problem becomes more prevalent as the molecular weight of the compressant increases.

The discussion above was limited to single-stage compressors. As the number of stages increases, performance maps tend to show a more sloping curve with less stable range as dictated by the particular application.

Variable Speed

A typical variable speed performance curve is shown in Figure 19-5. With variable speed, the compressor easily can deliver constant capacity at variable pressure, variable capacity at constant pressure, or a combination variable capacity and variable pressure.

Basically, the performance of the centrifugal compressor, at speeds other than design, are such that the capacity will vary directly as the speed, the head developed as the square of the speed, and the required power as the cube of the speed. By varying speed, the centrifugal compressor will meet any load and pressure condition demanded by the process within the operating limits of the compressor and the driver. It normally accomplishes this as efficiently as possible, since only the head required by the process is developed by the compressor. This compares to the essentially constant head developed by the constant speed compressor.

Factors Affecting Control

Most centrifugal compressors require some form of regulation. The type of control used depends first on the compressor drive. For turbine-driven compressors, normal control is accomplished by varying the speed. This method of controls permits a wide range of operation in a relatively efficient manner. Speed control is more efficient than throttling the flow at constant compressor speed; by creating resistance, an unrecoverable loss in power results. Variable speed and constant speed operation can be compared with Figures 19-6 and 19-7. Compared to constant speed operation, variable speed introduces some complications. Foremost is the selection of the critical speeds relative to the operating range. It may not be possible to select the critical speeds for compressor, driver, and the combined train to be above the maximum operating speed. In this case, the first critical speed should be well below the lowest operating speed. Operations at or near a critical speed shortens the life of the equipment at best and can lead to almost instant destruction at worst. It is generally recommended that the compressor manufacturer be given unit responsibility for variable speed equipment and for compressors with steam turbine drive. He is equipped to deal with the dynamics of the compressor train and can select a turbine com-

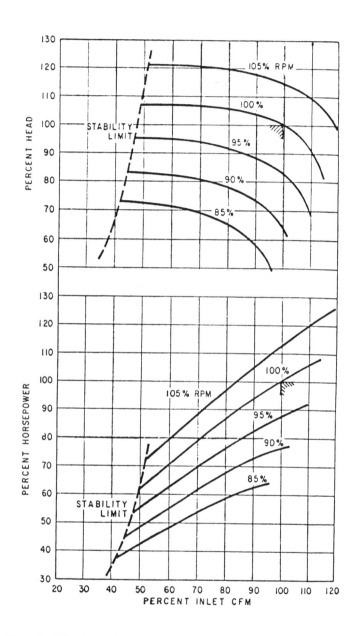

Figure 19-5: Typical variable speed performance curves for centrifugal compressor.

Mechanical Vapor Compression 197

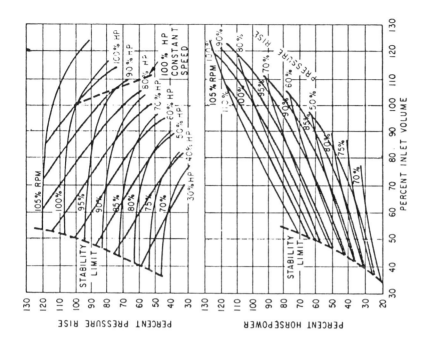

Figure 19-7: Typical variable speed performance curves. Compare the slope of the constant horsepower lines with Figure 19-6.

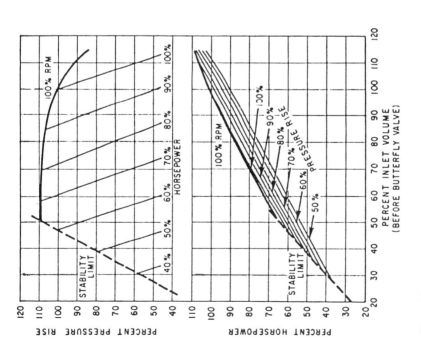

Figure 19-6: Typical constant speed performance curves. Compare the slope of the constant horsepower lines with Figure 19-7.

patible with the compressor. Overspeed protection and common lubrication systems will also be provided. Generally, speed variations in excess of 40% should be avoided. Speed control uses the minimum amount of energy to meet process considerations since there are no head losses from control valves.

For motor-driven compressors, the control can become more intricate, especially for the usual constant-speed motors. Motor-driven compressors can be controlled as follows:

(1) Solid-state AC speed controls are becoming more reliable and less costly. They should be considered in every application.

(2) Hydraulic or electric couplings can be used to obtain speed variations. This is not a popular method because of severe coupling efficiency penalties throughout the range of operation.

(3) Butterfly valves at the compressor inlet or discharge can be used to throttle the flow. Suction throttling is preferred because this reduces fluid density and results in a reduction in power required. Discharge throttling fixes the power consumption and is used only for small systems whose power efficiency can be ignored. Flow or pressure control can be achieved by either suction or discharge throttling; however the ability of the compressor to adjust to wide variations in load when controlled by throttling an in-line valve is inferior to speed control because the relationship between the valve position and weight flow to the process is not linear. The energy required to control compressor operation may differ significantly depending both on the location of the valve and on the compressor characteristics. When the compressor characteristic has a negative slope as is true of most centrifugal machines, especially multistage, the power required to meet the process demand is *less* when using suction throttling than when using discharge throttling. The difference in energy usage depends on the slope of the curve. If a rising compressor characteristic exists, discharge throttling would require a smaller energy use. (See Figure 19-8) This rising characteristic is usually found in low-speed applications ("squirrel-cage" blowers are an example) and has little process application. A throttling valve at the inlet does not change the head produced by the compressor; the reduction in power is solely due to the reduced mass flow.

(4) Adjustable inlet guide vanes can be used to adjust the characteristic curve of the centrifugal compressor. Adjustable inlet guide vanes are most efficient for conventional compressors developing less than 30,000 feet head or a multistage compressor with three stages or less. Inlet guide vanes are more expensive and complex than butterfly valves and are installed at the compressor inlet. Power savings by the use of inlet guide vanes are 10 to 15% for a single-stage unit, 5 to 10% for a two-stage unit, and 3 to 5% for a three-stage unit. The efficiency of inlet guide vanes is diminished

when multistage compressors are used because the guide vanes control the inlet to the first-stage rotor blades only. Adjustable guide vanes are available at the inlet of every stage but they are not widely used because of the expense. Guide vane control is much like suction valve throttling in terms of linearity of the control system. However, the energy required to throttle the suction flow is minimized since the inlet guide vanes change the angle of incidence between the inlet fluid and the rotor blades and thereby unloads or overloads the impeller. Inlet guide vanes operate by causing a pre-rotation to the vapor entering the impeller. Pre-rotation of the vapor ahead of the impeller in the direction of impeller rotation (co-rotation) restores the proper aerodynamic angle of attack. As a result, the head produced is reduced and with it the power required for a given mass flow. This is accomplished with a decrease in design efficiency. However the ratio of the head to efficiency decreases from its design value, resulting in lowered power. Rotation of the inlet guide vanes effectively shifts the compressor characteristic and this variable head-flow characteristic can readily be used for load control.

(5) A power wheel can be used at the inlet of the compressor upstream of the first-stage impeller. The fluid to be compressed expands through a set of movable guide vanes and is directed upon the turbine blades. The power wheel or turbine wheel is theoretically designed to more efficiently handle part load conditions than adjustable inlet guide vanes by itself. However, the power wheel has disadvantages and this control method is not popular. Major disadvantages are the inherent parasitic losses at the wheel at design conditions which lowers efficiency.

(6) Adjustable diffuser vanes can also be used, but this is not a popular control method because of the expense. They may also lead to heavy maintenance.

(7) A wound-rotor induction motor can be used for speed control. These motors are relatively expensive and inefficient and are not often used.

(8) A DC motor can be used for speed control. Again, this is a relatively expensive and inefficient drive.

SYSTEM CHARACTERISTICS

It is extremely important that the system resistance or characteristics be fully known before selecting a control system. Figure 19-9 shows a plot of pressure versus capacity for a constant speed compressor. Shown on this plot also are three different types of system characteristics. A compressor operating against a fixed head or pressure could have a system characteristic defined by

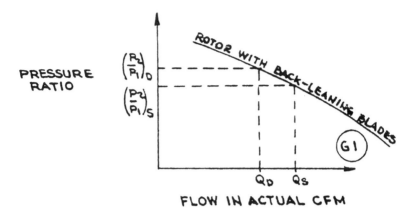

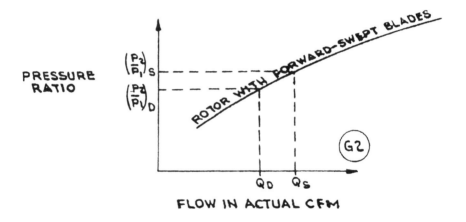

Figure 19-8: Comparison of power requirements when throttling the suction and discharge of a compressor.

AB. A compressor discharging into a large system through a long run of pipe would have a system characteristic defined by AD. A system which has essentially fixed top pressure with some pipe friction is represented by AC.

In Figure 19-9, the compressor head-capacity curve follows none of the system characteristics. Variation of the compressor outlet to meet the demand of the system, therefore, requires controls to regulate the volume, pressure, or a combination.

Mechanical Vapor Compression 201

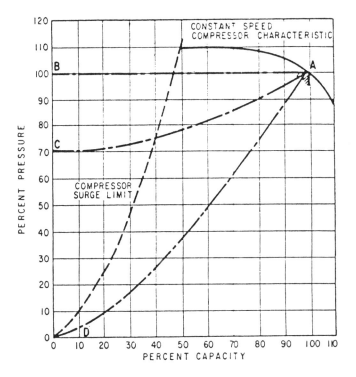

Figure 19-9: Pressure vs. capacity for constant speed compressor.

Vapor compression evaporator systems have characteristic curves that depend upon the type of calandria used and the amount of fouling experienced. For forced circulation and falling film calandrias, the capacity varies almost linearly with ΔT. For such systems, the evaporator operating lines start on the ordinate with an offset for the boiling point rise and are essentially straight lines. For natural circulation systems, the evaporator operating lines are not straight lines because boiling heat transfer rates are functions both of pressure and temperature difference.

Figure 19-10 indicates a performance map for an evaporator system having a linear relationship between capacity and ΔT using inlet guide vanes for compressor control. The lines have different slopes corresponding to different amounts of fouling. The compressor curves, each for a different inlet guide vane setting, show that compressor ΔT decreases with increasing flow. The intersection of the proper operating line with the proper compressor curve is the system operating point. The system should be designed with some margin, some extra capacity, and during normal operation the guide vanes would be slightly closed. The surge limit must be evaluated and operation must not occur in the surge region.

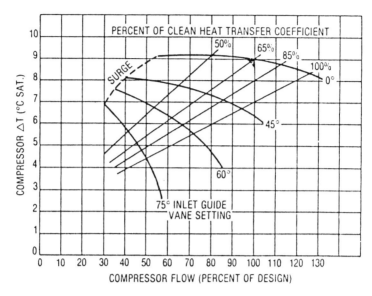

Figure 19-10: Performance curve of vapor compression evaporator with inlet guide vane control.

Speed control for the compressor of an evaporation system with linear characteristics is shown in Figure 19-11. Inlet guide vane control permits greater capacity turndown and allowance for evaporator fouling before the surge condition is reached.

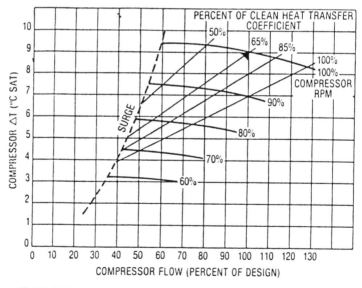

Figure 19-11: Performance curve of vapor compression evaporator with speed control.

RELIABILITY

Large centrifugal compressors have proven to be highly reliable in evaporator systems. Most operational problems have been caused by improper matching of compressor and evaporator and excessive entrainment. Several actions are recommended to maximize compressor reliability:

(1) minimize entrainment in the vapor by using suitable entrainment separators.

(2) recycle a portion of the superheated compressor discharge vapor to the inlet to dry any remaining entrainment. Water droplets can erode the compressor impeller.

(3) monitor solids content in the condensate. Solids in the vapor may build up on the impeller and break off in pieces to cause imbalance.

(4) examine the impeller periodically through a sight glass in the inlet line. Check for evidence of deposits or erosion. The impeller can be checked while running by using a strobe light.

(5) install vibration and temperature monitors on compressor, driver, and gear bearings to warn of imbalance or excessive wear. Automatic shutdown alarms will avoid catastrophic failures. Operators should not override alarms without determining the cause of abnormal vibration or temperature.

(6) open compressor casing drains during shutdowns and use a forced air dryer to keep the compressor dry to avoid corrosion.

(7) equipment must be properly aligned during installation and the alignment must be maintained during operation. A firm foundation and freedom from loads imposed by piping are required for proper alignment.

(8) flow uniformity at the compressor inlet must be provided. Adequate straight length of piping or properly designed turning and straightening vanes will satisfy this requirement. Vanes and other insertions can fail if vibration is excessive and lead to serious impeller damage.

Good operator training, regular inspection, and regular collection of operating data are important. A compressor is a piece of rotating equipment and it will require some maintenance. Equipment layout should facilitate necessary maintenance and reduce the danger of accidental damage.

EVAPORATOR DESIGN

Forced circulation and falling film evaporators offer characteristics that make them attractive for vapor compression systems. Falling film calandrias are particularly well-suited for vapor compression systems:

(1) they result in vapor with very little entrained liquid

(2) they provide high rates of heat transfer

(3) they require lower liquid circulation rates (smaller pumps)

(4) they are suitable for operation at low temperature differences.

Two factors are important when designing calandrias for vapor compression systems. Entrainment must be minimized to avoid both deposition on compressor impeller and to minimize fouling of the condensing side of the calandria. Venting of calandrias is also critical and has a greater effect on overall performance than for other systems.

APPLICATION

Vapor compression evaporation finds wide application when delta temperatures are low ($10°C$ to $15°C$), boiling point rise is low (less than $6°C$), vapors are not extremely corrosive, and fouling is not excessive or can be effectively handled. They are usually applied when compressor suction pressures are slightly above atmospheric or higher, although vacuum operation is also common. Generally, single effect systems are used but economics in many systems can easily justify multiple effects. Vapor compression can frequently be used for upgrading existing systems. They are generally more economical when the cost of electricity or its availability is favorable.

SUMMARY

The main advantage of vapor compression evaporation is significantly lower energy consumption. The energy-saving potential is founded in the thermodynamics of the process. The vapor compressor is crucial, but is basically a simple device. Attention to good design detail pays larger dividends with vapor compression than with conventional systems. Vapor compression has proven to be a reliable method of achieving low-energy evaporation when properly engineered and operated.

ECONOMICS

The costs for a mechanical compressor are primary based on the size of the drive and on the vapor rate and volume.

The theoretical mechanical adiabatic compressor horsepower can be estimated using the correlation below:

$$BHP = 1.89(R^{0.231} - 1)(CFM/100)(P_s/EFF) \qquad (19.1)$$

$$R = (P_D/P_s) \qquad (19.2)$$

where BHP = brake horsepower
R = compression ratio
P_D = absolute discharge pressure, psia

P_s = absolute suction pressure, psia
CFM = inlet volumetric flow, cubic feet per minute
EFF = compressor efficiency

Several published comparisons of vapor compression systems and multiple-effect systems have presented curves showing breakeven conditions based upon energy costs. These are presented in Figure 19-12.

Case A is a standard sextuple effect evaporator compared against mechanical compression for evaporating sulfate black liquor at a rate of 300,000 lb/hr (1978).

Case B is a comparison of a triple-effect evaporator and mechanical compression system for evaporating molasses-yeast residue at a rate of 33,680 lb/hr (1974).

Case C is a comparison of multiple-effect evaporator and mechanical compression for evaporating kraft black liquor at a rate of 300,000 lb/hr (1977).

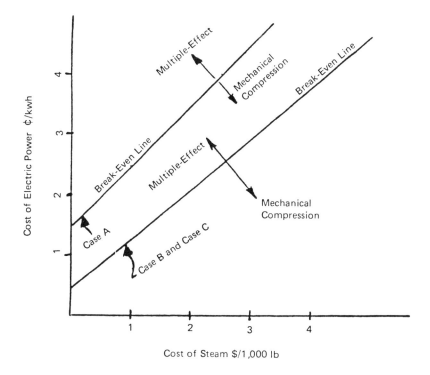

Figure 19-12: Comparison of vapor compression systems and multiple-effect systems.

20
Desalination

Seawater or brackish water is used for process applications or as potable water when fresh water is scarce. Six techniques are used for desalination. Five are evaporation processes: multiple-effect; thermocompression; mechanical vapor compression; once-through multistage flash; and multistage flash with brine recirculation. The sixth process, reverse osmosis, uses membrane technology for desalination.

Selection of the proper desalination process requires consideration of several criteria:

(1) operability and start-up

(2) maintenance and complexity

(3) capital and operating costs.

Final selection often depends upon factors which are site-specific.

STARTUP AND OPERABILITY

All six processes are easily operated once they have been started up and are under control. Time required for startup can vary from one hour or less for reverse osmosis units to four hours for multiple-effect evaporators. Ease of startup is not usually a major consideration because most desalination plants operate continuously with few shutdowns.

COMPLEXITY

All processes exhibit some degree of complexity based upon the need for controls, chemical addition, pumps, and other components. Reverse osmosis is the least complex process followed in order by once-through multistage flash, thermocompression, multistage flash with brine recirculation, multiple-effect, and mechanical vapor compression as the most complex.

Other factors that affect complexity are site-specific and are not included the the ranking above.

Seawater Supply: Seawater required per unit of water produced varies from 2 for mechanical vapor compression to 3 for reverse osmosis and 8-10 for the other processes. Seawater requirements may dictate more or larger equipment.

Seawater Quality: Evaporating processes are not greatly influenced by seawater quality if operating temperatures and concentrations are properly controlled. Seawater quality significantly affects the complexity and extent of pretreatment required for reverse osmosis.

Brine Disposal: The volume of brine to be disposed varies and is directly related to the required seawater supply.

MAINTENANCE

Maintenance requirements could be listed in the following order, from lowest to highest maintenance: thermocompression; once-through multistage flash; multistage flash with brine recirculation; reverse osmosis; multiple-effect evaporation; mechanical vapor compression. Several factors are important:

Thermocompression: Thermocompression systems operate at low temperatures with little scaling or corrosion. The process is simple with few pumps or controls.

Once-Through Multistage Flash: This system does not require acid addition and hot-end corrosion can be minimized with proper material selection and venting. Few pumps or controls are needed and frequent cleaning is not required with proper operation.

Multistage Flash With Brine Recirculation: More pumps and controls are needed. Two treatments are used for scale control: acid and non-acid. Non-acid systems require maintenance similar to once-through systems. Acid addition usually results in corrosion but better material of construction and proper operation can reduce corrosion related maintenance.

Reverse Osmosis: Membrane modules must be replaced periodically and high-pressure pumps require maintenance. Operation must be closely monitored.

Multiple-Effect Evaporation: Scaling can be increased because of maldistribution and concentration gradients. Acid treatment can result in corrosion. Additional pumps are needed.

Mechanical Vapor Compression: Scaling and acid treatment problems can also occur. The mechanical compressor and its motor, gearbox, and lubrication system require considerable maintenance.

ENERGY EFFICIENCY

Energy efficiency is usually measured by "economy" or performance ratio often expressed as pounds of water product per 1000 Btu energy input. However the six processes do not use the same type of energy and the relative costs of the type of energy required must be properly assessed.

Reverse Osmosis: Economies are 30–200 depending upon whether recovery turbines are provided. Electrical energy is used.

Meachanical Vapor Compression: Economies are typically 17–33 and electricity or high-pressure steam is required to drive the compressor.

Multiple-Effect Evaporation: These systems have maximum economies of 10–15. Low-pressure steam or hot water can be used.

Multistage Flash With Brine Recirculation: These systems have economies of 3–12. Greater economy is possible, but more stages are required. The practical maximum number of stages is about 50 where economy is about 12. Low-pressure steam or hot water can be used.

Once-Through Multistage Flash: These systems also have economies of 3–12 but are generally designed with economies lower than that of brine recirculation systems.

Thermocompression: Thermocompression systems typically have economies of 3–6 but 10 or greater can be achieved by using a large number of effects. High-pressure steam is required. Thermocompression systems operate at low temperatures which reduces corrosion and scaling.

CAPITAL COST

Capital costs are dictated primarily by capacity and economy; other factors that influence cost are operating temperature, materials of construction, pretreatment, chemicals, and auxiliary energy.

For small plants (less than 250,000 gallons per day) reverse osmosis plants have lower capital costs followed by multiple-effect (10% higher than reverse osmosis), once-through multistage flash (20% higher than reverse osmosis), multistage flash with brine recirculation (30% higher than reverse osmosis), and thermocompression and mechanical vapor compression (both 50% higher than reverse osmosis).

For large plants, multiple-effect and multistage flash with brine recirculation have lower costs followed by once-through multistage flash (10% higher), thermocompression (40% higher), and reverse osmosis and mechanical compression (both 50% higher).

OPERATING TEMPERATURE

Operating temperature affects capital cost, operation, and maintenance. Reverse osmosis systems operate at ambient temperatures (0–40°C); maximum operating temperatures for other processes are:

Thermocompression	65°C
Mechanical vapor compression	
Non-acid	75°C
Acid	100°C

Multistage flash, once-through
    Non-acid   85°C
    High-temperature additive   110°C

Multiple-effect evaporation
    Non-acid   85°C
    Acid   115°C

Multistage flash, brine recircualtion
    Non-acid   85°C
    High-temperature additive   110°C
    Acid   120°C

MATERIALS OF CONSTRUCTION

Materials of construction are selected to control corrosion and reduce maintenance at reasonable cost.

PRETREATMENT

Most processes require pretreatment but reverse osmosis often requires more complex pretreatment such as filtration and chlorination.

CHEMICALS AND AUXILIARY ENERGY

Process chemicals (such as non-acid treatment, acid treatment) and auxiliary electricity required for pumps vary for the six processes. Generally processes that require more auxiliary electricity require less chemical treatment. The evaporation processes generally require sulfamic or hydrochloric acid for cleaning; requirements depend upon the location. Reverse osmosis may require chlorine and cleaning agents such as detergents.

21
Evaporator Accessories

Various types of equipment, both major and minor, must be supplied for every evaporator in addition to the evaporator body itself. These include:

(1) condensers
(2) vacuum producing equipment
(3) condensate removal devices
(4) process pumps
(5) process piping
(6) instrumentation
(7) safety relief equipment
(8) thermal insulation
(9) equipment and pipeline tracing
(10) valves, manual and control
(11) process vessels
(12) electric motors and turbines
(13) refrigeration.

22
Condensers

Condensers are used in evaporation systems for two primary applications: to condense the vapor from the last effect and to serve as intercondensers in multistage vacuum producing systems. Two types of condenser are employed: direct contact and surface condensers. Both types can be of several designs.

DIRECT CONTACT CONDENSERS

Direct contact condensers condense vapors as they are contacted in some fashion with the cooling medium. Heat transfer surfaces are thus eliminated and vapor pressure drop is low. Because no surfaces are required, higher outlet cooling water temperatures can often be tolerated; however steam condensate is lost by mixing with the water. In some designs, more cooling water may be required than would be required for surface condensers. Direct contact condensers can generally provide lower condensing temperatures than surface condensers. They are also useful when entrained solids which would foul surface condensers are present.

Direct contact condensers are inexpensive and the penalty for oversizing is minimal if cooling-water pumping costs are not prohibitive. They can be built from a variety of materials. The primary consideration in applying these condensers is their environmental impact. If the cooling water will require waste treatment as a result of contact with process vapors, the waste treatment cost will usually be prohibitive and economics will dictate the use of surface condensers. Many existing direct contact condensers are being or have been replaced with surface condensers for this reason.

In some applications direct contact condensers use the fluid to be condensed as the cooling media. The coolant must be recirculated through an appropriate cooler as shown in Figure 22-1. These systems are used when low pressure drop is important, when the vapor has a high solid content, or when the process is very corrosive in the vapor phase. Such condensers are normally designed with vessel diameters sized as flash tanks. The spray nozzle is selected to provide the desired particle size. Knowing the spray rate and particle size, holdup time can be determined. Consequently the available heat transfer surface is known and heat transfer rates from the vapor to the particle can be calculated. Spray systems may result in considerable internal vapor circulation, especially if spray systems are small or numerous.

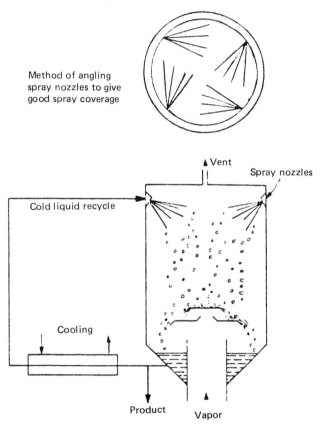

Figure 22-1: Spray condenser showing method of angling the spray nozzles to give good spray coverage, and recycling of the product.

Design of Direct Contact Condensers

Several factors must be considered in the design of direct contact condensers:

(1) location of vapor, water, and noncondensable connections

(2) water distribution arrangement

(3) direction of vapor and water flow

(4) vapor nozzle size

(5) water nozzle size

(6) noncondensable outlet nozzle size

(7) water discharge nozzle size

(8) body diameter and length

(9) mechanical design and corrosion allowance

(10) hot well size and capacity.

Three main factors affect the operation of any condenser:

(1) the quantity and temperature of the entering vapor

(2) quantity and temperature of the water supplied to the condenser

(3) the capacity of the pump provided to remove any noncondensables.

When designing a condenser for condensing process vapors from the last evaporator effect it is well to consider that the condenser is normally placed at the end of a train of operations which may themselves give variable and unpredictable results. Consequently a reasonable margin of safety must be provided for any condenser design. This does not mean that, in the face of a number of unpredictable variables, the condenser must be large enough to compensate for shortcomings in the design or operation of the equipment which precedes it.

SURFACE CONDENSERS

Surface condensers, as the name implies, provide a suitable heat transfer surface for condensing to occur. Consequently, cooling water and process fluids remain completely separated. However, the heat transfer surface is subject to fouling, corrosion, and plugging. Surface condensers are generally more expensive than direct contact condensers, but offer many advantages which frequently make them the most economical choice. Advantages include:

(1) Cooling water is not contaminated with the process. This is frequently the most important factor in selection of the condenser type.

(2) Condensate can be recovered. Steam condensate can be returned to the power house. Process condensate can be further processed.

(3) Cooling water can be used in series when multistage jets are used. Consequently, cooling water requirements may be reduced.

(4) Barometric legs or water removal pumps are not required.

(5) Lower headroom is required, generally.

(6) Corrosion may be reduced or eliminated when mixtures of process and water are highly corrosive.

(7) Vapor loads to secondary ejector stages may be reduced if water contains excessive amounts of dissolved gases.

The design of a surface condenser involves an optimization that considers total purchase, installation, and operating costs for both condenser and processing application. For instance, a condenser can be designed to cool the vapor-gas mixture leaving the condenser to within $3°C$ of the initial coolant temperature which may reduce the cost of a vacuum system. However the larger condenser surface required may not be economical. Generally, surface condensers will result in higher process pressure drop and lower temperature approaches to the coolant than will direct contact condensers.

Surface condensers can be of several types:

(1) shell-and-tube
(2) spiral plate
(3) spiral tube
(4) air-cooled
(5) evaporative cooled.

When water is used as a coolant, shell-and-tube units may be installed horizontally or vertically. Condensing may occur either on the tube side or the shell side. For air-cooled and evaporative-cooled condensers, the process almost always is condensed inside horizontal tubes, although air-cooled units may be vertical or inclined.

Shell-and-Tube Condensers

Vertical vapor-in-tube downdraft condensers offer several advantages over horizontal condensers. Many of the same advantages are offered by vertical vapor-in-shell condensers that use baffles designed to permit condensate to remain on the tube. The advantages of vertical condensers and differences between vertical and horizontal condensers are:

(1) Condensate subcooling is more efficiently accomplished in falling-film heat transfer. Horizontal tubeside subcooling uses only a small portion of the available area unless the tube is flooded. Flooded tubes lead to operational problems. Appreciable horizontal shellside subcooling can be accomplished only by flooding part of the shell, unless the condenser is designed so that the vapor passes vertically across the tube bundle only once (horizontal one-pass crossflow), with subcooling in the falling-film mode. Flooded shells usually result in higher vent losses.

(2) Vapors enter the tube at relatively high velocities affording good distribution and producing high heat transfer rates by reducing condensate thickness. Noncondensables are forced to the outlet and venting is more efficiently accomplished.

(3) Better heat transfer is obtained at high condensate loading. Vertical condensers are generally more economical than horizontal shellside condensers employing lowfinned tubes.

(4) Better heat transfer is achieved when condensing in the presence

of noncondensables. Condensate and uncondensed vapors and gases always coexist and are more nearly in equilibrium. In horizontal vapor-in-shell condensers, condensate drops from the tubes as it is formed and consequently, is not coexistent with vapor and noncondensables.

(5) Countercurrent flow is accomplished resulting in maximum possible subcooling of condensate. Noncondensable gases, which always exist in a condenser, are contacted with the lowest available temperature before removal. Vent losses are therefore at a minimum.

(6) Fluted tubes can be used.

(7) Pressure drop is generally lower for vertical vapor-in-tube condensers. However, for very low operating pressures (below 50 torr) shellside condensing will provide lower pressure drop if equipment is appropriate designed.

(8) More accurate control is afforded for vertical vapor-in-tube condensers.

(9) Shellside baffling can provide higher cooling-medium heat transfer.

(10) Tubeside condensing generally requires fewer components to be fabricated of more expensive materials of construction.

(11) Cooling water can not back up into the process preventing associated problems.

Updraft Versus Downdraft Condensers

With few exceptions, vertical downdraft condensers provide better heat transfer than updraft condensers. Occasionally, updraft condensers are used when minimum condensate subcooling is desired or when the remaining vapors will not condense in the presence of previously obtained condensate. Sometimes corrosion is minimized by preventing condensate and uncondensed vapors from contacting. Chemical reactions may also be avoided.

The primary disadvantages of updraft condensers are associated with flooding of the tube. In an updraft condenser, the condensate and incoming vapor flow counter current to each other at the tube inlet. If sufficient flow area is not available for both liquid and vapor, flooding or surging will occur, with slugs of liquid building up and escaping. This limits the capacity of the condenser and results in a serious control problem. To eliminate flooding, low vapor velocities are required, reducing the rate of heat transfer especially when condensing in the presence of inert gases.

Flooding in updraft condensers can be predicted from the following correlation:

$$v_v^{1/2} \rho_v^{1/4} (\sigma/\sigma_w)^{1/4} + v_l^{1/2} \rho_l^{1/4} = 0.75 [gD_i(\rho_l - \rho_v)]^{1/4} \qquad (22.1)$$

where v_v = upward superficial velocity of vapor phase
v_l = downward superficial velocity of liquid phase

ρ_v = vapor density
ρ_l = liquid density
g = acceleration of gravity
D_i = inside tube diameter
σ = fluid surface tension
σ_w = surface tension of water under atmospheric conditions, taken as 73 dynes/cm

For condensers in which the vapor and liquid flows are equal:

$$v_v = v_l, \text{ and}$$

$$v_v = \frac{0.5625 \, gD_i(\rho_l - \rho_v)^{1/2}}{\rho_v^{1/2}[(\sigma/\sigma_w)^{1/4} + (\rho_v/\rho_l)^{1/4}]^2} \tag{22.2}$$

The correlations above will predict the point at which flooding will occur. Generally, condensers are designed with a safety factor so that maximum velocities are approximately 85% of those resulting in flooding.

Flooding velocities can be increased in two practical manners:

(1) tilting the condenser 10° to 20° from the vertical can increase the flooding velocity considerably
(2) the ends of the tubes can be cut at an angle as shown in Figure 22-2.

Flooding velocities are increased as a function of the angle at which the tube end is cut. Improvement in performance varies as $(\cos \alpha)^{0.32}$:

angle α	improvement
30°	5%
60°	25%
75°	55%

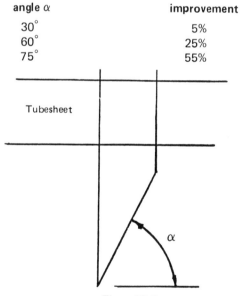

Figure 22-2.

Flooding in Horizontal Tubes

Condensate may partially or completely fill a horizontal tube as shown in Figure 22-3. If the tube is completely filled, venting can become difficult with resulting poor heat transfer and erratic control. Drainage of the tube can be greatly increased by sloping the tube a small angle from horizontal (3° to 5°).

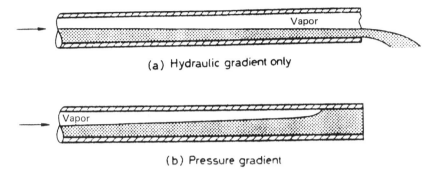

(a) Hydraulic gradient only

(b) Pressure gradient

Figure 22-3: Exit conditions in stratified condensation.

Figure 22-4 illustrates the relationship of the brink depth, l, to the inside diameter of the tube, d_i. The brink depth at the outlet of the horizontal tube can be calculated using the following equation:

$$q = 9.43(d_i/12)^{2.56}(l/d_i)^{1.84} \qquad (22.3)$$

where q = flow rate, cubic feet per second
       d_i = tube inside diameter, inches
       l = brink depth, inches

The ratio of l/d_i should not exceed 0.5 in order to assure proper venting. A value less than 0.3 is preferred.

Two-phase flow correlations should be used for a more rigorous approach to predict the actual flow regime.

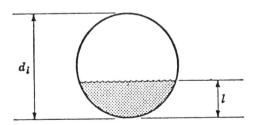

Figure 22-4: Relationship of brink depth l to the inside diameter d_i.

Shellside Flooding in Horizontal Condensers

Condensers in which condensate is formed on the outside of horizontal shell-and-tube equipment may be partially flooded with condensate. Shellside baffles must be provided and the flow path through the baffle cut is essentially a V-notch weir, as shown in Figure 22-5. The height of liquid in the shell can be estimated using the equation below:

$$W = 156[\tan(\alpha/2)]^{0.996} H^{2.47} \qquad (22.4)$$

where W = liquid flow rate, lb/sec
α = vertical angle of notch degrees
H = height of liquid above vertex, feet

For condensers with baffle cuts which are 20% of the shell diameter, α may be taken as 45°, and the equation above can be simplified to:

$$W = 65 H^{2.47} \qquad (22.5)$$

The height of liquid calculated using the correlations above is only approximate because it ignores any pressure drop effects and the fact that there are generally tubes in the flow area calculated. However, it can be used to give a reasonable approximation. If the height of condensate calculated is appreciable, then more than one condensate connection may be desired. The flow pattern of the condenser can be changed so that vapor and condensate make only one crossflow path across the tube bundle. Other baffle types may also be used.

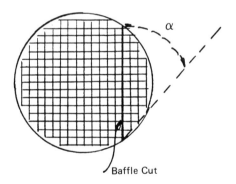

Baffle Cut

Figure 22-5.

Integral Condensers

Integral condensers are mounted directly upon the equipment in which vapors are produced which must be condensed. Integral condensers are frequently used in vacuum service because they eliminate large vapor lines and reflux pumps if reflux is required. Shell-and-tube, spiral plate, spiral tube, bayonet, and air-cooled condensers can be used as integral condensers. Condensing may occur either inside or outside of tubes. Figure 22-6 shows an integral tubeside condenser and Figures 22-7 and 22-8 show integral shellside condensers. The horizontal units may be designed with lowfin tubes and usually afford a single vapor pass across the tube bundle. This construction has the advantage of low pressure drop, good vapor distribution, and efficient subcooling of the condensate.

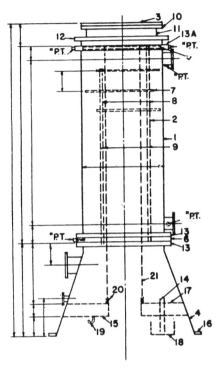

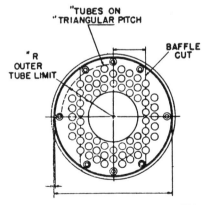

SCHEMATIC SECTION THRU SHELL
(number of tubes and spacers not correct)

Figure 22-6: Integral tubeside condenser.

220 Handbook of Evaporation Technology

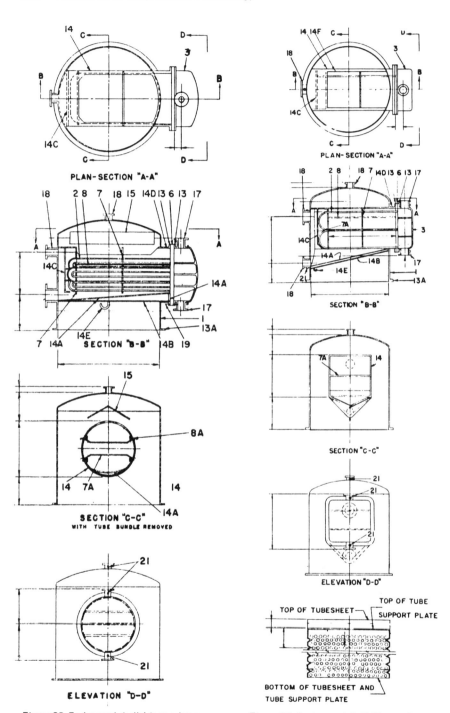

Figure 22-7: Integral shellside condenser.

Figure 22-8: Integral shellside condenser.

Integral condensers are not limited to vacuum service. They may be used when gravity drainage of condensate is required, when minimum system volume is desired, when expensive vapor piping is required, or when it is desirable to reduce structural steel requirements.

Condensate Connections

Condensate must often flow by gravity from the condenser. Hence, nozzles must be properly sized to ensure that the condensate line will be self-venting. Liquid in a self-venting line occupies a portion of the flow area, and gas moves freely with, or counter to, the liquid. Condensate nozzles should also be sized to prevent excessive liquid levels, especially in horizontal condensers. An equation for determining proper nozzle size is:

$$d_i = 0.92(Q_f)^{0.4} \tag{22.6}$$

where d_i = nozzle inside diameter, inches
Q_f = flowrate, gallons per minute

This equation predicts nozzles that act as circular weirs. No whirlpool will exist; the line will be self-venting; and the ratio of liquid height above the nozzle to nozzle diameter will be 0.25.

23

Vacuum Producing Equipment

Many evaporation systems operate at pressures less than atmospheric. Consequently, an understanding of vacuum producing systems is required of the evaporation technologist. Vacuum costs money, both in initial investment and in continuing operating costs. The first responsibility of the evaporation technologist is to determine whether vacuum is required, or whether there is a better method to achieve the desired result.

Vacuum technologists find it convenient to divide subatmospheric pressures into four regions:

Rough vacuum	760 to 1 torr
Medium vacuum	1 to 10^{-3} torr
High vacuum	10^{-3} to 10^{-7} torr
Ultra-high vacuum	10^{-7} torr and below

The rough vacuum pressure region is of greatest interest to the evaporation technologist because most evaporation processes requiring vacuum operate in this region.

Two types of vacuum producing devices are used: jet ejectors and mechanical pumps. Jet ejectors usually have lower initial costs, lower maintenance costs, but higher operating costs than equivalent mechanical systems. Mechanical pumps are more efficient users of energy. Jet ejectors require no moving parts and therefore are simplest of all vacuum producers.

Current economics favor the use of mechanical vacuum pumps in most applications. Steam jet ejectors are economically superior to other pumping devices when their initial cost and simplicity compensate for their inherently low efficiency.

JET EJECTORS

The operation of a stream jet ejector is illustrated in Figures 23-1 and 23-2.

Vacuum Producing Equipment 223

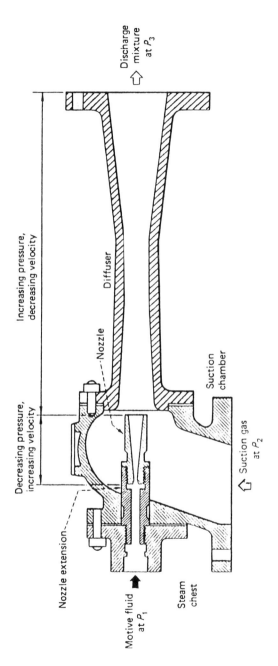

Figure 23-1: Steam jet ejectors have no moving parts and can be sized to handle high flowrates.

224 Handbook of Evaporation Technology

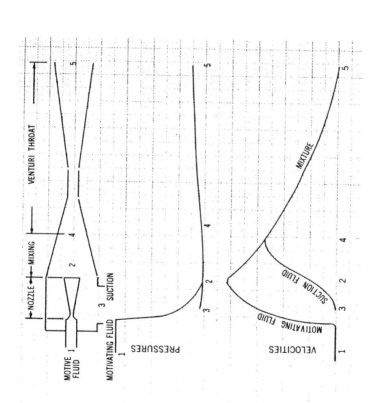

Figure 23-2.

A steam jet ejector consists of three parts: a motive steam nozzle, a mixing chamber, and a diffuser. Steam jets accomplish compression through momentum transfer. High-pressure steam, usually in the range of 100–200 psig, is accelerated to a high velocity in the steam nozzle as it expands through the converging and diverging section of the steam nozzle where potential energy in the form of pressure is converted into kinetic energy in the form of velocity. This high-velocity jet of steam enters the mixing chamber where the fluid to be pumped is entrained in the steam flow. The mixture of motive and suction fluids, now at a somewhat lower velocity, enters the diffuser in which the kinetic energy of the mixture in the form of velocity is converted back to potential energy in the form of pressure. The mixture is discharged at an intermediate pressure with a value somewhere between that of the original motive steam and the suction fluid. While the total energy in the process is conserved, the mixing process causes a substantial increase in the overall entropy; hence the steam jet has a low thermodynamic efficiency.

Multistage Jet Ejectors

The limit on the compression ratio for a single jet stage is about 15 and the optimum is often considerably less. A practical compression ratio for a single-stage unit is approximately 6 when discharging to atmospheric pressure. Jet ejectors frequently have multiple stages connected in series. The number of stages depends primarily on the overall compression ratio for the application. When multiple stages are used, the suction load for the first stage plus the motive steam to the first stage becomes the load to the second stage and so forth. Figure 23-3 shows a typical multistage jet installation.

Intercondensers

The total steam consumption in a multistage system can be considerably reduced if some of the steam can be condensed between stages. Multistage jet ejectors almost always have intercondensers to take advantage of this energy savings. Intercondensers are used whenever the interstage pressure results in a dewpoint of the discharge vapors from the previous stage high enough so that cooling water can be used for condensing. Generally this corresponds to an interstage pressure about 60 torr. When the interstage pressure is below 60 torr, intercondensers are not normally used unless the cost for refrigeration can be justified in reduced steam usuage for the jet ejector. The advantages of applying refrigeration to extend the operating range of intercondensers should not be overlooked. Possible advantages include: reduction in waste treatment costs; recovery of product; reduced capital and operating costs for the vacuum system; potential for debottlenecking existing systems.

Precondensers

Vent condensers and precondensers can usually be justified if they significantly reduce the load to the vacuum pump. Figure 23-4 indicates the function of the vent condenser and the precondenser. The removal of process vapors by a precondenser permits the use of smaller vacuum pumps and this may reduce both capital and operating costs. Precondensers may also increase reliability by

226 Handbook of Evaporation Technology

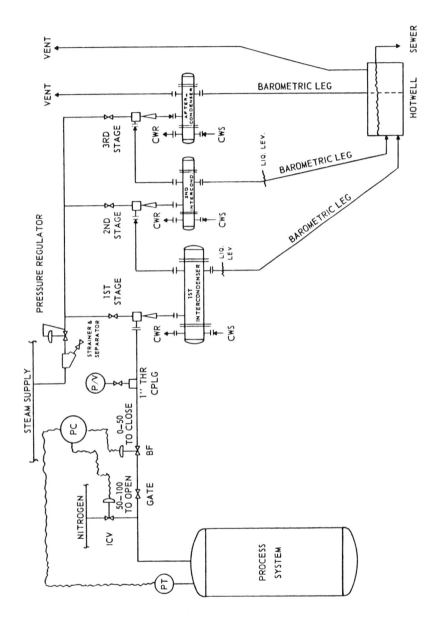

Figure 23-3: Schematic of a typical three stage steam jet ejector installation.

protecting the vacuum pump from solids and/or liquid carryover, and by reducing the concentration of corrosive vapors in the load to the pump. The advantages of refrigeration must also be considered. If precondensers are used, design must be optimized considering total purchase, installation, and operating costs for both the condenser and the vacuum unit.

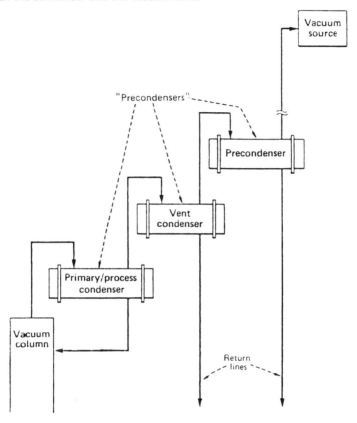

Figure 23-4: Condenser nomenclature. Usage assumes that condenser train is integral part of vacuum.

Aftercondensers

Aftercondensers are generally installed after the last stage of a steam jet ejector system. Aftercondensers are not necessary to the function of the jet as a compressor; their purpose is to minimize the steam discharge to the atmosphere. They also further reduce the discharge of any condensables from the process into the atmosphere. Aftercondensers also reduce the noise level of the discharge from the last effect.

A steam jet that discharges directly into the atmosphere produces a deafening roar. This problem is solved by installing a muffler or "silencer" at the stage discharge. An aftercondenser, if installed, provides sufficient noise reduction to eliminate the need for a silencer.

Barometric Legs and Hotwells

Because they operate under vacuum, the removal of condensate from an intercondenser can be difficult. A simple and economical solution results by installing the intercondensers high in a structure and running the condensate discharge line down to grade level where it is submerged below the liquid level of a tank of fluid which is open to the atmosphere. The condensate discharge line is referred to as the "barometric leg" and the tank is called a "hotwell". A design for hotwells is shown in Figure 23-5.

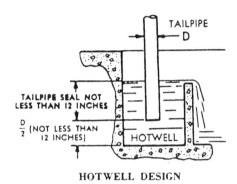

HOTWELL DESIGN

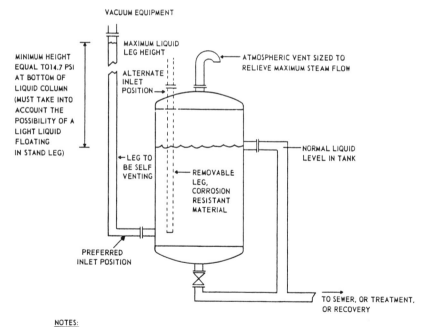

NOTES:
1) WILL NEED TO BE ABLE TO INITIALLY PROVIDE LIQUID LEVEL IN VESSEL.
2) 1 LEG PER JET STAGE REQUIRED.

Figure 23-5: Hotwell tank.

Vacuum Producing Equipment 229

When the system is operating, liquid from the hotwell rises in the barometric leg and reaches a level of steady state such that the interstage pressure plus the hydrostatic head in the barometric leg equals atmospheric pressure. Normally the intercondenser height is sufficient if the condenser drain nozzle is at least 34 feet above the hotwell liquid level. However, if the condensables in the load gas from the process system are not miscible with and are less dense than water, the barometric leg may fill with process condensable and greater height may be required. The volume of the hotwell must be adequate to hold the liquid in the barometric legs when the unit is shutdown. There must be enough liquid volume above the discharge of the barometric leg to assure that the discharge is submerged during startup.

If sufficient height is not available for a barometric leg, condensate may be collected in a small accumulator tank, the liquid level controlled, and condensate removed from the tank with a centrifugal pump. Such systems are called "low-level" hotwells and are illustrated in Figure 23-6. Each intercondenser must have its own accumulator, pump, and level control. Because this is costly, most jet systems are located high in the structure and use the barometric leg and hotwell system.

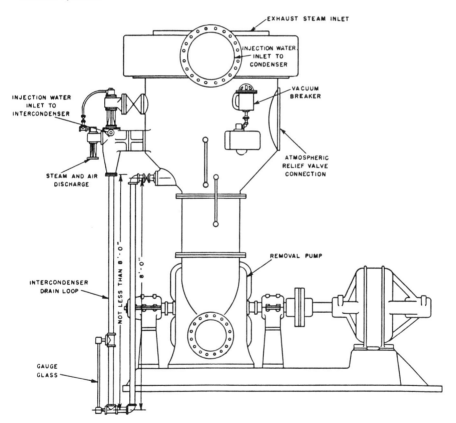

Figure 23-6: Two stage ejector with jet intercondenser serving a low level jet condenser.

Ejector Stage Chracteristics

Usually an ejector is designed for maximum efficiency at one operating condition. The quantitative performance characteristics of an ejector at conditions other than design cannot be accurately predicted. They must be determined by testing.

Ejector design may be either critical or noncritical. Critical design means that the vapor velocity in the diffuser is sonic and occurs at compression ratios above 1.8. In noncritical designs the vapor velocity in the diffuser is subsonic.

For critical design units, the relationship between motivating pressure and discharge pressure depends on the ejector design. An ejector is a one point design unit (optimum operation at a single set of operating variables). Once a unit is designed and built for a definite specifications of motive pressure, discharge pressure, and suction pressure, its load capacity cannot be increased without changing the internal physical dimensions of the unit. The ejector is a fixed orifice metering device and any change in motive pressure is accomplished by a proportional change in the quantity of motive fluid. Consequently ejectors designed in the critical range are sensitive to operating conditions other than those for which the unit was designed. The table below illustrates how these changes in operation can affect ejector performance.

Effect of Operational Changes on Critical Flow Ejector Performance

Motive Pressure	Discharge Pressure	Suction Pressure	Load Pressure
Decrease	Constant	Increase rapidly	Decrease rapidly
Constant	Increase	Increase rapidly	Decrease rapidly
Constant	Constant	Increase	Increase
Constant	Constant	Decrease	Decrease
Increase	Constant	Constant	Decrease gradually
Constant	Decrease	Constant	Unchanged

Ejector Efficiency

The usual concept of efficiency involves a comparison of energy output to energy input. For ejectors, overall efficiency is best expressed as a function of entrainment efficiency. The direct entrainment of the low velocity suction fluid by the motive fluid results in an unavoidable loss of kinetic energy. The fraction of the kinetic energy that is transmitted to the mixture through the exchange of momentum is called the entrainment efficiency. The kinetic energy which is lost is converted into heat which is absorbed by the mixture to produce a corresponding increase in enthalpy.

The capacity of an ejector handling other than saturated vapor is a function of the molecular weight and temperature of the fluid. If motivating quantities are equal, the ejector capacity increases as the molecular weight of the load gas increase. The capacity will be reduced as the temperature of the fluid is increased.

For ejectors with steam as a motive fluid, the quality of the high pressure steam has an effect on the operation of the unit. Most units are designed to operate with dry, saturated steam. If the quality decreases below 98%, a gradual decrease in both suction pressure and capacity is experienced, especially in units designed for high compression ratios and even more so for multistage units. Excessive steam superheat (greater than 30°C) can also adversely affect the capacity of any ejector. Not only is the energy level ratio decreased, but the volumetric increase tends to choke the diffuser. Diffuser choking can be avoided if the ejector is designed to use superheated steam as a motive fluid.

Basic Performance Curve—Critical Ejector

The basic performance curve for an ejector operating in the critical range is determined by fixing the motive steam pressure and the discharge pressure and varying the load to the ejector suction. Corresponding values of suction pressure are observed and the performance curve shown in Figure 23-7 can be developed.

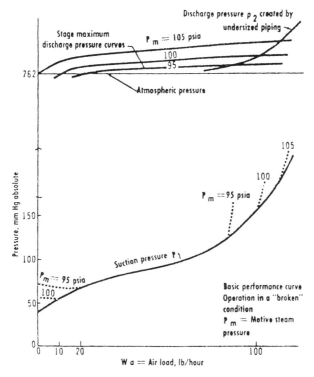

Figure 23-7: Basic ejector performance curves.

Next, at constant load, the discharge pressure is slowly increased. The suction pressure will remain constant until a limiting discharge pressure is reached.

At this condition, a small increase in the discharge pressure will produce a large increase in the suction pressure. The ejector is then said to be operating in a broken mode. The discharge pressure is then slowly decreased until the basic suction pressure is again attained. By definition, the discharge pressure which just permits a return to the basic suction pressure is called the maximum discharge or recovery pressure. Repeating this test with different loads will establish the maximum discharge pressure curve shown in Figure 23-7.

The maximum discharge pressure curve droops at small loads. If the ejector of Figure 23-7 discharges against atmospheric pressure, it will be on its basic performance curve at loads down to 10 lb/hr. At loads below this, the actual discharge pressure exceeds the maximum discharge pressure of the ejector, and the ejector will operate in a "broken" state with its suction pressure higher than the value on the basic performance curve.

If the motive steam pressure is next increased slightly, the mass flow of the motive steam will increase. This change will produce an insignificant rise in the suction-pressure curve, but it will raise the maximum discharge-pressure curve significantly. The ejector now will operate on its basic curve down to zero load because its new maximum discharge pressure at zero load exceeds the actual pressure aganst which it must discharge.

If the motive steam pressure were reduced slightly, there would be an insignificant lowering of the suction-pressure curve, but the maximum discharge-pressure curve would be shifted downward significantly making the ejector "broken" at loads below 20 lb/hr.

In summary, for a typical ejector stage operating in the critical range, the basic performance curve is essentially fixed. The maximum discharge-pressure curve is somewhat variable, and the ejector will operate on its basic performance curve if the maximum discharge pressure is not exceeded.

An ejector designed for noncritical operation has somewhat different performance characteristics and these are described in Chapter 18. The performance curve for noncritical ejectors is a function of the motive steam pressure and the actual discharge pressure. Thus, it has no broken mode of operation.

Unstable Ejectors

Performance characteristics for an unstable ejector are shown in Figure 23-8. Its performance curve terminates at small loads, below which it develops a cyclic mode of operation. It operates in a stable manner unless the load (and pressure) drop to the lower limit of its curve. Then the flow pattern abruptly breaks down and reverses. The reverse flow of load gas and steam out of the ejector suction connection continues until the pressure at the suction connection is somewhat higher than the pressure at which it becomes unstable. Then, it abruptly resumes stable operation and continues cycling. The reverse flow of vapors during the unstable portion of the cycle, called backfiring, forces water vapor into the suction system. When any stage in a multistage ejector becomes unstable, it may upset the preceding stages. Increasing the steam pressure will reduce or remove the instability as shown in Figure 23-8. Adding a fixed load to keep the ejector above its stable limit is sometimes useful.

In a few applications, an unstable design is desirable because of its low steam usage. If the load is accurately known and backfiring is not harmful, it is appropriate to permit instability at pressures below the useful range of the ejector. Another application for the unstable design is with multistage ejector stages between which there are no condensers. The second stage in a series of noncondensing stages may be an unstable design relying on the motive steam from the first stage to keep it operating in a stable manner. Similarly, the third stage may rely on motive steam from the first two stages for its stability, and so on. In using such a design, an operator should note that turning off the first one or two stages may make the ejector system unstable at operating pressure.

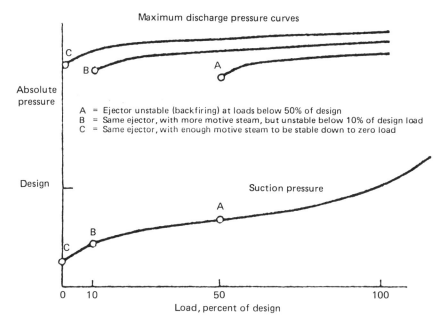

Figure 23-8: Performance curves for unstable ejector design.

Multistage Ejector Characteristics

The characteristics of a two-stage condensing ejector are shown in Figure 23-9. For simplicity, it is here assumed that the intercondenser removes all water vapor from the air load to the second stage.

The ejector shown is designed to maintain a suction pressure of 40 mm Hg absolute (40 torr) handling 100 lb/hr dry air. At a load of 120 lb/hr the maximum discharge pressure of the first stage is just equal to the suction pressure of the second stage. If the air load increases slightly above 120 lb/hr, the suction pressure of the second stage will exceed the maximum discharge pressure of the first stage, breaking the first stage and raising its suction pressure above that shown on the basic performance curve. Similarly, at a load above 140 lb/hr, the second stage breaks because its maximum discharge pressure is exceeded by the actual back pressure of the discharge piping. This ejector would be severely broken at loads more than 140 lb/hr of air.

Because the maximum discharge pressures of both stages at air loads below 120 lb/hr exceed the actual discharge pressures, the ejector operates on its basic performance curve from 0 to 120% of design load. If the motive steam pressure to this ejector is increased, the unbroken range of the ejector will be slightly extended. If the steam pressure is decreased, the unbroken range also will be slightly decreased and the ejector may become broken or unstable below small loads such as 10 to 20 lb/hr. If the steam pressure is reduced too much, the unbroken range may narrow quite rapidly to the point where the ejector is stable at only one air load and unstable or broken at all other load conditions. If it is reduced any more, the ejector may be completely unusable.

Examination of Figure 23-9 will show how the pressure of water vapor in the load to the second stage will reduced the maximum dry air capacity of the two-stage ejector.

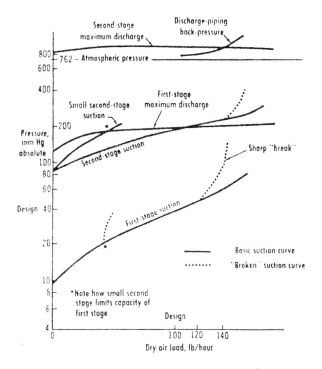

Figure 23-9: Two stage condensing ejector performance.

Troubleshooting Steam Jet Ejectors

Steam jet vacuum equipment is very reliable. However, unsatisfactory performance of an ejector can be caused by external or internal causes. Unsatisfactory performance can also be classified as sudden or gradual. The gradual loss of vacuum will normally suggest internal erosion or corrosion; a sudden loss of vacuum will normally suggest external causes. Since it is easier to check external causes of trouble, all possible external causes should be checked first.

External Causes of Trouble

(1) low steam pressure

(2) wet steam

(3) high water temperature or insufficient water flow

(4) entrained air in cooling water

(5) high discharge pressure

(6) fluctuating water pressure

(7) change in load — excessive air leakage

(8) corroded barometric legs

Internal Causes of Trouble

(1) eroded or corroded parts, particularly nozzles and diffusers

(2) clogged nozzles, diffusers, and strainers

(3) leaks in steam chests

(4) clogged or fouled water supply

(5) clogged water discharge

(6) excessive leakage — cracked or worn parts

(7) Intercondenser water nozzle eroded

Accurate instruments are necessary to check an ejector system. Steam pressure gages must be calibrated periodically; compound steam pressure gages are recommended because they will not be damaged if subjected to vacuum. Thermometers must be accurate. The type of vacuum gage required depends upon the degree of vacuum. A mercury absolute gage is usually satisfactory for systems with three or fewer stages; more sophisticated gages are required for systems with more stages.

If the problem is internal, the system must be checked by stages. Since earlier stages will not function unless the final stage is working properly, the final stage must be checked first. Also, the final stage has the smallest nozzle orifice and is more susceptible to erosion and plugging. The discharge of the last stage should be checked both visually and audibly. If water is coming out the discharge, the condenser ahead of the last stage is flooded. If a popping noise can be heard, the stage is not performing properly. Since the majority of ejector problems are found with the last stage, a simple performance check should be made. If corrosion or erosion is evident in the last stage, it is likely that similar problems exist in earlier stages and these should also be checked. If the final stage is found to be problem free, then the preceding stages should be checked in reverse order.

Test Procedure: If a jet operates steadily, but fails to produce the required vacuum:

(1) install a *reliable* vacuum gage (mercury manometer or McLeod gage) at the suction to the first stage

(2) close the gate valve isolating the jet from the system

(3) check the vacuum gage to see whether it reads the manufacturer's rated, no-load suction pressure. If it does, chances are the jet is not at fault and the system should be checked for leaks.

If the vacuum gage does not read the manufacturer's no-load suction pressure, the jet is at fault:

(1) check for wet steam

(2) check for low steam pressure

(3) check for blockage of atmospheric vent

(4) insure proper elevation of condensers

(5) check for inadequate cooling water flow to intercondensers

(6) check for corrosion of barometric legs

(7) check for worn, blocked, or corroded nozzles or diffusers

(8) check for icing (only if suction pressure is less than 5 torr)

If the jet backfires:

(1) check for low steam pressure

(2) check for blockage of vent

(3) check for worn, blocked, or corroded nozzles.

Wet steam is the most common cause of jet malfunction. The appearance of the plume of steam from a valve opened to the air will indicate the steam condition. If the steam is dry or slightly superheated, the plume will have a transparent portion near the valve. If the plume is completely opaque white, the steam is wet. Wet steam can cause erosion of both the nozzle and the diffuser. If these parts are badly eroded they should be replaced.

MECHANICAL PUMPS

Mechanical pumps can be of many types with the only common features being that they have moving parts and require rotating mechanical work to operate. Commonly used types include: rotary blower, rotary piston, rotary vane, trochoid, and liquid ring pumps. All these types, except for the liquid ring pump, are positive-displacement pumps, and, as such, cannot tolerate slugs of liquid. Any significant liquid slugging will destroy these machines immediately, often violently. All these types can be used singly, in stages of the same type, or in stages of different types combined. They can also be staged with steam jet ejectors.

Most of these pumps present certain degrees of unreliability and are not available in a wide range of materials of construction; as a result, their application is limited. The liquid ring pump is not subject to many of these limitations

Vacuum Producing Equipment

and offers other characteristics that make it useful for many applications.

Mechanical vacuum pumps are more efficient than steam jets. Typical calculated efficiencies of various pumps are shown in Figure 23-10. These curves represent estimates, because efficiency is a function of pump size and because there are significant variations in the efficiencies of pumps offered by different manufacturers. However, Figure 23-10 is generally representative of vacuum pump efficiencies.

For these curves, multistage ejectors are assumed to be provided with surface intercondensers (20°C cooling water); efficiencies for noncondensing multistage ejectors will be much lower.

All mechanical pumps are assumed to be electrically driven. The curve for two-stage liquid-ring pumps assumes that the sealing fluid will have a vapor pressure less than 1 torr at 20°C.

Efficiencies for the rotary-vane pump are based on a two-stage machine operated as a dry compressor.

Figure 23-10 assumes no condensables. Condensing effects are significant when liquid-ring pumps and condensing steam jets are used. When condensables constitute no more than 20% by weight of the total flow to the vacuum pump, condensing effects can nomally be ignored in estimating thermal efficiencies.

Multistage condensing steam jets can compete with mechanical pumps on the basis of thermal efficiency when condensables are the principal load from the vacuum source.

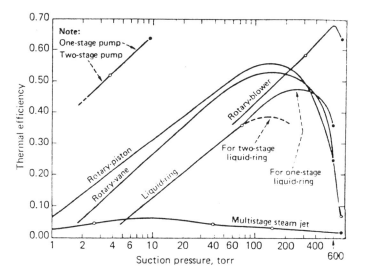

Figure 23-10: Rough estimates of thermal efficiency of various vacuum-producing systems.

Liquid-Ring Pumps

Liquid-ring pumps have only one moving part — the rotor. All of the functions normally performed by mechanical pistons or vanes are accomplished by the liquid ring.

The basic design is shown in Figure 23-11. There are a number of variations of this design, but all operate on the same principle. Prior to startup, the pump casing is partially filled with a sealant liquid, called a hurling fluid. When the pump is turned on, the rotating impeller throws (or hurls) the liquid to the periphery of the casing forming a liquid ring which conforms to the cylindrical pump body. The rotor axis is offset from the body axis; thus, a piston action is established as the liquid almost fills, then almost empties, each of the chambers between the rotor blades.

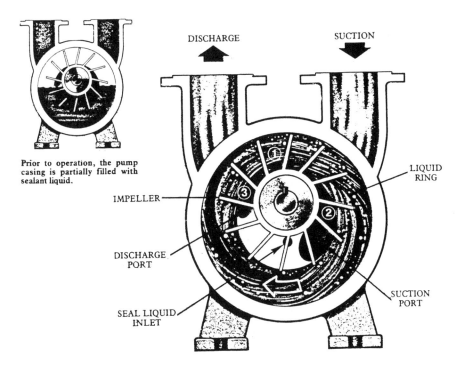

Figure 23-11: Single stage liquid ring pump cross-section.

The reciprocating liquid piston draws the process gas into the chamber through an inlet port, compresses it, and forces the gas, along with some of the hurling fluid, through a discharge port. Compression ratios for a single-stage pump can reach 10 when discharging to atmospheric pressure.

The shaping of the liquid ring depends on the speed of rotation, contours of the casing and rotor blades, and compression ratio across the pump. The compression chamber does not fit a pattern that can be readily defined or precisely measured; and the pump cannot be accurately rated on the basis of swept volume.

Thus the liquid-ring pump is not a true positive-displacement compressor. Instead, it is essentially an isothermal machine with the ring acting as a heat sink. Both evaporative and condensing effects must be considered in sizing liquid-ring pumps. Evaporative cooling takes place whenever dry gases are introduced at temperatures higher than those of the seal liquid. Condensation occurs when pumping gas that is saturated with vapor; in this situation the pump serves as a direct contact condenser. Condensation of process vapors in the pump actually improves the performance of liquid-ring pumps. Once-through use of sealing liquid to prevent concentration of condensate, and corrosion-resistant internals, are recommended when handling highly corrosive materials. If only small amounts of contaminants are present in the process vapor, the hurling fluid can be cooled and totally or partially recycled. This is especially true if the contaminants can be easily separated. If corrosive elements are present, the large dilution by hurling fluid when used once-through will prevent most corrosion problems. This can, however, result in a large load to the waste treatment facility.

The minimum useful pressure in a liquid-ring pump is limited by the vapor pressure of the hurling fluid. As the working pressure approaches the vapor pressure of the hurling fluid, a greater portion of the connected power is consumed in flashing, pumping, and recondensing the hurling fluid. For this reason, the temperature and vapor pressure of the hurling fluid must be as low as practical. Water, ethylene glycol, and Dowtherm A or Therminol VP-1 are commonly used as hurling fluids, although many other fluids can be used. Low density fluids require a higher speed of the impeller in order to prevent the liquid ring from collapsing. High viscosity fluids cause higher power losses in the pump. In some applications where the substance being pumped is condensed in the pump, this same substance can be used as the hurling fluid.

Most manufacturers rate their pumps with $60°F$ hurling water. For other conditions and other hurling fluids, the actual capacity must be corrected to rated condition as below:

$$V_{60} = V_a (P_s - P_{60})/(P_s - P_a) \tag{23.1}$$

where V_{60} = pumping capacity with water at $60°F$ as the hurling fluid, ACFM
V_a = actual pumping capacity required, ACFM
P_s = pump suction pressure
P_{60} = vapor pressure of water at $60°F$
(13.25 torr, 0.5216 inches Hg, 0.2562 psia
P_a = vapor pressure at actual temperature of hurling fluid

Application: The liquid-ring pump is simple and reliable with only one moving part and no metal-to-metal contact. Liquid slugs and entrained solids are easily handled. Most of the vapors that condense do so in the pump and are removed in the hurling fluid. The exhaust of a liquid-ring pump is relatively cool and is saturated because of the condensing effect of the pump. Liquid-ring pumps are often constructed as a two-stage unit in one common housing. Such units are efficient for suction pressures as low as 50 torr and can be used for suction pressures as low as 20 torr.

VACUUM SYSTEM RELIABILITY/MAINTENANCE

Two questions must be considered when assessing anticipated reliability and maintenance:

(1) What are the consequences of a complete and unexpected loss of vacuum?

(2) Will associated maintenance result in significant downtime with resultant loss of production?

Important considerations when projecting reliability include:

(1) response of the vacuum system to process upsets
(2) the ability of the system to withstand abuse
(3) skills required for field maintenance
(4) effect of difficult environments on performance.

Summary

There is no vacuum source that is the best choice for every application. The technology for protecting mechanical pumps from process upsets and abuse is highly developed. This technology can be used effectively in designed vacuum systems to meet specific requirements. In several applications, mechanical pumps have demonstrated reliability comparable with, or superior to, liquid-ring pumps or steam jet ejectors.

Generally, though, steam jet ejectors and liquid-ring pumps are used for evaporator applications.

MULTISTAGE COMBINATIONS

Often a single-stage vacuum device (pump or jet) cannot attain the necessary compression ratio. By connecting several devices in series, lower suction pressures become workable. This is commonly practiced, and the combinations are almost unlimited.

In many cases, multistage vacuum pumps consume less energy than single-stage pumps while doing the same job. Power requirements may be reduced as much as 50%. Rotary blowers are often called mechanical booster pumps because they are commonly used as a first stage in a multistage system.

A relatively new application is the so-called hybrid or combination system in which a steam jet first stage (s) (followed by an intercondenser) acts as a forepump for a liquid-ring pump (either single or double stage). This system may be especially desirable when significant condensing will occur at the conditions between the ejector (s) and the pump. In many cases the ejector stages of the system can be operated with steam pressures as low as 5 psig.

Air jet ejectors which use ambient air as a motive fluid are also used as booster pumps for a liquid-ring system. These may be expensive to operate because the motive air must be pumped back to atmospheric pressure in the backing pump.

The staging efficiency is particularly true for steam jet ejectors because the motive steam for one stage becomes the load for succeeding stages. Interstage condensers can dramatically reduce steam consumption; consequently all multistage jets should be provided with intercondensers. The reduction in operating cost completely overshadows the increase in capital investment.

Combination ejector/liquid-ring systems and mechanical systems that use rotary blowers backed with other mechanical pump types are versatile and energy efficient. Thermal efficiencies of combination pumping systems are shown in Figure 23-12. Ejector/liquid-ring systems offer the highest reliability and are versatile for both wet and dry loads over a wide range. Rotary-blower/rotary-piston systems are the most efficient and have been extensively applied.

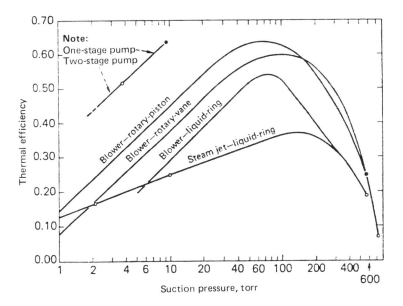

Figure 23-12: Integrated pumping systems are energy-efficient, and are available as complete packaged units.

SIZING INFORMATION

The most important parameters affecting final selection of the vacuum system are the suction pressure and the throughput.

Single-unit steam-jet ejectors can be designed for large volumetric flows (up to 10^6 ACFM). Liquid-ring pumps can handle flowrates to 10,000 ACFM in a single unit. Rotary-piston and rotary-vane pumps are limited, respectively, to 800 and 6000 ACFM; but both have relatively flat operating curves across the entire rough-vacuum region. Rotary blowers extend the range of integrated mechanical systems to 30,000 ACFM. Figure 23-13 indicates practical lower limits for suction pressures for various vacuum systems.

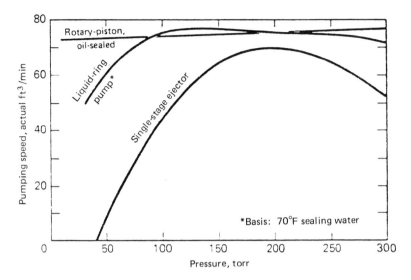

Figure 23-13: The range of operation of a vacuum producer varies, depending upon its type—rotary vane similar to rotary piston, rotary blower similar to steam jet.

Specifying the Load

The flow rate of gas or "load" to a vacuum system consists of noncondenseable gases and condensable process vapors. Noncondensables come from five sources:

(1) gases dissolved in the feed

(2) instrument blowback gases

(3) by-product gases from any chemical reactions in the process

(4) air leakage into the system

(5) gases liberated as cooling water is heated if direct contact condensed are used.

Air leakage is generally the largest component of the load gas.

After the required system pressure has been defined, flow rates of condensables and noncondensables must be estimated.

Air Leakage

Air can leak into a system around flanged joints, valve stems, pump and agitator shaft seals, instrument connections, and the like. Leakage can also occur through metal porosities and cracks along weld seams. Air leakage is unavoidable in commercial systems.

In almost all design situations, air leakage rates cannot be accurately predicted. The only real accuracy necessary is that extremes be avoided. The designer is in reality specifying a maintenance level for the system when he estimates air leakage. All vacuum systems will have a finite leakage which increases with time. As long as the design leakage rates result in reasonable maintenance requirements, the leakage estimate was a good one. Low estimates result in a frequent, costly "leak finding", while high estimates result in high energy usuage and a large air throughput in the system. Care should result in reasonable costs for both energy and maintenance.

Various methods have been proposed for estimating air-leakage rates:

(1) Leakage rates can be correlated with the volume and the pressure level of the system. This approach is more widely accepted and is easier to do. However this approach can lead to errors because it does not distinguish between complexities of various systems of equal volume.

(2) Leakage rates can be estimated by assigning nominal leakage rates to system components. This method can be cumbersome and is not readily adapted to predict leaks resulting from fabrication and assembly errors.

(3) Procedures which combine the two approaches above can also be used. This approach wil generally estimate air leakage rates less than the first approach above and will require a more leak-tight system.

A reasonable method for specifying air leakage in evaporation systems is to use Figure 23-14. For a given combination of system pressure and system volume, Figure 23-14 eatablishes an air-leakage rate below which a system is considered to be "commercially tight". The figure reflects in a relative manner:

(1) It is more costly to remove air from a high-vacuum system than from a rough-vacuum system.

(2) High-vacuum systems are usually designed and maintained such that the total air leakage is less than in rough-vacuum

(3) Systems having larger internal volumes will usually have more and larger air-leak paths.

Condensable Load

The largest portion of condensable vapor load is generally product since noncondensable gases are generally saturated with product vapors at process vent conditions. The amount of condensable that enters the vacuum system should be minimized for several reasons:

(1) the value of the product otherwise lost

(2) condensables may be an air pollutant in the vent from the vacuum pump

(3) condensable may be a water pollutant which must be treated after it is discharged with steam jet condensate from intercondensers

(4) condensables are an additional load for the vacuum system.

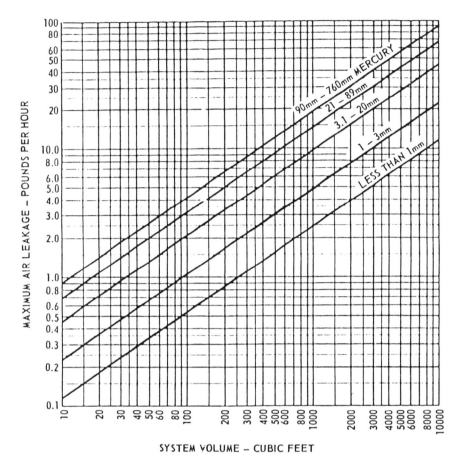

Figure 23-14.

The condensable load is minimized by cooling process vents as low as practical. Vent condensers or precondensers are often economical.

Significant amounts of condensables as load to the vacuum system can be an important consideration in the design and selection of the vacuum pump. Vacuum system suppliers must be provided condensable data in order to properly design the vacuum system. Information required include molecular weight, vapor pressure, specific heat, latent heat, transport properties, and solubility in water.

Free Dry Air Equivalent

When selecting a vacuum pump, the load at ambient conditions must be corrected to the standard conditions at which the manufacturers rated the pumps. Steam jets are usually rated on the basis of free dry air (FDA) equivalent. The actual load must be corrected to an equivalent load of air which is a function of the molecular weights of the load gas components. Each component can be corrected to equivalent FDA by using the molecular weight entrainment ratio from Figure 23-15.

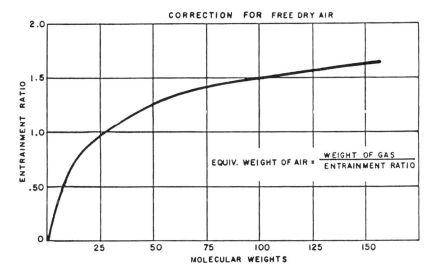

Figure 23-15: Molecular weight entrainment ratio.

Safety Factors

Safety factors are normally applied to the estimated air-leakage rates in order to improve reliability. Pump capacity gradually decreases as wear occurs. Good judgement must be used in establishing safety factors. Energy costs for oversized equipment can be staggering. Safety factors should be applied only to throughput. The penalty for applying a safety factor to both suction pressure and throughput can overwhelm the economics of vacuum systems and may dictate the type of pump required.

An overdesign factor in the range of 1.5 to 2.0 applied to the air-leakage rate is recommended. This factor should also be applied to the associated rates for saturated vapors in process streams. A factor of 2.0 is recommended for multistage steam jets with compression ratios above 6; 1.5 is adequate for single-stage jets with compression ratios less than 4 and for mechanical pumps. The larger safety factor is usually justified for multistage ejectors because they do not respond well to overloading.

If a steam jet ejector is used, an additional step is suggested: estimate the steam consumption and the number of stages. It is usually prudent to avoid jet stages which required less than 50 lb/hr because the resulting small nozzle orifice and diffuser bore will wear rapidly and lead to unsatisfactory operation. Also, the efficiency of very small jets is considerably lower than in large units. The orifice diameter for steam jets can be estimated as below:

$$W = 50 \, d^2 \, P_m^{0.96} \qquad (23.2)$$

$$d^2 = (W/50) \, P_m^{-0.96} \qquad (23.3)$$

where W = steam flow, lb/hr
        d = nozzle bore, inches
        P_m = motive steam absolute pressure, psia

For a steam rate of 50 lb/hr, the resulting relation for nozzle orifice diameter is:

$$d = P_m^{-0.48} \qquad (23.4)$$

For steam at 165 psia, the orifice size will be only 0.086 inch.

Multistage jet systems operating with suction pressures less than approximately 25 torr will normally be designed with one interstage pressure at approximately 60 torr. This criterion permits water-cooled intercondensers to be used as early as possible and greatly reduces the steam required for efficient operation.

Interstage pressures for multistage jet systems in which essentially all motive steam is condensed in intercondensers can be estimated using the correlation below:

$$P_i = P_s(R)^{(N_i/N_s)} \qquad (23.5)$$

where P_i = interstage absolute pressure
        P_s = suction absolute pressure
        N_i = interstage number with 1 representing the first stage (suction stage)
        N_s = total number of stages (estimate from Figure 23-16)
        R = overall compression ratio (equivalent to system absolute discharge pressure divided by the system absolute suction pressure)

For suction pressures less than approximately 25 torr, the first stages will not have intercondensers. For such systems, one interstage pressure can be assumed to be 60 torr. Subsequent interstage pressures can be established by assuming a total system with a suction pressure of 60 torr.

Once interstage pressures have been estimated, the steam requirements for each stage can be also estimated from Figure 23-17. If the estimates indicate steam flows less than 50 lb/hr, the design air load should be adjusted so that 50 lb/hr of steam is required as the minimum to any stage. Normally, the atmospheric pressure discharge stage will have a suction pressure of about 150-200 torr and will require the least amount of steam.

ESTIMATING ENERGY REQUIREMENTS

Energy requirements can be estimated once pressure levels have been established and loads have been estimated.

Vacuum Producing Equipment 247

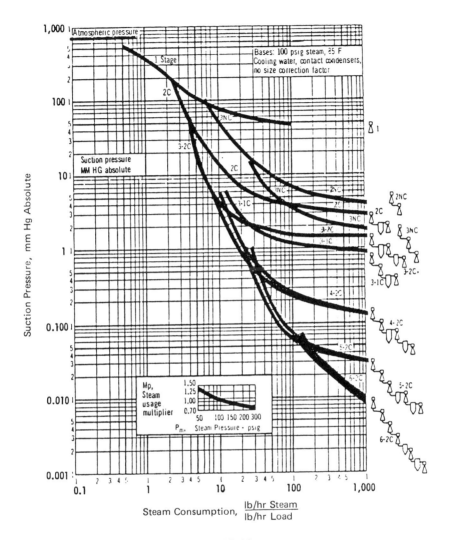

Figure 23-16.

Steam Jets

Steam requirements for steam jets handling only noncondensables can be estimated from Figure 23-16. This estimate can be corrected for motive-steam pressure as shown in the small chart included in the figure. Figure 23-16 is an estimating tool and manufacturers can frequently design more efficient jets that will use less steam than indicated, especially if interstage condensers can be used.

Steam requirements for steam jets with condensables in the load can be estimated using information from Figures 23-18 and 23-19. Again, these figures are intended for use as estimating tools.

248 Handbook of Evaporation Technology

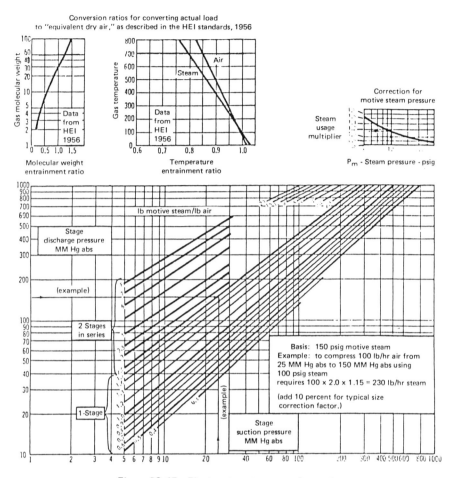

Figure 23-17: Ejector-stage steam requirements.

Steam jets with intercondensers require cooling water. Water requirements can be estimated as below:

	Number of Stages			
	2	3	4	5
Surface Condensers Water rate*	0.7	1	2	3
Direct Contact Condensers Once-Through Water Water rate*	1	2	3	4

*gpm/10 lb/hr steam

Cooling water costs are usually quite small compared to steam costs. The table above reflects the relative costs and their effect upon ejector efficiency.

Vacuum Producing Equipment 249

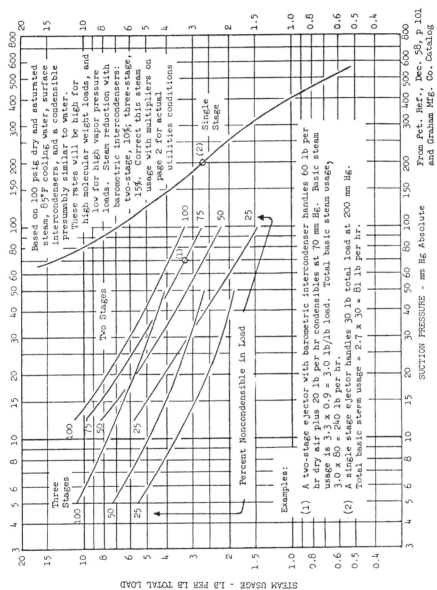

Figure 23-18: Steam jet ejector—basic steam usage—single stage, two stage condensing, and three stage condensing. From Pet. Ref., Dec. 58, p 101 and Graham Mfg. Co. Catalog

250 Handbook of Evaporation Technology

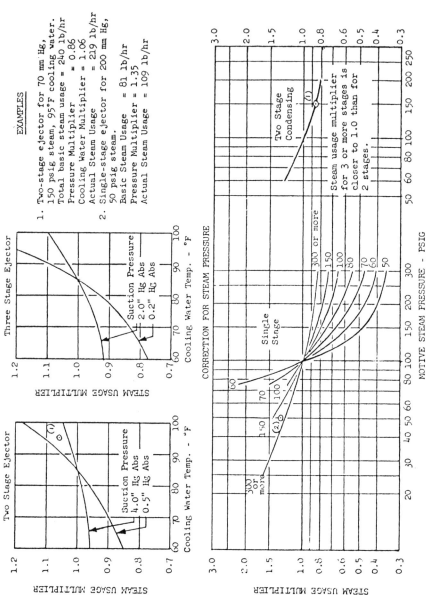

Figure 23-19: Steam jet ejector–steam usage correction factors.

Mechanical Pumps

Estimating the power requirements for mechanical pumps is more complicated. Estimates can be obtained from data presented in manufacturers' catalogs. When using catalogs, the load must be corrected from the actual conditions to the standard conditions at which the pump is rated by the manufacturer.

Power requirements can be estimated based on the flowrate (FDA equivalent) and the suction pressure. This method uses the equation below and is approximate only although it correlates well with test data especially for liquid-ring pumps. Suction pressures should be limited to the range of 10 to 500 torr.

$$BHP = a\,(FDA/P_s)^m \qquad (23.6)$$

where BHP = brake horsepower
FDA = flow rate, lb/hr (FDA equivalent)
P_s = suction absolute pressure, torr
a = constant from table below
m = constant from table below

Type Pump	a	m
reciprocating (oil sealed) pump	7.14	0.963
rotary piston (oil sealed) pump	7.63	1.088
liquid-ring pump	13.8	0.924
air ejector/liquid-ring pump	22.5	0.810

INITIAL SYSTEM EVACUATION

In addition to normal, steady state operating conditions, vacuum pumps must evacuate the process system at startup. Generally, a vacuum system must "pull down" the system pressure from atmospheric to operating level in a specified time which is referred to as the "draw down" time. Usually, a vacuum pump sized at steady state conditions gives an acceptable draw down time. When it does not, the steady state capacity must be increased to provide an acceptable draw down time or an additional vacuum pump must be installed solely for providing initial draw down. Because of energy savings, the latter is almost always favored. For jet systems, the additional device is usually a one or two stage ejector with no intercondensers and is called a "hogging jet". It is highly energy inefficient but is operated only for the short time during draw down. The counterpart for a mechanical system is called a "roughing pump".

The capacity to draw down the system in a specified time can be calculated:

$$S = (V/\theta)\ln(P_1/P_2) \qquad (23.7)$$

where S = average pumping rate (actual volume per unit time at suction)
V = volume of system
θ = draw down time
P_1 = initial absolute pressure
P_2 = final absolute pressure

Steam requirements for evacuation or draw down can be estimated from information presented in Figure 23-20. Correction factors of Figure 23-19 apply.

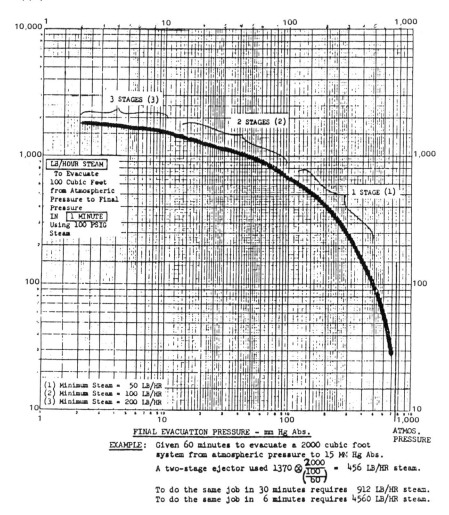

Figure 23-20: Noncondensing ejector—evacuation rate (approximate only).

CONTROL OF VACUUM SYSTEMS

The amount of gases that need to be removed from a process system will vary as leakage rates and production rates change and the capacity of the pump can vary with wear or clogging. Vacuum pumps are therefore generally provided with excess capacity above the immediate requirements. To prevent a pump from pulling the system down to a pressure lower than desired and to avoid pressure fluctuations, some sort of control must be provided.

Vacuum Producing Equipment 253

There are two basic methods for controlling vacuum pumps: suction throttling and using a load gas. Mechanical pumps may also use variable speed drive; single stage jets may also use a regulating spindle.

Suction Throttling

A simple feedback system can control the process pressure if a throttling valve is installed in the suction line between the process and the pump, as shown in Figure 23-21. The main problem with this system is that the reponse time is directly tied to the system leakage rate. If the leakage rate is high, response if fast; but if the leakage rate is low, the response time will be very slow. Unfortunately, low leakage rate and fast response time is the most desirable combination.

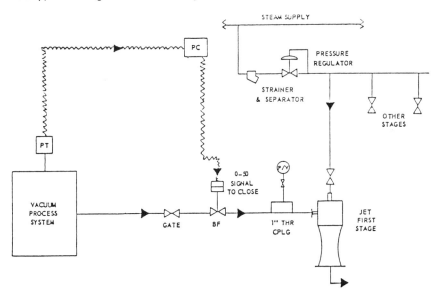

Figure 23-21: Controlling pressure in steam jet vacuum systems.

Load Gas

A load gas is a gas admitted to the vacuum system between the process and the vacuum pump. The feed rate of the load gas is controlled with a simple feedback loop to maintain constant system pressure. In real systems the load gas makes up the difference between pump capacity and leakage rate and provides good control. Load gas can be air (as shown in Figure 23-22), nitrogen or other inert gas (as shown in Figure 23-23), or discharge gases from the pump (as shown in Figure 23-24).

The ideal load gas is air, but some systems cannot tolerate air. The discharge gases from a steam jet will be wet and some systems cannot tolerate moisture. Discharge gases from a mechanical pump may contain oil or water, either of which may not be tolerated by the system. If the process requires, dry nitrogen or other suitable inert gas can be used but these may represent a significant operating expense.

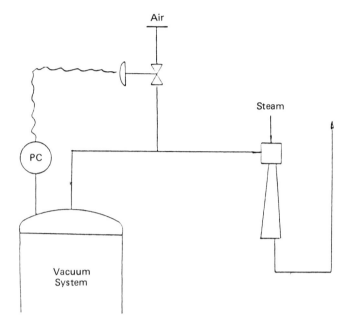

Figure 23-22: Control by atmospheric air load

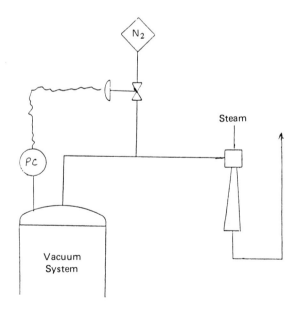

Figure 23-23: Control by nitrogen load.

Vacuum Producing Equipment 255

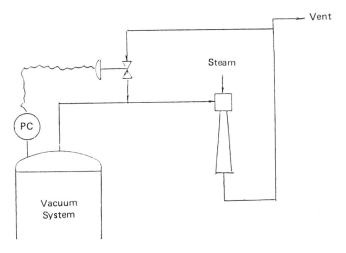

Figure 23-24: Control by recirculation of exhaust.

Combination Suction Throttling and Load Gas

By combining suction throttling with load gas as shown in Figure 23-25, the advantages of both can be retained. If the throttle valve is closed and the response time is too slow, load gas can be admitted to the system to bring the pressure back into equilibrium more rapidly. This method of control reduces the amount of any expensive load gas required.

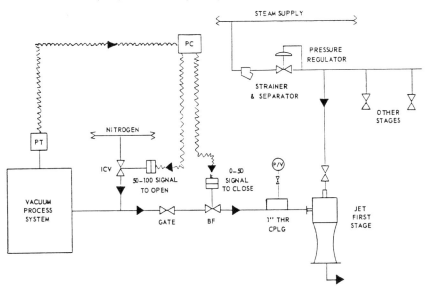

Figure 23-25: Controlling pressure in steam jet vacuum systems.

Variable Speed Drive

Mechanical pumps can be controlled by varying the operating speed of the pump. Some types, however, may not operate well at low speeds. Five types of adjustable speed drives are available: solid state AC, solid state DC, mechanical, electromechanical, and fluid. Many of these drives also result in substantial energy savings as speeds are reduced. Fluid and mechanical drives are relatively well known and have been long applied. In general, variable speed drives can only be justified economically for large power requirements. Two-speed electric motors used in conjunction with other controls are frequently cost effective.

Spindle Jets

Spindle regulated jets are usually not acceptable except for thermocompressors. A discussion of their application is given in Chapter 18. For critical vacuum pumps, regulating spindles are an ineffective means of control and can lead to unstable jet operation.

COSTS OF VACUUM SYSTEMS

The purchase price of vacuum systems depends upon the operating parameters and type of system. Single-stage jets and rotary blowers are among the least costly. The following equations can be used for making reasonable preliminary estimates of total installed costs. The equations are limited and should never be used as substitutes for detailed engineering.

Basis for the equations are:

(1) major modifications to utility supply systems will not be required.

(2) structures will not be significantly modified

(3) special supports will not be required

(4) an accuracy of plus or minus 30% is acceptable

(5) costs are mid-1981 and should be appropriately updated.

Steam Jets

$$\text{Installed Costs} = \$15,000(N_s + 2N_c)(W_s/1000)^{0.35} \tag{23.8}$$

where N_s = number of ejector stages
N_c = total number of intercondensers and aftercondensers
W_s = steam consumption based on 100 psig motive steam, lb/hr

Adjustments

Item	Multiplier
carbon steel/cast iron	1.0
stainless steel	1.1
Hastelloy	1.5
direct contact condensers	1.0
surface condensers	1.2

Liquid-Ring Pumps

$$\text{Installed Costs} = \$28{,}000(\text{BHP}/10)^{0.50} \qquad (23.9)$$

where BHP = motor horsepower of pump

Adjustments

Item	Multiplier
partial water recovery	1.0
total water recovery	1.1
stainless steel	1.6

COMPARISONS

Higher thermal efficiency is not synonymous with lower energy cost. The electrical equivalent of a unit of motive steam will usually cost 3 to 5 times more than the motive steam because of costs and inefficiencies associated with generation of electricity. Energy costs for mechanical vacuum pumps can be compared to energy costs for steam ejectors for a specified condition. Figure 23-26 shows such a comparison for mechanical pumps and steam jets. Figure 23-27 offers a comparison for liquid-ring pumps and steam jets.

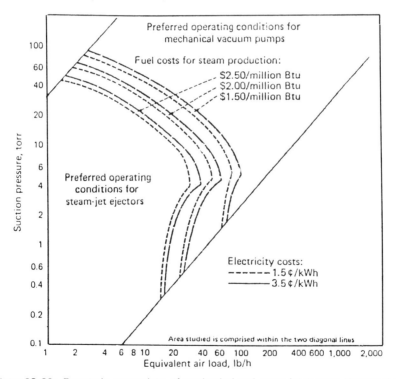

Figure 23-26: Economic comparison of mechanical and steam jet vacuum systems at two electricity costs.

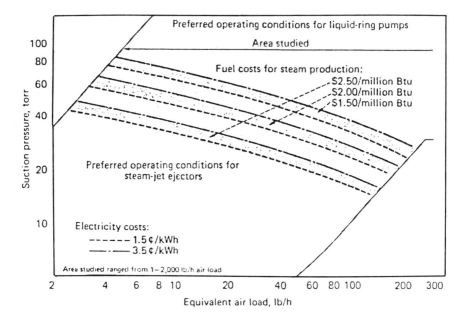

Figure 23-27: Economic comparison between liquid-ring-pump and steam jet vacuum systems.

ENERGY CONSERVATION

The greatest energy savings for vacuum systems is achieved through installing a vacuum pump that has an appropriately rated suction pressure and capacity. Things you can do to improve energy conservation:

(1) *Avoid the use of surplus equipment,* especially steam jet systems, that has more capacity or a lower rated suction pressure than needed. Surplus equipment is *rarely a bargain* in terms of energy cost.

(2) Do not arbitrarily specify higher design loads or lower suction pressure than is appropriate when purchasing new equipment.

(3) Select the appropriate type of vacuum system.

(4) Operate jets at their rated steam pressure.

(5) Keep steam dry.

(6) Periodically check pumps for wear or corrosion.

(7) Install vent condensers or precondensers.

24
Condensate Removal

Steam condensate must be removed from evaporation systems at three primary locations:

(1) calandria steam chest
(2) steam tracing systems
(3) steam header drip legs.

Three methods are used to remove condensate, primarily for calandria steam chests. Tracing and steam header systems can usually be handled by the first of these methods. The three methods are:

(1) steam traps
(2) liquid level control with condensate removed with gravity or pressure
(3) liquid level control with condensate removed by pumping.

Many evaporator calandrias condense steam at subatmospheric pressures. Consequently, condensate must be removed by pumping, either with mechanical pumps or pumping steam traps. Large condensate loads are usually best handled when liquid levels are controlled and condensate removed with the appropriate method.

LIQUID LEVEL CONTROL

Large condensate loads are frequently present in evaporator systems. Large quantities of condensate are best handled in an instrumented condensate pot or tank in which liquid level is controlled. Liquid-level control will reduce steam consumption by 2-17%, depending on the steam pressure and trap selected for service. Condensate can be removed by gravity or pressure in many cases. When the condensing pressure is below atmospheric, condensate must be removed with mechanical pumps or with pumping steam traps, which are described later. Pumping steam traps have a limited condensate handling capacity. Proprietary liquid-movers which have much larger capacities can also be used.

A typical instrumented condensate removal system is shown in Figure 24-1. The condensate tank must be located so that the controlled level does not create a liquid level in the calandria. Condensate line between calandria and tank must be self-venting. Controlling condensate levels in calandrias is not recommended as discussed in Chapter 31. If the condensate tank is adequality elevated, condensate can be removed by gravity even when condensing under vacuum. Pressure operation can use the pressure of the condensate to move it to the desired location. When the elevation does not permit condensate to be moved by gravity, pumping traps or proprietary liquid-movers can be used.

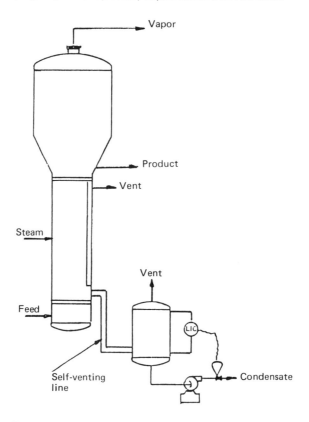

Figure 24.1: Typical instrumented condensate removal system.

Level instrumentation can be a compact pneumatic controller. Self-contained pneumatic instruments provide a positive level readout, a level sensor, a control point, and a pneumatic output control signal all in one small housing.

For most flashing-condensate services, a standard stainless steel control valve will provide long reliable life. Special trims and seats are available for high-pressure service. The control valve should always be sized for flashing service. In many instances, this will require a large valve-body with reduced trim to maintain valve-position control, because the flash steam causes high back pressure in the small valve-bodies.

Both pumping traps and liquid movers use steam as a motive force to pump condensate to the desired location. The steam must be vented between each cycle. For vacuum operation, a vacuum vent is required and usually represents an energy inefficiency unless the vent can be condensed in the process to recover the heat. The steam can be vented back to the calandria steam chest and condensed directly against the process, however, this may represent an operational instability unless properly designed. The vent rate must be controlled at a rate which does not create problems in the calandria or evaporation system. The receiver must be adequately sized to permit stable operation. Generally, it is better to vent to a separate condensing location.

STEAM TRAPS

It would be a rare process plant that has no steam traps, but many know little about these devices. Consequently, they are generally ignored, not maintained, and as a result, waste large amounts of energy. Steam traps waste energy because of two major causes—misapplication and neglect. Most plants can save 10-20% of fuel costs simply by having a formal active steam-trap program.

A steam trap is a device that permits condensate and noncondensavle gases to leave the system with minimum steam loss. The trap is required to first distinguish steam from condensate and noncondensable gases, and then to open a valve (or valves) that allows passage of condensate and noncondensable gases but closes to steam.

A steam trap handles only material that comes to it. Condensate can, in general, be expected to flow to the trap, but noncondensable gases cannot be guaranteed to flow to the trap. When a calandria is first placed onstream, the steam side is full of ambient air. Colder, denser air will collect at the bottom of the condensing surface remote from the steam inlet. At this stage, air is easily removed by the trap.

During operation, turbulence and diffusion mix air and steam together. Condensate forming on the heating surface is saturated with air present as a film on this surface. This assists in driving air to the condensate outlet leading to the trap. But there can be accumulation of air at locations remote from the condensate outlet nozzle.

In addition to being present throughout the steam system at startup, air enters whenever there is a shutdown that allows steam to condense and pull a vacuum. Also, it is constantly present if the deaerator is malfunctioning and feedwater conditioning-chemical dosage is inadequate. Carbon dioxide is also present in steam produced from water that is not completely demineralized. Besides the adverse effect of these gases on heat transfer, their presence leads to

262 Handbook of Evaporation Technology

corrosion in the steam-condensate system — another reason why steam traps are important for efficient plant operation.

Operating Principles

Traps distinguish between steam, condensate, and noncondensables by three principles: density difference, temperature differences, and phase change. These principles are used in the three major trap types: mechanical, thermostatic, and thermodynamic.

For a given operating condition, the density of condensate is considerably different from that of steam or air. The principle is used by mechanical-type traps. This density difference always exists, except at the critical pressure; however, the densities of steam and air are too close for distinction. It is also important to note that any flash steam produced in the line from the calandria is regarded as live steam. The trap will close even though condensate is following the pocket of flash steam.

The difference in temperature between steam and cooled condensate is the principle used by thermostatic-type traps. Note that condensate and the steam from which it is produced have the same temperature when in equilibrium. Therefore, to enable distinction by this type of trap, condensate must be cooled below the steam temperature. This may be achieved by locating the trap some distances from the calandria and/or leaving the inlet line uninsulated; or, by deliberately imposing a delay in condensate removal rate, allowing it to back up into the calandria or in the pipeline to the trap. Since temperature of condensate is sensed in this type of trap, it offers a method for controlling condensate discharge temperatures (e.g., in systems where it is worthwhile to extract some sensible heat from condensate). Thermostatic traps can discharge line-flash steam, provided its temperature is measureably below that of live steam. And assuming that air is cooler than steam (which is true at startup when air removal is of the greatest significance), a thermostatic trap will discharge the air.

The third principle used in trap design is the change of phase that occurs when condensate at any particular temperature experiences a drop in pressure, for example, when it flows through an orifice. This pressure drop results in formation of flash steam. The higher the condensate temperature upstream of the restriction, the greater the quantity of steam produced, for a fixed downsteam pressure. The flash steam is then used for closing a valve and/or as a means of choking off condensate flow, since steam occupies a much larger volume than condensate. Alternatively, the rise in pressure that occurs with increased steam velocity through a restriction is used to motivate a disk that closes off condensate flow. Traps that use a phase change are classed as thermodynamic steam traps. The quantity of flash steam produced is a function of inlet condensate temperature (and corresponding steam pressure) and downstream pressure. The pressure in the condensate return line (trap back pressure) is important. If it is too high, there will not be enough flash steam and the trap will not function. After closure of the valve (or disk) there must be a lapse of time for bleed-off or condensation of steam before the valve can open. This causes the trap to operate on a time cycle and thus it does not remove condensate continuously. Air is probably removed in the thermodynamic trap that operates by choking off flow, but heated air is likely to close off the disk trap in the same manner as steam.

MECHANICAL TRAPS

There are several basic types of traps in this class. However, they all have one principal in common: all are operated by response to the difference in density between steam and condensate. Density operated traps are purely mechanical in operation. They use some type of float to determine the condensate level in the body and thereby operate a valve mechanism. Each float-mechanism combination has fixed mechanical advantages requiring the valve seat orifice to be matched to a maximum operating pressure. Because the body always contains water, their use is generally confined to indoor applications unless provision is made to drain the body to prevent freeze-up damage when not in service. In some types the float serves only to release a counterweight which, as it falls, opens the discharge valve. As the float falls, it engages the counterweight for another discharge cycle. In some traps, the float itself serves as the valve.

THERMOSTATIC TRAPS

Thermostatic traps are temperature-sensitive traps; there are several basic types. They respond to either a temperature difference between the steam and the condensate or directly to a temperature of either steam or condensate. All thermostatic traps are operated and controlled by the temperature in the line upstream of the trap; time is necessary for the operating elements to either absorb heat to cause the valve to close or dissipate heat to cause the valve to open. They usually discharge condensate below the steam temperature and require a collecting leg before the trap to permit some subcooling of condensate. Therefore, sufficient piping length should be provided at the trap inlet to prevent the condensate backup from interfering with the heat transfer surface.

THERMODYNAMIC TRAPS

Thermodynamic traps utilize the heat energy in hot condensate and steam to control the opening and closing of the trap. They therefore respond to the difference in thermodynamic properties between steam and condensate.

STEAM TRAP SPECIFICATION

Steam traps are integral and important units in a process. Their design calls for the same analysis that other process equipment receives. There is less literature available on steam traps than for other equipment, but the References chapter lists several references that contain important information concerning steam traps. A trap that is undersized will cause condensate to interfere with heat transfer efficiency; while traps having too much excess capacity not only waste money, but act sluggishly and produce a high back pressure that may significantly reduce trap life.

The discharge capacity of the trap depends on the flow area of the valve orifice, the pressure drop across it, and the inlet temperature of the condensate. There is a considerable problem in measuring the pressure drop because hot condensate flashes as it passes through the valve orifice. Trap capacity is not truly defined by orifice size and pressure differential. Pressures upstream and downstream of the trap are also subject to variation, depending on calandria performance, flowrates, temperatures, and system back pressure. The orifice may never be fully open to flow because of the valve design. Nor can flow coefficients be measured with the same precision as for control valves, since the valve stem in the trap is often not definitely located with reference to the orifice.

In order to compensate for these variations in operating conditions, the designer uses a capacity safety factor to increase the calculated condensate load. This safety factor should be selected with care. One short-cut method that should be avoided: sizing steam traps to equal line size. This practice is never a substitute for analyzing process conditions; it invariably leads to specification of the wrong size steam trap.

COMMON TRAP PROBLEMS

Some of the more common problems associated with stea traps include:

(1) improper sizing
(2) poor performance of steam heated equipment
(3) high maintenance costs
(4) waste of steam.

SELECTION OF STEAM TRAPS

It is important that steam traps be properly sized. Since the various traps leak various amounts of live steam, substantial savings can be gained by selecting the most efficient trap for a given service. The most efficient trap can only be determined by actual use of the traps in plant service and testing the traps for efficiency of operation. Oversizing is one of the biggest problems in selecting steam traps.

The following factors should be considered when selecting steam traps:

(1) condensate load
(2) inlet pressure at the trap
(3) back pressure at trap outlet
(4) condensate temperature at trap inlet

(5) ratio of back pressure at trap outlets to pressure at trap inlet
(6) variation of condensate flow
(7) amount of air or noncondensables to be removed
(8) inlet and outlet piping
(9) cleanliness of outlet piping
(10) water hammer.

Process Steam Traps

Steam traps for large process equipment are required to handle high loads which justifies their selection on an individual basis. An accurate determination of the available differential pressure across the trap must be made for the various flow rates for minimum to maximum operating rates. The available trap inlet pressure is the saturation pressure that corresponds to the steam temperature required to heat the process fluid; it is not the full steam pressure at the inlet to the steam flow throttling valve. The trap back pressure is the condensate header pressure plus piping, valve, and elevation losses. Using accurate performance data allows the trap to be sized without using any "safety factor". "Safety factors" usually result in the selection of a trap that is too large. Oversized traps result in energy loss because of higher steam losses and shortened trap lifetime. Large condensate loads are best handled by controlling liquid level as discussed previously. Very large capacity steam traps are difficult to evaluate for steam loss. They are also subject to frequent failure.

INSTALLATION

The best steam trap will not function correctly if the piping to and from it is not properly designed and installed. Many factors are important and some will be discussed.

Double Trapping

Double trapping is the poor practice of placing two traps in series. Unless the second trap is an orifice type properly sized, flash steam from the saturated condensate entering the first trap will result in closing of the second trap. The net effect is poorer drainage than would have been achieved with only one of the traps. Undersized lines preceding a single trap is the equivalent of double trapping.

Multiple Trapping

Multiple trapping of equipment is often provided in the interest of space and investment economy. Generally, each of the heat exchanges connected to the common trap operate at sufficiently different conditions to result in poor drainage of condensate from one or more of the parallel equipment. A single trap should never be used to drain more than one piece of equipment. Even though the equipment may be identical in design, they will not operate identically and the steam flows will not divide equally.

Part-Load Operation

Part-load operating may reduce the ability of the trap to properly drain the condensate. This is especially true when low steam pressure is used and, when throttled at low loads, the pressure may fall below atmosphere or the pressure of the discharge header. Problems associated with this are discussed in Chapter 31.

Steam Locking

Steam locking may occur in systems where drainage lines from equipment to traps must run upward. Steam trapping consists basically of separating condensate from steam. This can occur naturally as a result of gravity. Whenever possible, drainage lines from equipment to traps should slope downward.

Inlet Piping

Inlet piping must be adequate to avoid water hammer and poor trap performance. They should not be undersized. To avoid sudden flashing of condensate, with its attendant problems, the line should be:

(1) adequately sized to minimize pressure drop
(2) sloped downward to the trap
(3) individually trapped to avoid bypassing
(4) provided with features to avoid steam-locking where gravity drainage is not possible
(5) sized for full trap capacity.

Discharge Piping

Discharge piping is as important as the inlet piping if the traps are to function properly and erosion is to be avoided. Condensate partially flashes when pressure is reduced by flowing through a trap. Flashed steam can not be avoided so it must be treated. Velocities in discharge piping must be low enough to avoid excessive eroison. Water hammer will generally not occur in flashing mixtures flowing in well-designed piping. Admitting water that is not at saturation temperature into a flashing mixture will invariably result in water hammer, with potentially destructive effects, and should be avoided. The merits of elevating condensate on the discharge side of steam traps has been much debated. Proper design will permit satisfactory operation whatever the choice. Proper design includes proper placement of check valves and considering the effects of the static head of condensate when sizing the trap.

Large Loads

Large loads can sometimes be best handled with multiple traps in parallel if level control is not justified. Multiple banks of traps for a single load have the advantage that the loss of a single unit does not result in a complete shutdown. One or more of the traps can be valved out of service at reduced loads also. Traps in parallel must be installed in the same horizontal plane.

Piping Details

Traps should be installed with no bypass valves. These valves tend to leak and waste energy. They also present an easy means to increase capacity by opening them. When the load later decreases, they will result in steam being wasted at a high rate. The proper way to solve the capacity problem is to install a trap having a higher capacity.

The process trap must be installed such that condensate does not back up into the heat transfer surface.

All process traps should have a strainer in the trap inlet line. An outlet shutoff valve should be provided to permit isolation of the trap for maintenance without having to drain the equipment or shutdown the condensate return header. The trap, strainer, and valves should be arranged to allow access for operation, maintenance, testing, and inspection. Modular hook-ups are recommended.

Traps should be installed below the equipment being drained. If this is not possible, a lift fitting or water seal should be provided at the low point. Discharge lines should be kept short to minimize freezeup damage. Traps should be installed to be self draining. Bucket traps must be protected from freezing and must be protected against loss of prime as a result of sudden or frequent drops in steam pressure.

Condensate Return Headers

Condensate return lines are more important today because significant amounts of energy can be conserved by collecting condensate and returning it for recycle. Condensate return lines must be adequate to avoid problems as a result of:

(1) accepting discharge from a greater number of items (usually steam traps) than originally planned. This is often the result of plant expansion or modification.
(2) a variety of pressures discharging into a common return header
(3) oversized steam traps.

Improperly sized condensate return headers result in overpressurization of the return system, venting of excessive amount of vapor, vapor binding of condensate pumps, faulty steam trap operation, and water hammer. The following factors should be considered:

(1) properly size traps
(2) properly size piping
(3) properly size flash tanks
(4) consider a separate return line for each header pressure
(5) properly insulate the line
(6) slope line toward flash tank
(7) expand the line on the bottom when pipe size increases are needed

(8) trap discharge lines should enter the top of the condensate header

(9) proper elevation of equipment to be drained.

EFFECT OF CARBON DIOXIDE

Carbon dioxide is often ignored in steam systems. However, when absorbed in water, it forms carbonic acid, which can be corrosive to all parts of the steam and condensate system. Its potential presence is frequently overlooked in the design of heat exchangers, steam traps, condensate systems, deaerators, and water-treating systems. Most steam systems require continual addition of make-up water to replace losses. Makeup water must be adequately treated, by demineralization or distillation, to remove carbonates and bicarbonates. If these are not removed, they can be thermally decomposed to carbon dioxide gas and carbonate and hydroxide ions. The ions will normally remain in the boiler water, but the caron dioxide will pass off with the steam as a gas. When the steam is condensed, the carbon dioxide will accumulate (since is is noncondensable); be passed as a gas by the steam trap; or if the condensate and carbon dioxide are not freely passed by the steam trap, become dissolved in the condensate and form carbonic acid. If carbonic acid is formed it can have a pH approaching 4 and be very corrosive to copper and steel. Even if both the gas and condensate are passed freely by the steam trap, the gas will become soluble in the condensate when subcooling occurs. If oxygen is present, the corrosion rate is accelerated.

When carbon dioxide is present, only those traps that discharge condensate and noncondensable gases at or near saturation temperature (within $3°C$) should be used. The introduction of new traps that claim to save energy by deliberately subcooling condensate has led to several problems. The new traps do not remove carbon dioxide and corrosion results. Steam leaks develop at threads, uhions, and welds.

The corrosion products (copper and iron carbonates) are soluble in the low pH system. When pressure drops across the small orifice, the corrosion products are redeposited as a solid. Plugged trap discharges quickly occur as a result, or solids cause trap malfunction.

Plants using makeup water that contains carbonates or bicarbonates can expect low-pH carbonic acid to cause difficulty in all portions of the steam systems if precautions are not taken. Possible precautions include:

(1) upgrading water-treating facilities

(2) properly venting heat exchangers

(3) using proper steam traps

(4) providing increased condensate recovery

(5) keeping condensate hot

(6) deaerating condensate

(7) separating condensate and makeup to deaerators

(8) discarding vent-condenser drain liquids

(9) applying chemical treatment.

Proper observance of these precautions can minimize carbon dioxide effects with resulting energy savings and reduced maintenance costs for steam systems.

STEAM TRAP MAINTENANCE

A malfunctioning steam trap is a great waster of steam. Almost any large plant has some (often many) energy-wasting traps. Traps waste energy because of two major causes—neglect and misapplication. Most plants can save 10-20% of fuel costs simply by having a formal, active steam-trap program.

The road to an effective steam-trap energy-saving program can be made much easier by following six steps:

(1) Select a steam-plant energy coordinator. The right person is the key to the success or failure of an effective program.

(2) Develop a plant standard of only a few approved steamtrap types.

(3) Monitor steam consumption for all products. Establish a cost basis for comparison and evaluation.

(4) Change out all steam traps to the plant standard and monitor steam savings.

(5) Develop and implement a steam-trap checking procedure to detect bad traps and change them out using the plant standard.

(6) Standardize on a set trap-checking fequency, with plant followup to make sure the program is working.

The first four steps are designed to solve misapplication; the last two are designed to eliminate neglect.

There are four essentials to an effective trap program:

(1) Select the right person to head the program.

(2) Set up a plant standard.

(3) Establish a routine trap-checking procedure.

(4) Set up a permanent crew for trap maintenance.

25
Process Pumps

Pumps, motors included, normally represent 5-10% of the total major equipment cost for new plant construction. They represent a much larger share of the total cost for electrical hardware and power wiring and this cost is usually several times that of the equipment cost for pumps and drivers. Pumps are significant consumers of process energy.

Pumps should be given proper attention when designing evaporator systems in order to avoid problems at startup and continuing maintenance costs. An understanding of pumps and pumping principles enables the evaporator technologist to optimize investment and operating costs for pumping systems and to avoid many problems which result from misapplication of pumps.

GENERAL TYPES OF PUMP DESIGNS

An evaporator designer is generally in error if the type of pump is selected before pumping conditions have been defined. Action steps for successful pump application could be listed as follows:

(1) determine pumping conditions and evaluate complete possible range of these conditions

(2) outline the entire pumping installation (elevations, physical layout, line sizes, pressure profiles)

(3) select a pumping unit

(4) determine proper installation, startup, and operation procedures

(5) establish proper preventative and corrective maintenance procedures.

When selecting a pump, many problems are encountered for proper handling of raw materials, products, and by-products for an evaporation system. Important factors that must be considered include:

(1) corrosion
(2) erosion
(3) mechanical reliability
(4) feasibility in process application
(5) process contamination
(6) energy costs
(7) effect of range of process viscosity
(8) process temperature ranges
(9) process vapor pressure ranges
(10) effect of solids in process
(11) minimum and maximum flow rates
(12) extremes of suction and discharge pressures
(13) effect of shear upon process fluid
(14) effect of dissolved gases
(15) toxicity of the process.

Many types of pumps are available, not all of which should be considered for a specific application.

NET POSITIVE SUCTION HEAD (NPSH)

The net positive suction head (NPSH) is the absolute pressure *in excess of the liquid vapor pressure* that is available at the pump suction nozzle to move the liquid into the eye of the impeller. NPSH must *always* be calculated using units of absolute pressure and then expressed as head. NPSH is a concept *entirely different* from a pump's suction pressure. The actual NPSH must exceed the required NPSH for a given pump for adequate pump performance. Without adequate NPSH, cavitation and mechanical damage to the pump can occur. NPSH is an important consideration when selecting a pump required to pump liquids from systems under vacuum or to handle near-boiling liquids or liquids with high vapor pressures. It is usually not practical to specify values of NPSH less than two feet.

Factors that must be considered in evaluation of NPSH requirements include:

(1) Minimum pressure in the suction vessel
(2) pressure drop in suction line and all equipment in the suction line
(3) density of pumped fluid
(4) minimum elevation of the liquid level in the suction vessel

(5) safety factor to allow for operation at capacities greater than maximum

(6) vapor pressure of the liquid being pumped at the maximum anticipated temperature under normal operating conditions. When the fluid contains dissolved gases, the vapor pressure should be that of the solution. Impurities such as dissolved gases and traces of volatile liquids produce appreciable differences in vapor pressure from that of the pure liquid.

CAVITATION

Cavitation is caused by the formation of vapor bubbles in a high-velocity, low-pressure region and by the subsequent collapse when the bubbles move to a higher pressure region. Cavitation can cause excessive erosion and vibration. A liquid moving through a pump will vaporize rapidly whenever the local pressure falls below the liquid vapor pressure.

With moderate cavitation in a centrifugal pump, the pump will sound as though it is pumping gravel or a slurry of sand and gravel. Severe cavitation will cause the discharge pressure to fall and become highly erratic and produce both flow and pressure pulsation. Pump cavitation can result from insufficient available NPSH, from a high suction piping velocity, and from a poor suction piping configuration.

Pumps often cavitate in spite of a designer's best efforts to meet NPSH requirements of pumps. This behavior is not well understood, and could occur for a number of reasons, including:

(1) high pump-suction velocities and long piping increase pressure fluctuations in the pump. Velocities less than six feet-per-second work best when NPSH requirements are difficult to meet.

(2) Vacuum systems seem more prone to unpredictable cavitation than pressure systems.

(3) Vapor pressure is used in NPSH calculations when actually saturation pressure might be more appropriate. A better definition of "vapor pressure" is needed. Be wary of dissolved gases.

(4) A suction-piping elbow next to the pump aggravates cavitation, especially at high flow velocities. About five diameters or more of straight pipe between pump and elbow will minimize pressure disturbances near the elbow.

PRINCIPLES OF PUMPS AND PUMPING SYSTEMS

The principles of designing good pumping systems are not difficult to comprehend. Once mastered, the designer can contribute to both energy conservation and process reliability. Understanding pumping principles will permit a prompt response to plant operating problems or modifications.

Pumping is the addition of energy to a fluid in order to move it from one point to another and not, as is frequently thought, the addition of pressure. Since energy is capacity to do work, the addition of energy to a fluid causes it to do work, such as flowing through a pipe or rising to a higher level. In the closely allied field of centrifugal compressors, "pumping" describes an unstable surging that occurs in many compressors at partial capacity.

Centrifugal Pumps

Every centrifugal pump operates on a predetermined "pump curve", as shown in Figure 25-1. The pump curve indicates the fixed relationship between total head and volumetric flow rate. The range and shape of the pump curve is determined by:

(1) impeller diameter
(2) impeller speed
(3) geometry of impeller, casing, pump inlet, and pump outlet.

Brake horsepower is also plotted on the curve. Note that brake horsepower decreases with decreasing flow, but does not reach zero at zero flow. There will be a noticeable and sometimes troublesome "heat effect" in cetrifugal pumps operating with the discharge throttled to abnormally low flow rates. Pump efficiency is also indicated on the curve.

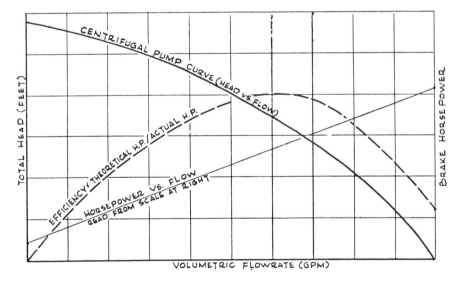

Figure 25-1: Typical performance characteristics.

For all but a few special considerations, it is possible for a centrifugal pump to be valved-in at the discharge without damage. A centrifugal pump will pro-

274 Handbook of Evaporation Technology

vide a wide range of flow rates, depending on the downstream resistance to flow. Relationships are as indicated below:

	Minimum Flow	Maximum Flow
Discharge pressure	maximum	minimum
Brake horsepower	minimum	maximum

The exact operating point on any pump curve is determined by the intersection of the pump curve with the "system curve" which shows pipeline and equipment head loss as a function of volumetric flow rates as snown in Figure 25-2. "System curves" must intersect a given pump curve at a definite *operating point.*

Graphical techniques such as Figure 25-2 are extremely useful in helping to visualize dynamic situations in a pumping operation and aid in selecting "optimum" process configurations or in proposing modifications to existing systems. The static head of liquid as a result of vessel elevation must be included when plotting system curves. If there is no elevation head, the system curve will pass through the origin.

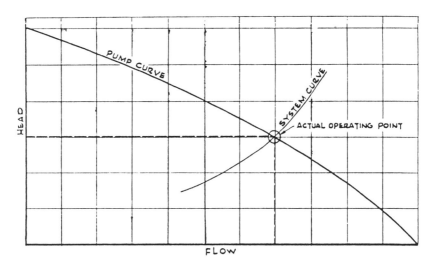

Figure 25-2: Intersection of pump curve and system curve.

Pump curves should *never* be extrapolated beyond the limits shown by the equipment supplier. Pump curves often terminate abruptly, with very little capacity increase at discharge heads lower than those shown by the vendor.

Safety Factors for Performance: When specifying pump performance, safety factors should not be applied to both capacity and head without careful graphical analysis of pump and process dynamics. It is normally adequate to specify pumps as below:

(1) use maximum value for total head

(2) use an appropriate safety factor for volumetric capacity only.

Figure 25-3 can be used to illustrate the effects of safety factors. Pump A operates exactly at the specified normal capacity and head. Adding a volumetric capacity safety factor without changing the specified head results in purchase of Pump B. Adding simultaneous safety factors to both head and capacity results in purchase of Pump C. The excessive head must be throttled across a control valve to restrain the flow.

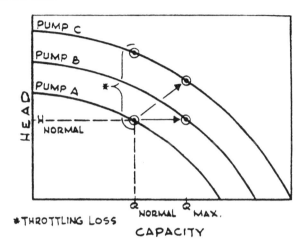

Figure 25-3: Safety factors in performance specifications.

Net Positive Suction Head (NPSH) Available: The available NPSH is the amount of absolute head available to move the liquid into the eye of the impeller. This must exceed the required NPSH for a given pump for adequate pump performance. Every pump has its own NPSH characteristics and is usually given in the form of performance curves furnished by the pump manufacturers. If the required NPSH for a given pump exceeds the available NPSH, cavitation will result with eventual mechanical damage to the pump.

Some effort should be made to avoid low available NPSH values (2 to 5 feet) because this may result in the pump manufacturer offering an oversized pump that will operate at part capacity and low efficiency. Whenever possible, excess NPSH over the requirements of the pump should be available. When a pump is handling a boiling liquid, the available NPSH is almost entirely dependent upon the elevation of the liquid level in the suction vessel.

A pump should have a throttling device in the discharge piping, either a valve or an orifice plate, to prevent the pump from operating with the equivalent of an open discharge. This could result in the pump operating in the high-capacity, low-head region of the pump where NPSH requirements become excessive.

Propeller Pumps

Propeller pumps are axial flow pumps and develop most of the head by the propelling or lifting action of the vanes on the liquid. Propeller pumps are usually applied for high-volume, low-head services and can be installed both vertically and horizontally. Propeller pumps are not generally used when developed head must exceed 30 feet.

Characteristics of propeller (and mixed-flow) pumps vary considerably from those of radial flow centrifugal pumps. The head curve of a centrifugal pump is fairly flat, while that of the propeller pump rises sharply when its flow is restricted. A comparison of a centrifugal and a propeller pump rated for the same conditions is shown in Figure 25-4.

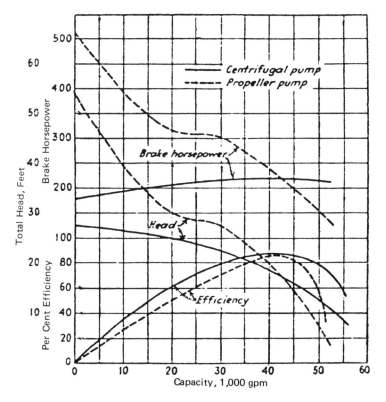

Figure 25-4: Characteristic curves of a centrifugal- and a propeller-type pump, each rated 42,000 gpm at 17 ft. head.

Propeller pumps require higher shutoff power than centrifugal pumps of the same rating, but they do not require oversize motors because shutoff operation can be avoided in several methods.

To overcome some of the disadvantages of the fixed-blade propeller pumps, pumps have been developed with adjustable-blade impellers. Maximum efficiency of a fixed-blade impeller occurs over a rather narrow discharge and head range, where that of the adjustable-blade impeller is maintained over a wide range. Blades can be adjusted manually or automatically while the pump is in service. This type of pump can be used to advantage where the head varies over a considerable range or where it is necessary to adjust the discharge from the pump.

AVOIDING COMMON ERRORS

The designer should be aware of some of the common errors for pumping systems.

Temperature Rise and Minimum Flow

A pump may overheat if it is required to operate at very low or zero flow for any significant period. Overheated pumps can create serious suction problems, mechanical problems, and safety problems. The time a pump may operate at zero flow without overheating should be established. The minimum safe flowrate for continuous heat removal during extended operation at low or zero flow should be determined.

If heat removal problems cannot be avoided in the pump system, circulation through an external cooler may be required.

Oversizing Pumps

Oversizing a pump can often create NPSH problems as well as waste energy and capital cost. A centrifugal pump is capable of operating over a wide capacity range, and the capacity at which it actually operates is determined by the matching of the system pressure drop and the pump head-capacity performance. Large calandria circulating pumps are often installed in systems having no control valve. If the pump has significantly greater head than is required by the system, the pump's capacity will increase until a match is reached as shown in Figure 25-5.

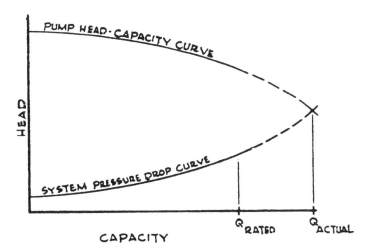

Figure 25-5: Pump head exceeds needs of system.

278 Handbook of Evaporation Technology

As the actual flowrate increases beyond the design point, a decrease occurs in the NPSH available to the pump because of additional pressure drop in the suction piping. At the same time, the NPSH required by the pump is increasing with flowrate so that the NPSH-available and NPSH-required curves intersect as shown in Figure 25-6. This point of intersection represents the onset of cavitation because of formation of vapor bubbles. Motor overload may also occur at unexpectedly high flow rates.

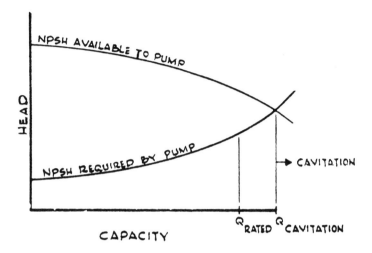

Figure 25-6: NPSH available and required.

One of the following steps should be taken when pump oversizing is experienced:

(1) trim the pump impeller
(2) reduce the pumping speed
(3) install an orifice in the discharge line to increase system pressure drop
(4) install a valve in the discharge line to enable system pressure drop to be varied.

The last two options waste energy.

26
Process Piping

Typically, process piping contributes 15-20% of the total investment for an evaporation system. Energy usage plus continued maintenance add additional costs. Clearly, sound sizing practices can have a substantial impact on overall system economics.

The literature on pipe sizing is massive reflecting its importance. Piping-flow phenomena are frequently complex and often not well understood. Too often, piping is sized using rules of thumb and sound sizing principles are ignored. Before undertaking any sizing calculation, the designer first decides what basis to use, often a difficult decision.

Most often, economic sizing is considered to be a balance between energy and investment costs. However, it is better to consider economical sizing in its broader context. Frequently there is one best and most economical size and that size may have no effect whatever on energy consumption. For example, pipe sizes to and from fixed-pressure utility headers do not affect energy consumption. Yet in a given application there is a minimum size that will give satisfactory operation, and is therefore the most economical size.

One rational approach to pipe sizing involves first determining if pipe size affects plant energy consumption. If so, optimum sizing should be given further consideration. If not, it should be ignored and another rational basis selected. In any case, the following sizing limitations should be considered:

(1) pressure-drop availability
(2) two-phase flow patterns
(3) drain-piping design
(4) erosion corrosion
(5) noise and cavitation
(6) flow distribution
(7) transient and vibration phenomena

(8) piping expansion and support.

Economical line size is affected by several factors:

(1) economic analysis
(2) piping configuration
(3) materials of construction
(4) physical properties
(5) control valves. The control valve for a controlled line requires a large fraction of total frictional pressure drop.
(6) insulation. Insulation tends to increase capital investment and decrease optimum pipe size.
(7) location
(8) inflation
(9) operating schedules.

Rules of thumb should be viewed only as guidelines or selection aids. The following guidelines should be considered:

(1) select a sound sizing basis; avoid rules of thumb
(2) consider special piping configurations or flow phenomena that might alter the selection
(3) use short-cut pressure-drop estimating techniques for small piping only (less than six-inch in diameter), or when pressure drop is unimportant
(4) avoid sizes not stocked in plant stores
(5) with turbulent flow, recognize that straight-pipe friction will usually dominate total frictional losses in small piping, and fitting losses will usually dominate the total in large piping.
(6) avoid the "eqivalent" length concept for turbulent flow through large-diameter fittings and for all laminar-flow fittings.
(7) with small piping (one-inch, e.g.), consider the need for larger piping to ensure adequate support
(8) avoid the use of valves (except control valves) that differ from line size
(9) use a design frictional pressure-drop approximately 1.3 times greater than the value calculated for new pipe to allow for pipe aging and uncertainties in the estimate.

DESIGNING DRAIN PIPING

Drain piping usually connects to an upstream pressure vessel and, even if the liquid level is controlled, gases and vapors can break through into piping. Whirlpools, low liquid levels, flashing, and turbulence can contribute to the gas entrainment.

Two flow regimes are possible for drain piping: self-venting and siphon flow. Liquid in a self-venting line occupies a portion of the pipe cross-section, and gas moves freely with, or counter to, the liquid. Siphon flow refers to a liquid or two-phase downflow where pressure increases in the direction of flow.

Lines will be self-venting if the following correlation is used to determine drain nozzle sizes:

$$D = 0.92 \, Q_F^{0.4} \tag{26.1}$$

where D = nozzle diameter, inches
Q_F = flow rate, gallons per minute

Short, horizontal pipe will also become self-venting at this transition. However, the transition is very sensitive to pipe slope and somewhat sensitive to length. Open-channel-hydraulics technology must be used for analysis of horizontal pipe.

COMPRESSIBLE FLUIDS

When the line pressure drop in gas or vapor flow exceeds roughly 40 percent of the absolute upstream pressure, density variations must be considered. The NASA charts can be used to solve compressible flow problems. Adiabatic, subsonic, compressible flow for two common cases are given:

(1) conditions are known at the upstream reservoir
(2) conditions are known at a downstream point in the pipe.

TWO-PHASE FLOW

The flow of a gas-liquid mixture in a pipeline is common. The flow patterns present should be recognized. For vertical piping, the pattern alone often determines the proper line size. Two-phase pressure drop correlations have a high uncertainty associated with them. Typically, the estimated pressure drop can be ±40% of the true value.

Two correlations are widely used: the Lockhart-Martinelli correlation and the homogeneous model of Dukler and others. Dukler's correlation will be slightly better for most applications. Both correlations are expected to give better accuracy for horizontal flow than for vertical flow. Only when the two-phase pressure drop is high will the estimate be relatively independent of pipe orientation.

Lockhart-Martinelli

To estimate the two-phase pressure drop with the Lockhart-Martinelli correlation, the procedure outlined below is followed:

(1) estimate the liquid phase friction pressure drop, ΔP_L, assuming it is flowing alone in the pipeline

(2) estimate gas phase friction pressure drop, ΔP_G, assuming it is flowing by itself in the pipe

(3) Estimate the two-phase pressure drop, ΔP_{TP}, from:

$$\Delta P_{TP} = [(\Delta P_L)^{1/n} + (\Delta P_G)^{1/n}]^n \qquad (26.2)$$

use n = 4.0 when both gas and liquid phases are in the turbulent flow regime
use n = 3.5 when one or both phases are in laminar flow.

(4) Equation 26.2 predicts total head loss; elevation effects must be taken into account.

Dukler Homogeneous Model

Dukler proposed a correlation based on the assumption that the gas and liquid flow at the same velocity (no slip) and that the properties of the fluid can be suitably averaged.

(1) Calculate average fluid properties:

$$\rho_{ns} = \rho_L \lambda_L + \rho_G (1 - \lambda_L) \quad \text{for density} \qquad (26.3)$$
$$\mu_{ns} = \mu_L \lambda_L + \mu_G (1 - \lambda_L) \quad \text{for viscosity} \qquad (26.4)$$

where $\lambda_L = Q_L/(Q_L + Q_G)$
Q = volumetric flow: Q_L = liquid; Q_G = gas

(2) Calculate an average no-slip Reynolds number:

$$Re_{ns} = (D_i G_T)/(\mu_{ns}): \quad G_T = G_L + G_G \qquad (26.5)$$

(3) Calculate frictional pressure drop:

$$\Delta P_f = (4 f G_T^2 L)/(2 g \rho_{ns} D_i) \qquad (26.6)$$

where f is the conventional Fanning friction factor

(4) In addition to the friction pressure drop, there will be a loss associated with expansion of the gas. This is sometimes termed an acceleration loss, A_{cc}; it will be important only when the pressure drop is a substantial fraction of the upstream absolute pressure.

$$A_{cc} = (G_T G_G P_{avg})/(g P_1 P_2 \rho_{G\,avg}) \qquad (26.7)$$

where P_1 = upstream pressure (absolute)
P_2 = downstream pressure (absolute)
P_{avg} = $(P_1 + P_2)/2$
$\rho_{G\,avg}$ = $(\rho_{G1} + \rho_{G2})/2$

ρ_{G1} = upstream gas density
ρ_{G2} = downstream gas density

(5) The two-phase pressure drop is given by:

$$\Delta P_{TP} = (\Delta P_f)/(1 - A_{cc}) \qquad (26.8)$$

Flashing Liquids

Dukler's analysis can be suitable modified for flashing flow. The two-phase pressure drop is:

$$\Delta P_{TP} = (G_T^2/2g)[2[(x/\rho_G)_2 - (x/\rho_G)_1] + (4fL/D_i\rho_{ns\ avg})] \qquad (26.9)$$

where x is the mass fraction gas
1 and 2 refer to upstream and downstream conditions respectively
$\rho_{ns\ avg}$ is the average no-slip density.

Elevation Effects

To estimate the two-phase pressure change ΔP associated with elevation difference ΔZ, Chisholm's holdup correlation is recommended:

(1) Estimate the gas holdup, R_G, which is the fraction of the pipe cross-section occupied by the gas:

$$1/R_G = 1 + [1 - x + (x\rho_L/\rho_G)]^{1/2}[\rho_G(1-x)/(x\rho_L)] \qquad (26.10)$$

where x is the mass fraction gas

(2) Estimate an average density, ρ_a, from:

$$\rho_a = R_G\rho_G + (1 - R_G)\rho_L \qquad (26.11)$$

(3) Estimate the two-phase pressure change associated with elevation difference ΔZ from:

$$\Delta P = (\rho_a \Delta Z g)/(144 g_c) \qquad (26.12)$$

where ΔP = pressure change, psi
ΔZ = elevation change, feet
g = gravitational acceleration
g_c = gravitational constant (4.18 × 10^8 ft/hr^2)

Two-Phase Flow Valve and Fitting Losses

Two-phase friction loss in valves and fittings can be estimated using the no-slip velocity, v_{ns}, the single-phase head coefficients, K, and the no-slip density:

$$v_{ns} = G_T/\rho_{ns} \qquad (26.13)$$

$$\Delta P = K\rho_{ns}(v_{ns})^2/9266(1 - A_{cc}) \text{ (psi)} \qquad (26.14)$$

where A_{cc} is the acceleration term estimated from Equation 26.7.

Siphon Flashing

Lifting a column of water more than 34 feet causes it to boil or flash. This same phenomena frequently occurs in water piping and in process lines, and it can lead to inoperable piping. When velocities are sufficient to create a siphon, liquids may flash and the vapors collapse violently in the downleg. This violent collapsing can shake the piping and associated support. The vacuum can also interfere with sampling and venting.

SLURRY FLOW

The flow of solids suspended in a liquid stream is encountered in some evaporator applications. The characteristics of the flow depend upon the physical properties and relative amounts of the solid and liquid and upon the shape and particle size distribution of the solid being transported.

Flow of solid-liquid mixtures (slurries) in pipes differs from the flow of homogeneous liquid in several ways. With liquids, the flow regime is defined by the physical properties of the fluid and piping; the complete range of velocities is possible. With slurries, two additional distinct flow regimes and physical properties are superimposed on the liquid system.

(1) *Homogeneous slurries.* The solid particles are homogeneously distributed in the liquid media, and the slurries are characterized by high solids concentrations and fine particle sizes.

(2) *Heterogeneous slurries.* Concentration gradients exist along the vertical axis of a horizontal pipe even at high flow rates; the fluid phase and the solid phase retain their separate identities. These slurries tend to be of lower solids concentration and have larger particle sizes than homogeneous slurries.

(3) *Mixed slurries.* Fine particles can join with the liquid media to form a homogeneous vehicle, while the coarser sizes are heterogeneous.

Critical velocity of a slurry in horizontal pipes is the fluid velocity below which the solids begin to deposit at the bottom of the pipe. In vertical pipes, solids that would be deposited if the pipe were horizontal are easily transported because their settling velocity is usually much lower than normal flow velocities. In general:

(1) For vertical pipelines, the normal flow velocity will prevent settling out of solids.

(2) For horizontal pipelines, turbulence must be sufficient to prevent settling out of solids.

The concentration of solids in a slurry system is usually expressed as the ratio of the weight of solids per weight of liquid (also weight percent). Concentration may also be expressed as volume percent, the two concentrations being related by solid and liquid specific gravities. Solids concentration in the slurry is an important parameter; it may affect the critical velocity and pressure drop.

The abrasive nature of slurries is a major consideration in design and selection of equipment for the slurry system. Proper selection and layout of major components of the system — pumps, pipe, valves and fittings, and instrumentation — are important for the proper functioning of the system.

Centrifugal Slurry Pumps: These pumps are characterized by limited casing pressure and efficiency. Efficiency is low because of the robust nature of the impeller design and relatively wide impeller throat clearance. Multiple pumps in series can develop final-stage discharge pressures up to about 600 psig.

Positive Displacement Pumps: These are used when pumping pressures above 600 psig are required. Several designs are available. For very abrasive slurries, the plunger-type pump is used with a flushing arrangement injecting clear liquid to keep solids away from the plunger packing. Piston pumps have application in less abrasive service; they have the advantage of displacing slurry on both forward and reverse strokes, but depend upon sealing the full differential pressure across the piston.

Piping Systems: The following considerations are important when laying out a slurry piping system:

(1) Flushing or draining the piping during normal or emergency shutdown.
(2) Replacement of wear points near pump discharge, sharp bends, and downstream of restrictions.
(3) Access for unplugging.
(4) Elimination of dead spaces at tees and tappings.
(5) Effect of erosion corrosion on pipe wall thickness.

The primary objective in designing a slurry system is to select a practical velocity/diameter combination which carries the solids and results in reasonable pressure losses. A practical design velocity depends not only upon the physical properties of the solid and liquid, but also on the solids concentration in the stream. A velocity in the range of 4 to 7 feet per second is usually practical and economical; however velocities above 7 feet per second may be necessary for strongly heterogeneous slurries. Pipe abrasion is a consideration at about 8 to 10 feet per second, and can be serious at higher velocities.

The effect of the presence of solids in a liquid stream on frictional pressure drop depends, among other things, on the rheological characteristics of the slurry. Some slurries are highly non-Newtonian and must be treated as such.

PIPING LAYOUT

Six related problems must be considered in the layout of piping for process plants:

(1) process requirements

(2) transmission of excessive stresses from piping to equipment, and the transfer of vibration from equipment into the piping system

(3) economy

(4) operational accessibility

(5) maintenance and replacement

(6) excessive stresses in the piping system.

Process Requirements

Except for some particular process pecularity, the best arrangement is not always the most direct method. All equipment is subject to some kind of maintenance, inspection, and cleaning: it must therefore be accessible. Safety may demand certain equipment be located at a specific location.

The layout of equipment can permit the simplest piping arrangement.

Transmission of Stresses and Vibration

The transmission of stresses from piping to equipment can be eliminated by installation of manufactured expansion joints. Corrosion, erosion, and cyclic stresses, however, limit the use of mechanical expansion joints. Piping stresses on equipment is usually controlled by proper piping arrangement and use of supports and anchors.

There are two reasons for eliminating stresses at piping connections:

(1) to prevent overstressing of the body of a machine or valve to which the piping is attached

(2) to prevent misalignment of equipment parts caused by differential movements.

Vibration in piping is usually encountered when piping is connected to moving machinery. Proper use of supports and hangers or vibration dampeners will eliminate most cyclic vibration. High-frequency vibration caused by high-speed machinery can be relieved by applying fixed supports. The correct location for these supports may be determined by experimental application of weights while the machine is in operation.

Economy

Cost considerations are often overshadowed by process requirements. Each system will vary, but the best economies in piping are usually obtained by simplification of the piping specification.

Accessibility

Valves and other attachments which must be used during operation and/or which require periodic maintenance must be located in a convenient position.

Maintenance and Replacement

The problem of maintenance and replacement of piping is usually not important except for piping conveying extremely corrosive or erosive fluids. Any particular piece of piping subject to extreme conditions of erosion or corrosion should be arranged so that replacement is simplified.

Piping Stresses

All piping stresses are essentially the result of temperature changes. There are ordinarily three conditions which cause stresses in piping: first, stress caused by internal or external pressure on the pipe wall; second, stresses remaining in the pipe wall after fabrication or erection; and third, stresses caused by temperature changes produced by the fluid flowing or by external conditions. Selection of proper pipe specifications and wall thickness will produce a safe design that will withstand the first condition. The second condition can be eliminated by stress-relieving and proper fabrication. The third condition is more intangible; although it can be anticipated, it cannot be precisely evaluated except by rigourous analysis.

Some method must be provided for eliminating or reducing the forces that are produced by changes in temperature of piping systems. The piping must be made flexible so that it can relieve itself of the major portion of these forces. This can be accomplished by providing flexibility in the piping layout or by the use of expansion joints or cold springing.

Flexibility Through Layout: Piping is rarely run in a straight line from one fixed connection point to another. Each time piping changes direction and is left free to move at the point of change, the system becomes more flexible. By making bends in more than one plane the flexibility is further enhanced.

Expansion Joints: The installation of an expansion joint is a straight run of pipe firmly anchored at both ends will alleviate the stress developed by expansion. Manufactured expansion joints are not extensively used in process piping. They are used for utility piping and in process applications where piping size or space limitations would make changes inconfiguration impractical or costly.

Cold Springing: Forces generated by expansion are caused by movement of piping as its dimensions change from ambient to operating temperatures. One method for reducing final stress levels in the hot condition is to cut the pipe short and spring it into place while cold. This springing produces a tensile stress on the system which changes to a compressive stress in the hot condition. Both of these stresses will be lower than that which would occur if the pipe had been cut to exact cold dimensions. This technique is called cold springing. When practiced, actual field measurements for any given piece of piping should be made.

27
Thermal Insulation

Thermal insulation, properly applied, assures effective operation of process equipment and conserves valuable heat. There is always a wide variety of insulating materials for a given situation and careful consideration must be given to each type so that an intelligent decision may be made. Thermal insulation should be the simplest, most generally accepted, cost-effective method of saving energy immediately available to a plant owner.

If we would take a casual look at a typical evaporation system, we would have little reason to expect that energy, being an invisible commodity, is being wasted by the misuse of thermal insulation. Taking a closer look at this evaporation system, we would see insulated lines with valves and flanges left bare. Of course, these are a source of energy loss. Looking a little closer, we would find still another source of energy loss due to the lack of adequate maintenance. This common source of energy loss is deteriorated weather barriers. As weather barriers deteriorate, they permit the insulation to get wet. As insulation becomes wet, its thermal conductivity increases and its efficiency decreases thus allowing a greater energy loss.

Energy loss due to the lack of insulation and improperly installed or maintained insulation may easily be visualized by monitoring steam usage before and after a rainstorm. During a recent 15 minute sudden rainstorm at a typical chemical plant, the steam load increased from 638,000 lb/hr to 677,000 lb/hr within a 15 minute period. After the rain stops, the steam load does not immediately decrease because all water that enters the insulation must be vaporized and driven out. We can easily justify the insulation of all flanges, valves, manways, etc.

Bare flanges, valves, manways, etc., and deteriorated insulation may be visually detected by an inspection of the system. However a much more subtle source of energy loss, one that is much harder to detect, are those lines and equipment that appear to be adequately insulated, but with a less than appropriate thickness of insulation.

Energy conservation is the most common use for thermal insulation. In most cases, it is not too difficult to recognize when and where to use insulation. Insulation is commonly used to accomplish the following:

(1) conserve energy
(2) control heat transfer
(3) control temperature
(4) retard freezing
(5) protection from burns
(6) control fire
(7) prevent surface condensation or ice formation.

The amount of insulation that should be used when insulating for energy conservation must be based on economic factors. Many factors affect the cost of energy:

(1) energy investment costs
(2) energy operating costs
(3) maintenance costs
(4) depreciation period
(5) return on investment

Many factors also affect the installed cost of insulation:

(1) type of insulation
(2) material cost
(3) labor rates
(4) thermal conductivity
(5) average ambient temperature
(6) operating temperature

The optimum or economic thickness of insulation is determined by optimizing the cost of energy and the costs of insulation. A discounted-cash flow analysis of insulation economics must be conducted using appropriate costs for investment and amortization periods. Insulation for energy conservation is an excellent investment and when optimized, a balance between investment and savings will be obtained.

28

Pipeline and Equipment Heat Tracing

The transfer and storage of materials that have high freezing points and that become viscous or solid at normal ambient temperatures have long been a difficult problem. Consequently, piping and equipment often need to be heated and this is usually accomplished with some type of heat tracing system. Three types of tracing systems are employed:

(1) steam (or vapor) tracing
(2) liquid tracing
(3) electrical tracing

The choice of a particular tracing method depends upon several factors:

(1) characteristics of the fluid in the pipeline or equipment
(2) heat load required
(3) heating fluid availability
(4) pipeline or equipment geometry
(5) whether equipment or pipeline to be heated is yet to be installed or is already in place
(6) costs of various forms of energy.

Tracing systems are generally not designed to raise the temperature of process fluids; they are designed simply to maintain fluid temperatures at some suitable value. Tracing systems should be properly thermally insulated.

Tracing systems can be classified as below:

Fluid Systems
    Liquid or Vapor Tracing
        Internal
            Straight or spiraled tubes
            Integral
        External
            Straight or spiraled tubes
            Plate heaters
            Jackets
    Electrical Heating
        Impedance
        Resistance
            Internal
                Cables
            External
                Cables
                Tapes
                Strips
                Panels
        Induction

Energy savings can be achieved by the proper design, installation, operation, and maintenance of tracing systems. The most commonly used heating medium is steam; however, liquid systems and electrical systems have many advantages and their application is increasing. An evaluation of all possible methods is required for any specific application.

29
Process Vessels

Every evaporator system requires several process vessels for operation. The designer must determine the need and size for any process vessel from energy and material balance information and any special requirements for operating flexibility, storage capacity, or future expansion needs. The design basis for each vessel should be clearly established.

The responsibility to select the number of process vessels required for a system lies with the process engineer. To determine the number of vessels to be used, the process engineer usually must evaluate several factors for each vessel under consideration. The cost of installing, operating, and maintaining a vessel must be balanced against any detrimental effect that might be incurred if the vessel is not installed. Whenever possible the process engineer should strive to minimize the investment required for process vessels without sacrificing the quality and safety of the design.

The true need for a process vessel should be determined after considering the following factors:

(1) Operating Flexibility: Will the omission of this vessel create a bottleneck in the processing unit or in the distribution facilities?

(2) Material Compatibility: Could the material in this vessel be stored elsewhere without adversely affecting its quality?

(3) Physical Property Constraints: Does the material in this vessel have to be stored under special provisions for safety or environmental reasons?

(4) Future Need: Could this vessel be installed at a future date without causing insurmountable problems?

(5) Existing Facility Utilization: Will an existing vessel meet the needs of the system?

Some vessels commonly specified for operating units are raw material storage and feed tanks, process surge tanks, mixing and decantation tanks, and product storage tanks. Other vessels should be considered for:

(1) Off-quality material produced during startup, shutdown, or upset conditions and during changeovers.

(2) Material returned from customers.

(3) Small volume, special grade products not made on a routine basis.

Whatever the reason, a clear understanding of the *need* for every vessel specified is needed.

30
Refrigeration

It would be very difficult for many evaporation systems to function without process refrigeration. The three basic types of refrigeration units are mechanical, steam jet, and absorption systems. Mechanical refrigeration systems are typically used for process cooling applications requiring temperature levels between 15°C and −100°C. Steam jet refrigeration systems may be specified for process cooling applications requiring temperature levels between 2°C and 25°C. Absorption refrigeration systems are typically specified for process cooling applications requiring temperature levels between 5°C and 10°C when excess steam or hot water is available.

Several basic system design considerations should be evaluated when selecting and specifying a refrigeration system. Considerations include:

(1) purchase cost
(2) installation cost
(3) operating costs
(4) downtime costs: equipment reliability and spares required
(5) maintenance costs
(6) space limitations
(7) safety considerations
(8) delivery requirements
(9) environmental concerns: noise and vents/flares
(10) laws and regulations.

Other specific items to be considered early in the design stage include:

(1) *peak load:* avoid causing excessively high peak loads that can otherwise be handled

(2) *turndown:* full turndown is not always available on packaged refrigeration units

(3) *rate of load variations:* this is important for instrumentation and control

(4) *location of installation:* outdoor versus indoor installation is an important factor.

Refrigerants absorb heat not wanted or needed and reject it elsewhere. Heat is removed from the system by evaporation of a liquid refrigerant and is rejected by condensation of the refrigerant vapor. This evaporation-condensation process occurs in absorption, mechanical and steam jet refrigeration systems.

MECHANICAL REFRIGERATION

Mechanical refrigeration systems are the most versatile units available. Mechanical refrigeration systems offer wide application flexibility and adaptability for process temperature requirements. Two basic classifications are direct and indirect systems. The "direct" refrigeration system is applicable to many industrial applications. The term "direct" is used because the process stream is cooled directly within the refrigeration system (chiller or evaporator). The "indirect" refrigeration system is applicable when it is more practical to pump the cooling medium to the process. This is normally accomplished by cooling a brine or intermediate fluid in the refrigeration system and circulating it to various process heat exchangers.

The equipment in a refrigeration system serves only to provide the refrigerant in the liquid state at the place where cooling is desired. Since evaporation of the liquid is the only step in the refrigeration cycle that produces cooling, the properties of the refrigerant should permit high rates of heat transfer and minimize the volume of vapor to be compressed. High rates of heat transfer should also be afforded at the condenser, where heat must be rejected. Generally, the selection of a refrigerant is a compromise between conflicting requirements and often is influenced by properties not directly related to its ability to transfer heat. For example, flammability, toxicity, density, molecular weight, availability, cost, corrosion, electrical characteristics, freezing point, and the critical properties are often important factors in selection of a refrigerant.

Several types of fluids are used as refrigerants in mechanical systems; a wide temperature range therefore is afforded.

STEAM JET REFRIGERATION

Steam jet refrigeration may be used for process cooling between 2 and 25°C when excess steam is readily available. Steam jet systems generally provide between 10 and 100 tons of refrigeration capacity per unit. Steam jet refrigeration systems are similar to mechanical systems in that an evaporator, compression device, condenser, and refrigerant are the basic system components. A steam jet ejector serves as the compressor. High steam usage makes this system uneconomical in many applications. However, the system is simple, rugged, highly reliable, and has low first cost and maintenance costs.

ABSORPTION REFRIGERATION

Absorption refrigeration cycles employ a secondary fluid, the absorbent, to absorb the primary fluid, refrigerant vapor, which has been vaporized in the evaporator. The two materials that serve as the refrigerant-absorbent pair must meet a number of requirements; however, only two have found extensive commercial use: ammonia-water and water-lithium bromide.

Water-lithium bromide systems cannot be used for low temperature refrigeration because the refrigerant turns to ice at $0°C$. Lithium bromide crystallizes at moderate concentrations and therefore is usually limited to applications in which the absorber is cooled with cooling water. Other disadvantages are associated with the low pressure required and with the high fiscosities of the lithium bromide solution. The pair does offer the advantage of safety and stability and affords a high latent heat of vaporization. The system has wide application in air-conditioning and fluid chilling applications.

Ammonia-water systems are more complex than water-lithium bromide systems, but can be used at temperatures down to $-40°C$.

31
Control

Many books have been written concerning the methods of process control for specific unit operations. Basically, automatic control is relatively easy to comprehend. In many ways it is like manual control. However, the automatic controller does not necessarily duplicate what the human operator does by hand. Automatic equipment gives continuous, minute attention to the control application. Automatic controllers can compute and remember, but they cannot reason from new conditions, nor can they forecast beyond the data which are built into them.

The instruments required for a modern unit normally represent 5-8% of the total unit investment. In some cases, especially if a process computer is required, this cost may be significantly higher. The average installed cost for a full control loop is approximately $10,000.

The complexities of modern units make it virtually impossible to operate them without an adequately designed process control system. Instrumentation is installed to reduce operating labor and make the most effective use of process raw materials and energy. Proper instrumentation also increases process safety and workability. There are, however, many opportunities for wasting money on instrumentation that won't be used or won't work as intended. The benefits of properly designed control can justify many times over the cost.

The basic process control requirements can be established only after an operating philosophy has been developed. The process variables to be controlled must be established. The desired location of measuring elements, control valves, and controllers must be determined. The safety aspects and operational information must be evaluated.

MANUAL CONTROL

Early processes depended almost completely upon manual or hand control for operation. Control was based on human experience, sight, sound, and memory. Experience is the primary requirement of manual control. A number of processes are manually operated because:

(1) no instrumentation is available to meet the requirements
(2) cost of available instrumentation is excessive
(3) labor costs for maintenance of elaborate instrumentation are high.

In general, there are relatively few processes which cannot be made completely automatic. There are few manually operated processes which do not use instrumentation for indicating and recording. Processes may be started, stopped, or adjusted manually from a central control station by the use of remotely controlled equipment.

EVAPORATOR CONTROL SYSTEMS

Any control system which results in the controlled operation of the evaporation process must maintain both an energy and material balance across the evaporator boundaries. Consequently, a control system designed to maintain a specified evaporation must simultaneously manipulate product withdrawal and energy supplied to the evaporator, if efficient operation is to be achieved. The control system must be able to treat changes in feed flowrate or composition. The control system should function to reduce heat input with a reduction in feed rate. It should also function to change the overhead vapor rate (amount evaporated) as changes in the feed composition occur.

The objective of the evaporation system must be established before processing equipment can be evaluated for dynamic control. Specifications for processing equipment are always influenced by cost, maintainability, and operability. Of equal importance is the effect equipment selection has upon control system performance. Several factors need consideration when establishing objectives:

(1) Is control needed? Some simple systems may not need control. Some systems may operate at maximum evaporation rates because the product is further processed downstream; consequently only a minimum of control is needed.
(2) Which streams require close composition control?
(3) Which streams should have minimum flow changes?
(4) Must the evaporator be able to process feed streams with fluctuating volume and/or composition?
(5) Is the heat source dependable? If it is a condensing heat source, such as steam, does it have a constant supply pressure?
(6) What cooling medium is used on the condenser? Is it vaporizing? Is excess refrigeration available, if refrigeration is required for condensing?
(7) What turndown is required?

(8) Is fouling expected?

(9) What type of calandria will be used: forced or natural circulation?

The choice of a control system should be based on the needs and characteristics of the process. Evaporators tend to have relatively large mass and energy storage capacity and have significant dead time. Material balance and energy control should be used whenever possible to improve performance and reduce operating costs. Material balance control uses a measure of performance (temperature, pressure, composition) to directly control a process flow rate. Energy balance control uses a measure of performance to directly control an energy input rate, such as steam to a calandria. Energy balance and material balance are interacting and measures of performance must be chosen to minimize this interaction.

The control systems to be considered include: feedback, cascade, and feedforward. If the major process loads (feed rate and composition) are reasonably constant and the only corrections required are for variations in heat losses or fouling, feedback control is adequate. If steam flow varies because of demands elsewhere in the plant, a cascade system may be the proper choice. If, however, the major load variables change rapidly and frequently, feed forward or feed forward in conjunction with feedback should be considered.

Feedback Control

Feedback control systems monitor final product quality by manipulating heat input. The internal material balance is maintained by level control on each effect. In feedback control a correction for a disturbance cannot be made until the effect of the disturbance is detected. Some time is also needed to measure the deviation and make the correction. Process time lags require again that time pass before the effect of the correction can be known. Meanwhile, the controlled variable continues to deviate from the desired value for some time. In feedback control, one manipulated variable is controlled from one measured value.

Feedback control requires that a deviation exist between the measured variable and the set point before control action can take place. The feedback controller changes its output until the deviation (error signal) between the set point and the measured variable is negligible, which is essentially a trial-and-error process characterized by the oscillatory nature that feedback control often exhibits.

Multielement Control

Multielement control occurs when two or more input signals jointly affect the action of the control system. Examples of multielement control are: cascade control, ratio control, and override control.

Cascade Control: In cascade control, the output of one controller is the set point for another controller. Each controller has its own measured variable with only the primary controller having an independent set point and only the secondary controller providing an output to the process. The addition of a secondary controller permits variations to be corrected immediately before they can affect the primary variable.

In cascade control, at least two control loops exist: the primary (slow or outer loop) and a secondary (fast or inner loop). The process is separated into two parts; one part contains the external disturbances while the other part contains the long time constants. The slave or secondary loop is used to reduce the effects of supply-side disturbances. The master or primary controller senses the desired controlled variable.

When a cascade system is placed in operation, both controllers are initially set in manual during startup. After the process stabilizes, the secondary controller is placed in automatic and the correct action settings are determined. When the secondary controller operation in automatic is satisfactory, the primary controller is then placed on automatic and its optimum control action settings are determined.

Ratio Control: In ratio control, a predetermined ratio is maintained between two or more variables. Each controller has its own measured variable and output to a separate final control system. However, all set points are from a master primary signal that is modified by individual ratio settings.

Override Control: In override control, either the highest or the lowest signal from two or more input signals is automatically selected. Common examples include:

(1) suction and discharge pressure compressor control

(2) temperature or vacuum and differential pressure control for distillation columns.

Feedforward Control

Feedforward control is a control action in which information concerning one or more conditions that can disturb the controlled variable is converted into corrective action to minimize deviations of the controlled variable. In a true feedforward control system, the controlled variable is not used as one of the inputs. However, it is usually difficult, as well as not practical, for the controller-computer to include corrections for all variables. Therefore it is customary to include a feedback loop to prevent inaccuracies in the feedforward loop from adversely affecting the process.

In most evaporator applications the control of product quality is constantly affected by variations in feed rate and composition to the evaporator. In order to counter these load variations, the manipulated variable must attain a new operating level. In pure feedback or cascade arrangements, this new level is achieved by trial-and-error as performed by the feedback controller.

A control system able to react to these load variations when they occur rather than wait for them to pass through the process before initiating a corrective action would be ideal. The technique is called feedforward control. There are two types of load signals are inputs to the feedforward control system where they compute the set point of the manipulated variable control loop as a function of the measured load variations. The unmeasured load variables pass through the process, undetected by the feedforward system, and cause an upset in the controlled variable. The output of the feedback loop then trims the calculated value of the set point to the correct operating level. In the limit, feedforward control would be capable of perfect control if all load variables could be defined, measured, and incorporated in the forward set point computation. Practically, the expense of accomplishing this goal is usually not justified.

At a practical level, the load variables are classified either as major or minor, and the effort is directed at developing a relationship which incorporates the major load variables, the manipulated variables, and the controlled variable. Such a relationship is termed the steady-state model of the process. Minor load variables are usually very slow to materialize and are hard to measure. Minor load variables are easily handled by a feedback loop. The purpose of the feedback loop is to trim the forward calculation to compensate for the minor or unmeasured load variations.

The third ingredient of a feedforward system is dynamic compensation. A change in one of the major loads to the process also modifies the operating level of the manipulated variable. If these two inputs to the process enter at different locations, these usually exists an unbalance or inequality between the effect of the load variable and the effect of the manipulated variable on the controlled variable. This imbalance manifests itself as a transient excursion of the controlled variable from set point. If the forward calculation is accurate, the controlled variable returns to set point once the new steady-state operating level is reached.

In terms of a concurrent flow evaporator, an increase in feed rate will call for an increase in steam flow. Assuming that the level controls on each effect are properly tuned, the increased feed rate will rapidly appear at the other end of the train while the increased steam flow is still overcoming the thermal inertia of the process. This sequence results in a transient decrease of the controlled variable, and the load variable passes through the process faster than the manipulated variable. In a countercurrent evaporator operation, the manipulatted variable passes through the process faster than the load variable. This dynamic imbalance is normally corrected by inserting a dynamic element (lag, lead-lag, or combination) in at least one of the load measurements to the feedforward control system. Usually, dynamic compensation of that major load variable, which can change in the severest manner (usually a step change) is all that is required. For evaporators this is usually the feed flowrate to the evaporator. Feed composition changes, although frequent, are usually more gradual and the inclusion of a dynamic element for this variable is not warranted.

The three ingredients of a feedforward system are:

(1) the steady-state model
(2) process dyanmics
(3) addition of feedback trim.

Development of the steady-state model for an evaporator involves material and energy balance. A relationship between feed composition and product composition is also required. The process dynamics require that a lead-lag dynamic element be incorporated in the system to compensate for any dynamic imbalance. In some applications evaporators are operated with waste steam, in which case the feed rate is proportionally adjusted to the available steam, making the feed the manipulated variable and steam the load variable. Generally, steam is the manipulated variable. The final consideration is feedback trim. As a general rule feedback trim is incorporated into the control system at the point at which the set point of the controlled variable appears.

One additional load variable which if allowed to pass through the process would upset the controlled variable is steam enthalpy. In some applications the steam supply may be carefully controlled so that its energy is uniform; in other applications substantial variations in steam enthalpy may experienced. The objective is to consider the factors that influence the energy level of the steam supplied and to design a system that will protect the process from these load variables.

The feedforward system imposes an external material balance as well as a an internal material balance on the process. The internal balance is maintained by liquid level control on the discharge of each effect. Analysis of a level loop indicates that a narrow proportional band (less than 10%) can achieve stable control. However, because of the resonant nature of the level loop which causes the process to oscillate at its natural frequency, a much lower controller gain must be used (proportional bands 50-100%). A valve positioner is recommended to overcome the nonlinear nature of valve hysteresis.

The heat to evaporate water from the feed material is directly related to the boiling pressure of the material. Consequently, the absolute pressure is normally controlled at the last effect. Many methods are available for controlling the pressure. Ambient conditions can profoundly affect the operation of an evaporator since all the energy used is ultimately rejected to the atmosphere. Most processes are designed to operate at a certain production rate under one set of ambient conditions. Since that rate is rarely maintained and ambient conditions are constantly changing, control is applied to establish an equilibrium representative of design specifications. The principal failing in this mode of operation is that process efficiency can actually be increased at reduced loads and under favorable conditions for most processes. Controls should be designed to cooperate with, rather than fight, the environment. For these savings to be realized, evaporator pressure must be allowed to float usually as the condenser floats. To operate at minimum pressure successfully, however, requires additional controls, rather than the elimination of pressure control. Pressure must be permitted to change only gradually to prevent upsets from flashing, flooding, splashing, or foaming. If evaporator temperature is used as an index of composition, changing pressure will effect that relationship and appropriate compensation is required. To save energy, evaporator pressure measurement should also be used to modulate heat input.

CONTROL OF EVAPORATORS

The two primary variables in evaporator operation are usually product concentration and evaporation rate. The method of control depends upon the evaporator type and method of operation. When evaporation rate is to be constant, a steam flow controller is generally used. Steam flow control usually is accomplished by throttling steam which results in lowering of available temperature levels. Steam may therefore be controlled to achieve maximum evaporation capacity. Steam pressure controllers may be used to protect the equipment or to assume substantially constant temperatures in the early effects of the system. Constant or slowly-changing boiling temperatures greatly reduce entrainment losses as a result of flashing. Constant temperatures in the later effects of the evaporator can be maintained with a pressure controller on the last effect.

When a product fails to meet specification, it is either rejected as a waste or low-value product, burned as fuel, reprocessed, or blended with feed-stock or overpure product. The result is the use of more energy per unit of production. To avoid these penalties, operators typically make their products much purer than necessary, which also uses energy. Substantial savings can be achieved by being able to control product closer to specifications. This is the principal justification for advance control systems for evaporators.

Having developed the ability to control closer to specification, the time has come to question the specifications themselves. The first matter of business is to estimate the energy savings possible by relaxing specifications. In some cases, specifications are established by law; in others they are in the terms of contracts between buyer and seller. But many transfers are within a company, or between its divisions, and arbitrary specifications are still applied. Making specification product is not enough — they must be made profitably. So there is more than a single objective; there are generally two more objectives than there are specifications on the products:

(1) to minimize manufacturing costs

(2) to set production at its most desirable value.

It must be recognized, however, that variations in product quality incur a twofold penalty:

(1) The average purity must exceed specifications by the amplitude of variation, thereby increasing operating costs.

(2) Variations about an average purity further increase the operating cost above that required to generate the average purity, in relation to the amplitude of these variations.

The best control system should be used in the design of evaporator systems. Products which are off-specification require additional time, expense, and energy in reprocessing. A properly designed control system can do much to reduce these wastes. The proper control system will result in near optimum energy usage during normal operation.

The structure or organization of the control system is the most important aspect of instrument systems in achieving efficient operation. Significant reductions in the operating costs of evaporation systems can be achieved by providing control systems which are designed to minimize energy usage.

A control system can force a plant to operate efficiently; but the control system cannot change the plant. No expense for instrumentation can make an eight-cylinder Cadillac operate for the same energy costs as a four-cylinder Chevette. Energy-sensitive control systems can and should be used as we strive to operate evaporation systems more efficiently.

AUTO-SELECT CONTROL SYSTEM

In many processes the final product is the result of a two-step operation. The first step produces an intermediate product which serves as the feed to a final concentrator. The aim is to insure that the process is run at the maximum throughput consistent with the process limitations. In this two-step evaporation there are three limitations to the process including:

(1) the steam supply to the intermediate concentrator can be reduced due to demands in other parts of the plant

(2) the steam supply to the final concentrator can be reduced due to demands in other parts of the plant

(3) the final concentrator can accept feed only at or below a certain rate, and it is desired to run this part of the process at its limit.

Each part of the system has its own feedforward-feedback system. The final concentrating process does not require feed concentration compensaton so that it is only necessary to establish the ratio of steam to feed and adjust the ratio by feedback. Ordinarily, the system is paced by the final concentrator feed signal which is fed to the auto-select system. The final concentrator controller output manipulates the feed to the intermediate process so as to maintain the final concentrator feed at the set point limit.

Should either steam supply become deficient, its output would be selected to adjust the feed to the intermediate process. Thus, the process is always run at the maximum permissible rate consistent with any process limitations.

PRODUCT CONCENTRATION

Product concentration can be controlled by measuring a number of physical properties. On-stream composition analyzers are often used. Commonly used physical properties include: density, boiling point rise, temperature/pressure combinations, temperature difference, conductivity, differential pressure, refractive index, buoyancy float, and viscosity. Each method has certain advantages as well as limitations. In all cases, however, a representative measurement location must be carefully selected to eliminate entrained air bubbles or excessive vibration, and the instrument must be mounted in an accessible location for cleaning and calibration. The location of the product quality transmitter with respect to the final effect should be considered also. Long piping runs between the product and the instrument increase deadtime, which in turn reduces the effectiveness of the control loop.

Control is usually effected by controlling the discharge of product from the evaporator. Feed rate and flowrates between effects are then adjusted to maintain constant liquid levels. When this is not possible, product concentration may be controlled by monitoring the feed flow. Liquid level control in evaporators may be important for product concentration, to maintain heat transfer rates, and to prevent scaling, salting, and fouling. Level control may also reduce splashing and entrainment.

In modern process plants many control laboratory functions have been displaced by automatic instruments designed to measure continuously the properties of a flowing process stream. Processes require a physical measurements or a chemical analysis of the product are burdened with an excessive time lag when operators must await results from a control laboratory before changing conditions. Even though the initial cost may be high, the use of stream analyzers has proven profitable in processes requiring rigid quality control. The continuous record eliminates time lag and prevents production of large quantities of product which do not meet specifications.

Boiling Point Rise

Boiling point rise expresses the difference between the boiling point of a constant composition solution and the boiling point of pure water at the same pressure. For example, pure water boils at $100°C$ at one atmosphere and a 35% sodium hydroxide solution boils at about $121°C$ at one atmosphere. The boiling point rise is therefore $21°C$. However, the vapor from the solution is superheated steam at $121°C$, but the steam will condense at $100°C$ only. Consequently, boiling point rise represents a loss of total available temperature difference.

Duhring's rule states that a linear relationships exists between the boiling point of a solution and the boiling point of pure water at the same pressure. Thus, the temperature difference between the boiling point in an evaporator and the boiling point of water at the same pressure is a direct measurement of the concentration of the solution. Two problems in making this measurement are location of the temperature sensors and control of absolute pressure.

The temperature sensors must be located so that the measured values are truly representative of the actual conditions. Ideally, the sensor measuring liquor temperature should be just at the surface of the boiling liquid. This location can change, unfortunately, if the operator decides to adjust liquid levels in a particular effect. Many times the liquor sensor is installed near the bottom of the evaporator flash body where it will always be covered. This creates an error due to hydrostatic head which must be compensated for in the calibration.

The vapor temperature sensor is installed in a condensing chamber in the vapor line. Hot condensate flashes over the sensor at an equilibrium temperature dictated by the pressure in the system. This temperature less the liquid boiling temperature (compensated for head effects) is the temperature difference reflecting product concentration.

Changes in the absolute pressure of the system alter not only the boiling point of the liquor, but the flashing temperature of the condensate in the condensing chamber as well. Unfortunately, the latter effect occurs much more rapidly than the former, resulting in transient errors which may take a long time to resolve. It is therefore critical that absolute pressure be closely controlled if temperature difference is to be a successful measure of product concentration.

Conductivity

Electrolytic conductivity is a convenient measurement to use in relationships between specific conductance and product quality (concentration) such as in a caustic evaporator. Problem areas include location of the conductivity cell so that product is not stagnant but is flowing past the electrodes; temperature limitations on the cell; cell plugging; and temperature compensation for variations in product temperature.

Differential Pressure

Measuring density by differential pressure is a frequently used technique. This method is frequently used in the pulp and paper industry to measure and control density of green liquor, heavy black liquor, clay or starch slurries, lime slurries, and lime-mud slurries. The flanged differential pressure transmitter is preferred for direct connection to the process; otherwise lead lines to the transmitter can become plugged by process material solidifying in the lines. Differential pressure transmitters are more frequently used on feed density than on final density measurements.

Gamma Gage

This measurement is popular in the food industry because the measuring and sensing elements are not in contact with the process. It is very sensitive and not subject to plugging. Periodic calibration may be required due to the half-life of source materials. Occasionally, air is entrained especially in extremely viscous solutions. Therefore, the best sensor location is a flooded low point in the process piping.

U-Tube Densitometer

The U-tube densitometer, a beam-balance device, is also a final product density sensor. Solids can settle out in the measuring tube, causing calibration shifts or plugging.

Buoyancy Floats

Primarily used for feed density detection, the buoyancy float can also be applied to product density if a suitable mounting location near the evaporator can be found. Because flow will affect the measurement, the float must be located where the fluid is almost stagnant or where flow can be controlled (by recycle) and its effects zeroed out. A Teflon-coated float helps reduce drag effects.

CONDENSER CONTROL

Several methods have been used to control the amount of heat removed in a given condenser. Some of these include controlling the cooling medium flowrate or temperature, changing the amount of surface available for heat transfer, and introducing inert gases into the condensing vapor.

Constant-Pressure Vent Systems

The condensate from a condenser is subcooled; there is no way to avoid subcooling even when it is not desired because of the temperature gradient across the condensate film required for heat transfer. In some cases the condensate is purposely subcooled in order to reduce product losses with the vent.

For total condensers with essentially isothermal condensation, subcooling results in a pressure reduction, unless something is done to prevent it. This results because the subcooled condensate has a vapor pressure lower than the operating pressure. The condenser outlet pressure will be the vapor pressure of the condensate when no inert gases are present. Permitting the system pressure to vary as the degree of subcooling changes is not usually desirable.

A constant-pressure vent system is normally provided to prevent this pressure variation. Inert gases are provided or removed as required to ensure that the system pressure drop is reflected only by the friction drop and not by changes in vapor pressure. Figure 31-1 indicates such a system. The inert gases should be introduced downstream of the condenser. Introducing inerts upstream of the condenser will reduce the rate of heat transfer requiring more heat transfer surface. This action, however, is sometimes used to control a condenser.

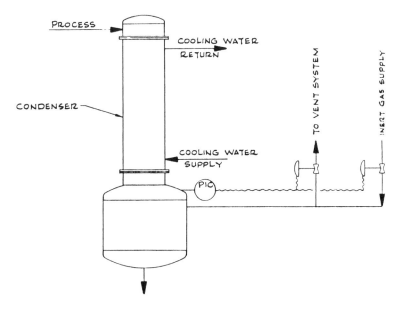

Figure 31-1: Constant pressure vent system for condensers.

Condensate Flooding

Control of condensers by flooding part of the area with condensate is not recommended for several reasons:

(1) Venting of inert gases is relatively difficult. In many services, inert gases must be vented at the liquid-vapor interface. Flooding of condensate provides a variable liquid level making it difficult to vent this interface. Failure to remove inert gases may result in corrosion and will certainly result in lowered rates of heat transfer.

308 *Handbook of Evaporation Technology*

(2) Control is relatively sluggish because liquid levels cannot be changed rapidly unless provisions are made to pump condensate to and from the condenser. This requires a separate condensate source. Generally, with flooded systems, condensate build-up occurs only at the rate at which it can be condensed, while condensate removal can be effected fairly rapidly.

(3) Product losses with the vent stream is increased because the condenser is vented before the condensate is subcooled.

Vertical vapor-in-shell condensers can be designed to properly vent inerts when condensate flooding is desired. Other types of condensers cannot be easily designed to be properly vented as flooding by condensate occurs.

When the condenser and condensate tank are closely connected, the condensate tank must be properly sized in order to permit the condensate liquid level to be controlled somewhere in the condensate tank. If the condensate tank is too small, liquid level control can be achieved only by flooding part of the condenser, especially when the condensate is pumped from the tank. Liquid level must be maintained in the condensate tank and not in the condenser.

Direct-Contact Condensers

Many types of direct-contact condensers can be controlled by reducing the flow of cooling medium, usually water. Some types cannot be controlled in this manner, however.

Water-Cooled Condensers

Throttling cooling water flow to water-cooled condensers is generally not recommended because of fouling and silting which frequently occur at reduced water velocities. Consequently, water-cooled condensers should be provided with tempered-water systems as described in Chapter 13. Tempered-water systems result in better control, reduced maintenance, and reduced pumping costs.

Air-Cooled Condensers

The preferred way to control air-cooled condensers is to control the amount of air flowing across the condenser with controllable-pitch fans. Controllable-pitch fans produce good control and also result in reduction of energy required for driving the fans. Typical energy reductions achieved with air-cooled condensers are shown in Figure 31-2.

Air-cooled condensers are normally sloped toward the outlet to provide for better condensate drainage in the tubes. It is extremely important that air-cooled condensers be installed so that the outlet of the tubes in a pass is at least at the same elevation as the inlet to avoid condensate flooding; preferably it should be a few inches lower. This is even more important when the freezing point of the condensate is higher than the lowest possible air temperature. Multipass condensers may be sloped in more than one direction at a fairly nominal cost. In some instances, the condensate outlet header is provided with a "sump" by making the outlet header deeper. This prevents the bottom row of tubes from being flooded by condensate and avoids problems associated with improper venting.

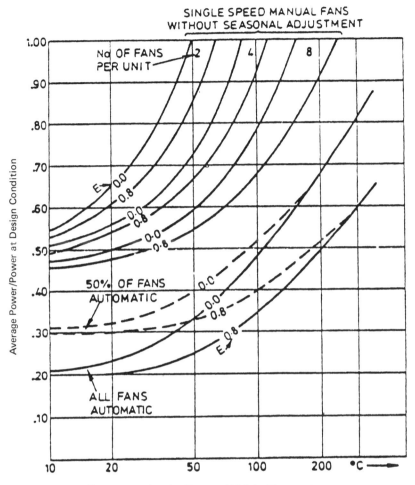

Figure 31-2: Average correlation of power saving by fan control (for approximate use only).

Temperature Level: (Process-Air) Inlet Temperature

$$T_1 - t_1$$

$$E = \frac{T_1 - T_2}{T_1 - t_1}$$

Where T_1 = process fluid inlet temperature °C
T_2 = process fluid outlet temperature °C
t_1 = design ambient air temperature °C

Single-pass air-cooled condensers must usually be somewhat oversized resulting in subcooling of condensate. This results because the air temperature at each row of tubes is different; consequently, each row condenses a different amount of vapor. In order to insure that all vapor entering the top row (highest air temperature) is condensed, it may be necessary to provide more area than required for all other rows. Multipass condensers avoid this problem, but pressure drop considerations do not always permit multipass designs.

Refrigerated Condensers

Refrigerated condensers can be cooled directly by boiling of refrigerant or with a circulating brine. Brine systems can be controlled by throttling the flow of brine to the condenser. Boiling systems can be either flooded or direct-expansion. Either can be controlled with a back-pressure controller on the boiling side. Flooded systems also require liquid level control in order to avoid excessive entrainment.

Three problems are common in boiling refrigerant systems:

(1) plugged leads on liquid level transmitters. Oil from the compressor may collect in the connections and solidify at the cold temperatures. Purge gas should be used.

(2) At high boiling rates, liquid carryover may result. A flow indicator in the vapor line will display the carry-over and provide a warning before damage to the compressor can occur if compressor suction scrubbers are not provided.

(3) Sometimes inerts can be charged into the refrigeration system. Inerts must be purged from the system in order to avoid reduced rates of heat transfer.

CALANDRIA CONTROL

Control of natural circulation calandrias presents some characteristics not found in other equipment. When heating with condensing vapors, changes in condensing pressure affect four variables: heat transfer coefficients, driving force (temperature difference), liquid composition, and circulation rate. The same four variables are affected when throttling the flow of a liquid heating medium. The whole mode of operation is changed when one variable is altered; and it's not always possible to predict from experience which direction the change will take.

Control by flooding of condensate introduces another variable — change in heat transfer area. Condensate flooding also reduces the available pressure drop because the hydrostatic head is reduced. Again, the entire mode of operation is altered. Also, condensate flooding often results in improper venting — which adds another variable to a complex system. Failure to vent inerts may lead to corrosion, especially with steam.

Kettle-Type Reboilers

The liquid level on the shellside of kettle-type reboilers should be sufficient to ensure that all the tubes are covered with the boiling fluid. Controlling by varying process liquid levels may result in fouling of the heat transfer surface as part of that surface is deprived of liquid. In addition, the temperature difference may be affected as the hydrostatic head (which affects boiling temperature) is reduced. If product is not removed from the kettle, an adequate purge may be required to reduce the effects of concentration.

Falling Film Vaporizers

The temperature gradient across the liquid film must be kept relatively low (less than $15°C$). Excessive temperature differences between the process and utility fluids may result in boiling of the fluid on the heat transfer surface with resultant fouling. Film boiling may also be effected with subsequent reduction in the rate of heat transfer.

Inert gases are sometimes injected into a falling film evaporator in order to reduce the partial pressure required to vaporize the volatile component. This technique will often eliminate the need for vacuum operation. Enough inert gas must be injected to achieve the desired results, but too much can produce flooding and entrainment with resultant poor control.

Steam Heated Calandrias

Steam-heated calandrias with process boiling temperatures less than $100°C$ can present control problems, especially at reduced rates and during startup. In most such cases, low pressure steam is used for heating. Control is usually achieved by throttling the entering steam in order to reduce the pressure at which it is condensed. At reduced rates, or before fouling has occurred, this often results in steam pressures less than atmospheric or less than the steam condensate return system pressure. The steam is usually removed through steam traps which require a positive pressure to function. In order for the trap to function, steam condensate floods part of the steam chamber until the steam pressure is sufficient to operate the trap. This leads to poor control and all the problems associated with condensate flooding.

There are several alternatives for the solution of this problem. The best of those should be selected for any given application. No solution is best for all cases. These alternatives include:

(1) Provide a pumping steam trap which can work against a vacuum in the steam chamber. This solution may require a vacuum vent system.

(2) Install a vacuum breaker to permit air to enter the steam chamber at pressures less than atmospheric. This would require corrosion-resistant materials of construction.

(3) Install a vacuum breaker and add a non-corrosive gas (such as nitrogen) source. Inert gas would be admitted when the steam pressure was throttled below atmospheric.

(4) Provide a condensate tank with controlled liquid level in the tank. This permits pressure control of the steam; the condensate tank would provide for adequate liquid level control without requiring

a condensate level in the steam chamber. A pump would be required to remove condensate from the tank. Inerts would be vented from the tank, possibly requiring a vacuum vent sources.

(5) Install a tempered-water system. The necessary equipment would be provided to heat water which in turn would heat the process.

(6) Design the calandria with two or more shells so that only a part of the heat transfer surface may be used during startup or at low rates.

A similar problem sometimes occurs when heating with high pressure steam. In these cases there is not enough pressure differential between the steam supply header and the condensate return header to allow the steam condensate pressure to be properly controlled. This problem usually occurs when the process boiling temperature is lower than the saturated steam temperature at a pressure equal to that of the condensate return header.

Liquid Heated Calandrias

Calandrias heated with sensible heat from a hot liquid are normally controlled by throttling the liquid flow. Sometimes control may be achieved by controlling the temperature of the heating medium. Forced circulation calandrias heated with hot liquid may present control problems.

For single-pass forced-circulation calandrias either of these modes will result in adequate control. When the calandria is multipass either of these modes may result in problems with the control of the system. Throttling flows or changing temperatures of the heating medium may result in a temperature cross between the process and utility fluids. If this situation is approached or realized, precise control may be difficult because a small change of flow or temperature may result in a large change in driving force (temperature difference). This possibility can be minimized by designing for high process fluid pumping rates in order to approach an isothermal situation on the process side of the calandria.

EVAPORATOR BASE SECTIONS AND ACCUMULATORS

Evaporator base sections and accumulators are generally too large for good dynamic control. Evaporator and accumulator liquid levels should be proportional-only control. This permits the volumes to be used as surge capacity. Valve positioners should be used on the control valves in level control loops to avoid cycling due to hysteresis in the valve.

All base level controllers in a train should have proportional-only control. In some plants, these controllers are difficult to find because they have been considered inferior. Proportional control is the best, if you can permit the measurement to move around the set point. Surge capacity is frequently provided in the design of accumulators, but then negated by specifying proportional-plut-reset controllers. Proportional-only control *cannot* be achieved with most instruments by turning the reset adjustment to the maximum reset time. A slow cycle with high amplitude will result.

GUIDELINES FOR INSTRUMENTS

New evaporation systems are designed to be controlled and supervised from modern control centers that assure safe, efficient, operation. The basic design philosophy is that all control systems be designed to fail safe in the event of instrument air failure, power failure, combination of both, or other utility service failure. Further, the control equipment should be designed to prevent release of flammable or toxic materials, and should be designed to permit safe maintenance of instrumentation.

Display Information

Sufficient information should be displayed in the control room to permit the control room operator to take action in the event of an upset or to identify the problem for direct attention of the outside operator. Typical information required includes alarms, variable indication, critical running lights, etc. Location of controllers (local or panel) should be determined by how often the set point of that controller requires attention. Prime candidates for local control are non-critical level controllers (condensate flash tanks, surge tanks), tank pressure controllers, non-critical heater temperature controllers. etc. Local indication should be provided for evaluation by the outside operator.

Measurement Locations

Measurement locations should not introduce unnecessary dead time into the control system. Sufficient thermowells and analyzer sample connections should be provided to permit troubleshooting. Provisions for all data needed for testing or troubleshooting should be included in the control system design. Sometimes the instruments are not provided until needed, but the connections should be available at the proper location. Frequently overlooked is the need for good flow measurement. Increased use of computers provide a tremendous tool for gathering data for material and energy balance. The tool is useless, however, if inaccurate flow measurement is installed. Good flow metering is relatively easy to achieve during design; it is very expensive to correct problems after the equipment has been installed.

Control Valve Stations

Control valve stations should be designed considering the following:

(1) The shutoff valves around the control valve should be of the same specification as the lines and the same size as the control valve.

(2) The control valve bypass valve should be a globe valve of the same size as the control valve and should meet line specifications.

(3) The bypass valve and line should be located above the control valve to reduce the possibility of plugging while the bypass valve is closed.

(4) Consider equipping the control valve with a manual actuator to set the valve at a fixed position if a bypass line is not installed.

(5) Vents and drains should be installed between the shutoff valve and the control valve.

(6) The control valve station should be located to facilitate operation and maintenance.

Primary Flow Elements

Primary flow elements should be installed properly to assure good measurements. The following types of measuring devices are commonly used:

(1) Orifice runs are the mainstay for flow measuement. Proper attention must be given upstream and downstream piping. Sharp-edged orifice plates are generally used for smaller lines (6 inch and less); square-edged plates are generally used for larger lines.

(2) Turbine meters are used for liquid measurement only and need adequate upstream and downstream piping. If solids are present in the measured stream, a filter should be installed upstream of the turbine meter to prevent damage to the meter. Air eliminators may be required to protect the meter from shock as a result of gas or air pockets.

(3) The magnetic flow meter is used for liquid measurement, but generally requires only a minimum straight run upstream piping equal to the pipe diameter. The magnetic flow meter will not operate effectively in a fluid that has an electrical conductivity less than 10 micromho.

(4) Weir plates, flumes, elbow taps, target flow meters, ultrasonic flow meters, and positive displacement meters may be preferred under certain conditions.

Primary Level Measurements

Primary level measurements are made using different devices depending upon process conditions and the accuracy required. General types include:

(1) Differential pressure transmitters should be used to measure liquid level except where impractical, such as large storage tanks, extreme temperature variations, high pressures. For level measurement of steam drums, compensating datum columns should be used with differential pressure transmitters.

(2) Load cell transmitters are used to measure bulk solids level when accuracy is important.

(3) External displacer type transmitters or controllers may be used when differential pressure devices are not practical. Adequate valving should be provided for maintenance.

(4) Float and tape type level gages should be used to measure liquid level in large field storage tanks. The gage installation kit must be fitted inside the tank by the tank fabricator.

(5) External level switches for alarms and shutdowns, and for batch control, should be float type. Internal displacers or probe-type level sensors are not recommended except in atmospheric tanks where they can be removed with the tanks in service.

(6) Sight glass gages should be used only where necessary. Gage glasses are used for direct observation of liquid levels. Only armored gage glasses should be used in process service. Flat glasses, reflex or transparent, should be used when gage glasses are required. Reflex glasses should be used for clean, relatively colorless fluids. Transparent glass is used for interfaces, viscous fluids, and dirty fluids. Illuminators are used on transparent glasses where manual illumination is impractical. Protective mica shields should be used when liquid attacks the glass and for steam service. Frost shields are required if the operatng temperatures are below $0°C$. Gage cocks should be furnished for all gage glasses.

Primary Pressure Measurements

Three types of process measurements are common:

(1) Absolute pressure is usually measured with absolute pressure transmitters; direct connected pneumatic controllers may be used for local control. Absolute pressure transmitters with suppressed ranges may be used for maximum accuracy.

(2) Gage pressure is normally measured with pressure transmitters for remote readouts. Direct connected pneumatic controllers may be used for local control.

(3) Differential pressure is normally measured with a standard pressure transmitter. Pressure transmitters with suppressed range may be used for maximum accuracy.

Pressure or vacuum is measured and indicated for several reasons:

(1) provide an operating guide
(2) indicate abnormal pressure
(3) provide test data
(4) troubleshooting.

Criteria for pressure gage installation:

(1) make sure a pressure gage is needed
(2) could a pressure test connection be supplied rather than a permanent gage?
(3) make sure gage is mounted where it can be observed
(4) what is the condition of the flowing stream; will pulsation or vibration be present?

(5) what accuracy is required for the service?

Pressure gage accessories include:

(1) chemical attachments prevent corrosive or solidifying fluids from entering the pressure gage. A chemical attachment is a flexible diaphragm held between a pair of flanges. Pressure gages are usually mounted on the chemical attachment by the manufacturer.
(2) Siphons are used to prevent hot vapors from entering the elastic chamber of a pressure gage. Siphons are used to reduce errors as a result of temperature and to prevent damage to the gage.

Liquid-filled gages should be used on reciprocating equipment where vibration or pulsation is expected. Gages should have pressure-relieving cases.

Pressure Regulators

Pressure regulators are used primarily for gas pressuring systems for tanks.

Control System Utilities

Control systems require instrument air and blowback systems. Clean, dry instrument air at adequate pressure is required. The blowback gas or purge liquid must be at adequate pressure and must be compatible with the process. In some systems, blowback is not permitted and this must be recognized when designing the control system.

PROCESS COMPUTERS

Integration of a computer system into a process unit requires clear definition of the functions to be performed prior to design of the system. Most processing units built today require more sophisticated control and reliable information obtained on a consistent basis. Computer installations are often essential.

Process computers provide information, alarms, data records, and on-line control and optimization. They can have a pronounced effect upon performance. Process computer costs are highly dependent upon the application.

Process computers frequently used are called minicomputers and microcomputers. The computers can be used in an integrated system or separately to perform various functions. The equipment is designed to perform the following tasks:

(1) record process variables
(2) on-line analyzer scanning
(3) process control
(4) optimization
(5) data logging
(6) alarm scanning

(7) engineering calculations

(8) machinery monitoring.

Process computers are frequently minicomputers and are capable of read-write functions, foreground-background tasks, scheduling of computer programs, and have programming capability. The peripherals associated with minicomputers include:

(1) alarm typers

(2) message typers

(3) log printers

(4) cathode ray tubes (CRT)

(5) programmers input/output typers

(6) high speed line printers

(7) bulk storage on disk units

(8) card readers

(9) paper tape reader and punch.

Minicomputers are flexible in their application in that large machines or several small machines may be installed.

Microcomputers are small dedicated computers which in general are single purpose computers. Some of the uses include analyzer control, controllers, flow measurement, data logging, and special mathematical functions. These systems are replacing some of the simpler minicomputer tasks which can be done more reliably and economically by microcomputers. They are not as flexible because they are designed for specific requirements.

Programmable controllers are specific microcomputers that are designed to replace electromechanical relay logic systems. These instruments can be connected to computers for data logging or the computer may act in a supervisory capacity over the programmable controller.

The decision to include a process control computer must be made when the project scope is defined. Measurements are required for production loop, operator guidance, material balances, energy balances, and closed loop control of process variables. These measurements must be completely defined and specified.

32

Thermal Design Considerations

Many factors must be carefully considered when designing evaporator systems. The type of evaporator or heat exchangers, forced or natural circulation, fouling, and intended service are all important and have already been discussed. Many other factors are also important and some are presented below.

TUBE SIZE AND ARRANGEMENT

These factors are important in determining the performance of a tubular exchange.

Tube Diameter

Smaller diameter tubes permit more compact and economical designs than do larger diameter tubes. Smaller tubes, however, may require more pressure drop. Larger diameter tubes may be required if frequent cleaning is expected. Larger tubes result in higher circulation rates when natural circulation is used.

Tube Length

Since the investment per unit area of heat transfer surface is less for long exchangers with relatively small shell diameter, minimum restrictions on length should be observed.

Arrangement

Tubes are arranged in triangular, square, or rotated-square pitch. Triangular tube-layouts result in better shellside heat transfer coefficients and provide more surface area in a given shell diameter. Square pitch or rotated-square pitch layouts are generally used when mechanical cleaning of the outside of tubes is required; sometimes widely spaced triangular pitch facilitate cleaning. Both types of square pitches offer lower pressure drop — but lower heat transfer — than triangular pitch. Square pitch arrangements are also more susceptible to flow-induced tube vibration and to acoustic resonance.

EXTENDED SURFACES

The ratio of outside to inside surface for bare tubes is usually in the range of 1.1 to 1.5 depending on the tube diameter and tube wall thickness. When the thermal resistance on the outside of the tube is much greater than that on the inside, more heat can be transferred (per unit length) through tubes that have a greater ratio of outside to inside area. Such tubes are referred to as extended surface (finned) tubes and are available with external to internal surface ratios ranging from 3 to 40.

Internal extended surfaces are less commonly used, but they are available in some materials of construction. Even less frequently encountered are tubes with extended surfaces on both inside and outside of tubes.

Extended surfaces are normally produced with fins either longitudinal or transverse to the tube axis. Other forms are also available, but not widely used. Fins may be attached to the prime surface by some mechanical means; hence there is a bond resistance to heat transfer which increases with the temperature of the application. Fins may also be manufactured by extrusion from the prime tube or by welding fins to a tube; both of these techniques eliminate any bond resistance.

Extended surfaces are used when the heat transfer coefficient on one side is much less than on the other side. A rule of thumb states that extended surfaces are usually economical when the heat transfer coefficient on the outside of the tube is less than one-third that on the inside. Heat is transferred through a fin by conduction resulting in a temperature gradient across the fin; consequently high termal conductivity is preferred for fin materials. Fin efficiency is defined as the ratio of heat transferred by a fin to what would be transferred if the fin were uniformly at the base temperature (the temperature of the unfinned portion of the tube). Fin efficiencies are usually in the range of 65-100%. Under such conditions, increasing the outside surface by adding extended surface is more economical than increasing the number of tubes.

Lowfinned tubing is formed by extruding fins from the base metal. The fin diameter is maintained the same as the outside diameter of the tube ends, which are standard tube sizes. This type of tubing can, therefore, be used interchangeably with bare tubes of the same diameter and is readily incorporated into shell-and-tube equipment. Area ratios are in the range of 2.5 to 5.

Lowfinned tubing is often used when fouling is expected. Lowfinned tubes often are more easily cleaned than plain tubes.

SHELLSIDE IMPINGEMENT PROTECTION

Impingement devices are often used in tubular equipment to protect the tubes from erosion due to fluid impingement at the shell inlet nozzle. One approach to the design of impingement plates is to reduce the entering velocities. The most common inpingement design is to expand the nozzle size to twice the pipe diameter in a standard reducer (Figure 32-1). A plate somewhat larger in diameter than the pipe is installed in the enlarged section. The high veolocity jet, including any entrained particles, strikes the plate. The average velocity at the bundle inlet is reduced to one-third its former velocity and the entrainment is reduced even more.

This type of design is suitable for the smaller nozzle sizes, but is seldom used for nozzles greater than 8 inches in diameter. Large nozzles require a large concentric reducer size; consequently, a distributor belt (vapor belt) is often the better choice to achieve low inlet velocities. A distributor belt is an enlarged portion of the shell which permits the fluid to enter the bundle at more than one point. Distrbutor belts can also be designed to provide inpingement protection. Vapor belts (Figure 32-1) provide the lowest entrance velocity, the lowest entrance pressure drop, and the best fluid distribution in addition to better impingement protection.

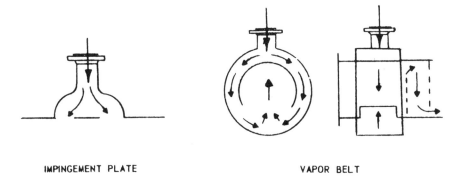

IMPINGEMENT PLATE   VAPOR BELT

Figure 32-1: Impingement protection.

Sometimes tubes are omitted from the tube bundle in order to provide space for an impingement plate. The impingement plate in this case is generally a circular or rectangular plate located a short distance from the row of tubes and secured by various means. A modification of this type is to form the plate and surround the bundle with it as a partial shroud. Omitting tubes from the tube bundle has many advantages, especially when inlet nozzle velocities are very high or when tubes must be omitted from the baffle windows in order to prevent problems associated with tube vibration.

Two other methods are sometimes used to provide impingement protection. One method utilizes the first two rows of tubes adjacent to the nozzle for impingement protection. The tubes are plugged and rods are sometimes inserted through the tubes to prevent excessive tube movement in the event of a tube being cut sufficiently to allow movement. The other method replaces the first two rows of tubes adjacent to the nozzles with rods of the same diameter at the tubes. Both of these methods are effective and inexpensive in the common materials of construction.

Generally impingement plates should not be perforated. There may be times when a flow distribution plate also serves as an impingement plate. There is seldom sufficient space to install a plate to serve both as an impingement plate and as a flow distribution plate unless all tubes have been omitted from the baffle windows.

FLOW DISTRIBUTION

The location of nozzles is important, especially on the tube side. There are basically two ways that nozzles can be installed on the channels; parallel to the axis and perpendicular to the axis. These are called axial and radial respectively. The radial nozzle (Figure 32-2) is much preferred over the axial nozzle because it provides a more uniform distribution of the fluid to the tubes. Axial flow nozzles can afford good distribution if some type of diffuser is provided.

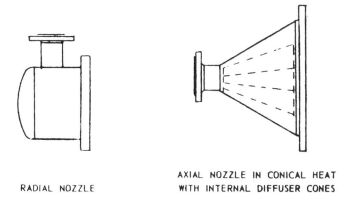

RADIAL NOZZLE

AXIAL NOZZLE IN CONICAL HEAT WITH INTERNAL DIFFUSER CONES

Figure 32-2: Nozzle orientation.

Flow distribution is also important on the shellside, especially in cases where parallel flow baffles or RodBaffles are required. Vapor belts, or distribution belts, serve to distribute shellside fluids in many cases. Vapor belts may be used to permit fluid to enter the shell over the entire circumference or they may be designed such that the fluid enters only a specified area of the shell. Vapor belts reduce entrance velocities and entrance pressure drops while effecting good distribution.

Shellside distribution may also be effected by installing perforated plates downstream of the inlet nozzle. This approach is often used for condensers when the shellside fluid makes only one pass across the tube bundle. The perforated plate must be adequately positioned above the tube bundle to prevent excessive jetting and impingement.

33
Installation

Many details must be considered when installing heat transfer equipment. Many of these may seem of minor importance but it is precisely these small details that often lead to poor performance and increased maintenance.

VENTING

Vent connections are normally provided. In many cases they must be permanently installed. In other cases intermittent venting may be acceptable. See Chapter 13 on Evaporator Performance for a discussion of evaporator and evaporator accessory venting.

SIPHONS IN COOLING WATER PIPING

The water side of equipment should operate under a positive pressure whenever possible. Frequently, water-cooled units are mounted high in a structure and the available pressure is barely adequate to deliver water to its intended use. Often a vacuum exists and boiling of the water may occur. This causes corrosion and hydraulic problems, both in the exchanger and the outlet water piping. Any control valve should be placed in the outlet piping so that the maximum available pressure is realized at the exchanger. No design should be complete without considering what the pressure at the outlet of the water side of the exchange will be when operating. If a decision is made to operate under a vacuum, the following should be considered:

(1) The pressure should not be so low that the water can boil at the process temperature. Boiling the water will increase the venting load and result in fouling because of scale formation. Increased corrosion also often results.

(2) A pressure gage should be provided near the water outlet. A pressure gage will permit the pressure to be observed and the vacuum created by the siphon can be controlled.

(3) An adequate means to check the vent flow should be provided and the vent flow should be checked regularly in order to ensure the exchanger is properly vented.

Caution and good judgment should be exercised when a decision is made to operate the water side under a vacuum.

U-BEND EXCHANGERS

Generally, U-bend exchangers should not be installed in a vertical position. Vertical U-bends are difficult to vent or drain on the tubeside because connections cannot be provided at the U-bend end.

Multipass exchangers with straight tubes preferrably should not be installed vertically unless tubeside velocities are relatively high and provisions are made to ensure that the tubes operated completely flooded.

EQUIPMENT LAYOUT

Sufficient space should be allowed in the equipment layout to remove tube bundles from removable bundle units. Consideration should be given to units expected to require periodic maintenance or cleaning. Early recognition should be given to space requirements of air-cooled and evaporative-cooled exchangers. Equipment layouts should be recognizing that longer tube lengths result in more economical heat exchangers. Care should be exercised to avoid forcing the equipment to fit the layout rather than providing a layout to match the equipment.

PIPING

The manner is which an exchanger is piped up influences its performance. Horizontal units should have inlet and outlet nozzles on the top and bottom of the shell or channel. Nozzles should not be on the horizontal centerline of the unit. In general, fluids should enter the bottom of the exchanger and exit at the top, except when condensation occurs. Units are almost always designed to be counterflow and the piping must reflect this. When concurrent flow has been specified, it is equally important that equipment be piped to suit. For multipass units, the shellside fluid can be admitted at either end of the shell without affecting the thermal performance. However, for some cases better distribution or mechanical reliability can be achieved with the inlet nozzle at a specific location.

34
Design Practices for Maintenance

By anticipating potential problems, the designer can avoid high maintenance or cleaning expenses and costly shutdowns. To anticipate maintenance problems, the designer needs to be familiar with the plant location, process flowsheet, and anticipated plant operation. Some of the questions that must be considered include:

(1) Will the equipment need to be cleaned? How often? What method will be used? Can cleaning be performed in place, or must the equipment be removed for cleaning?

(2) What penalty will the plant pay for leakage between the fluids handled?

(3) What kind or production upsets can occur that could affect the equipment? Will cycling occur?

(4) How will the equipment be started up and shut down?

(5) Will the equipment be likely to require repairs? If so, will the repairs present any special problems?

(6) Should the equipment be situated in the plant layout to facilitate maintenance?

(7) Would a different type of evaporator or different materials of construction reduce maintenance?

(8) Has adequate attention been given to the piping and auxiliary equipment associated with the evaporator?

(9) Have personnel been properly trained? If not, is a training program scheduled?

STANDARD PRACTICES

Standard practices must be established to achieve an economical balance between the cost of the equipment and the cost of maintenance and production losses. These practices may vary in different locations of the same plant as the penalty for eqipment failures, and the severity of the service, vary.

REPAIR FEATURES

If the equipment will be used in corrosive service and if the components will have a short life expectancy, the designer should select units that are easy to repair. Removable bundle units may be required. In addition, the plant equipment layout should be arranged to facilitate removal and repair. Maintenance costs and production losses can often be reduced by specifying equipment with standard components or designing equipment to be interchangeable among several different services.

CHEMICAL CLEANING EQUIPMENT

Removal of various scales and deposits from either the outside or inside of tubes by chemical cleaning is a common cleaning technique. Plant maintenance crews are often able to provide chemical cleaning or washing for many applications. Qualified organizations are frequently employed to determine the nature of deposits, furnish proper cleaning fluids with inhibitors, and provide all necessary equipment and personnel for a complete cleaning job. Chemical cleaning, however, is not always effective and some types of fouling can be removed only with physical or mechanical means.

MECHANICAL CLEANING EQUIPMENT

Mechanical cleaning must be used when an effective chemical cleaning method cannot be found. The most common method is to blast the surfaces with high pressure water streams (hydroblasting). Other means include scraping, wire brushing, drilling, cutting, and sandblasting. Mechanical cleaning cannot usually be easily accomplished for the inside of U-bend tubes. To mechanically clean the outside of tubes requires that the tube bundle be removable and that the tubes can be on square pitch or wide triangular pitch to enable the tool or water jet to penetrate the bundle.

BACKWASHING

Silt is often deposited in heat exchangers cooled with cooling water, especially when cooling water is on the shellside. "Silting" often occurs during-rainy periods, when low water flow rates are maintained, or because water strainers have been bypassed. This silt must be removed to prevent excessive corrosion and loss of effective heat transfer surface. Once silt has been deposited it tends to pack and agglomerate, making removal difficult. Chelating agents or polyelectrolytes are sometimes added to cooling water to remove silt, but their use may not be economically justified in all services. Often to reduce or eliminate silting, backwashing is provided by periodically reversing the water flow direction. To be effective, backwashing must be performed on a routine schedule. Backwashing techniques vary from plant to plant depending upon the quality of the water. In some plants, proper maintenance of water strainers eliminates the need for backwashing. Special valves are sometimes used to facilitate backwashing and minimize process upsets resulting from the need for backwashing. Automatic backwashing valves are also available. Brushes or sponge balls can be shuttled back and forth through tubes as water flow direction is reversed to clean tubes mechanically. These methods are often effective.

AIR INJECTION

Another method to reduce silting is to periodically inject air into the cooling water. Air injection must normally be more frequent to achieve the same results as with backwashing. This method is also more expensive and adds to the venting problem. It has the advantage, however, that no process upsets occur as experienced occasionally when backwashing. Air injection, to be effective, must be performed on a routine schedule. A timing mechanism may be used to inject air for short time intervals over a period of several days.

Sometimes a combination of backwashing and air injection is used with good results. During backwashing, silt deposits may remain in the space between the back face of the tubesheet and the nozzle. Air can be injected through the lower tubesheet vent or drain connections to remove these deposits.

35
Mechanical Design

Many factors affect the mechanical design of evaporator systems, particularly of the calandrias. The two most important are the temperatures and pressures to which the equipment will be subjected. Not only are the temperatures and pressure during normal operation important, but upset, startup, shutdown, dryout, cycling, pulsating pressure, and safety relief requirements are equally important. Other considerations include external loadings from supports or piping and vibrations transmitted from external sources. Wind loadings and earthquake loads must also be considered. Anticipated life expectancy and future service should be considered.

An analysis of the process should be carefully conducted to determine the maximum fluid pressures and the degree of vacuum that can be anticipated. It is important to determine both minimum and maximum anticipated operating temperatures in order to obtain the design temperature. This combination of temperatures and pressures will determine the mechanical design of the equipment.

MAXIMUM ALLOWABLE WORKING PRESSURE AND TEMPERATURE

Several special terms are used to describe factors which affect the mechanical design of equipment. Some of these are discussed:

Operating Pressure and Temperature: The operating pressure is the pressure at the top of a pressure vessel at which it normally operates. The operating temperature is the temperature that will be maintained in the metal of the vessel for the specified operation of the vessel.

Design Pressure and Temperature: The design pressure is the pressure used to determine the physical characteristics of the different components of the vessel. It usually is the sum of the maximum allowable pressure and the static head of the fluid in the vessel. The design temperature is the metal temperature coincident with the design pressure.

Maximum Allowable Working Pressure (MAWP): The MAWP is the maximum gage pressure permissable at the top of a completed vessel in its operating position for a designated temperature. This pressure is the basis for the pressure setting of the pressure relieving devices protecting the vessel.

The maximum allowable working pressure (MAWP) is the maximum pressure to which the equipment can be safely subjected. This MAWP will appear on the nameplate of the vessel. For calandrias, the tubeside and shellside will have separate nameplates; the MAWP for each side must be determined and may not necessarily be the same value. The MAWP should generally not be less than the maximum anticipated operating pressure divided by 0.90. The maximum temperature and which the MAWP will be realized must also be specified. In addition, a minimum design temperature should also be specified. The mechanical design of the equipment is governed by these values of pressure and temperature. Normally, these maximum design requirements are determined by the pressue and temperature conditions that can be realized if the equipment is "shut-off" or valved in or by a safety valve or relief device limitation. In some cases, upset, startup, or dryout conditions determine the maximum allowable working pressure and temperature and hence govern the mechanical design of the equipment.

UPSET CONDITIONS

Upset, startup, or dryout conditions may present temperature-pressure combinations drastically different from the normal operating conditions. These upset conditions may dictate the choice of heat transfer equipment and may determine the mechanical design of the selected type. It is therefore imperative that these upset conditions be carefully considered and evaluated. Upset conditions may include, in addition to pressure and temperature changes, variations in flow rates, possible additional services, pressure and thermal cycling, surging, batch operation, cleaning procedures, and composition or physical property variations. In some cases, full understanding of any reaction kinetics involved in the system may be important. The equipment may need to be designed for total containment of any explosions.

THERMAL EXPANSION

Differential thermal expansion between various components of calandrias, especially shell-and-tube types, has an important effect on the mechanical design of the equipment. Some types of tubular exchangers incorporate into the basic design and fabrication means to provide for thermal expansion. Other types, specifically fixed tubesheet units, must often be provided with expansion joints in the shell to meet specified differential thermal expansion between the tubes and shell. Another critical area affected by thermal expansion in fixed tubesheet units is the shell-to-tubesheet juncture. Gasketed joints may also be affected by thermal expansion; therefore both the flange design and the gasket selection in some cases may be determined only after evaluating the effect of any expected thermal expansion. In some cases, the manner in which an exchanger is piped up with the respect to the two fluids may be critical to avoid problems resulting from thermal expansion. Thermal expansion may also determine the manner in which tubes are attached to the tubesheet. Thermal expansion problems may also be caused by frequently shutting down and starting up the equipment.

TUBE-TO-TUBESHEET JOINTS

The main function of tube-to-tubesheet joints is to seal the tubes tightly to the tubesheet. And for most equipment, an additional major function is to support the tubesheet against pressure-induced loads.

Tube-to-tubesheet joints are made by:

(1) stuffing the space between the tube and the hole with packing

(2) sealing the tube to the hole with an interference fit

(3) welding or brazing the tube to the tubesheets

(4) using a combination of methods.

For metallic tubular exchangers, two methods are commonly used to secure the tube to the tubesheet. Most units have tubes expanded, or rolled, against the tubesheets; tube holes are normally grooved to provide strength. This method has proven to be reliable for many applications. Tubes may be fusion-welded to the front face of the tubesheets. Welds may be strength welds; or seal welds may be sufficient. Strength welds rely upon the welded joint for strength; tubes are usually only lightly rolled to eliminate the crevice. Seal welds rely upon a fully expanded tube-to-tubesheet joint for strength; tubes are welded only to minimize leakage. Seal welded joints are acceptable for most applications.

Welded tube-to-tubesheet joints are used for high pressure services, for high temperature services, when clad tubesheets are employed, for services which cycle frequently, for lethal service, and when more reliability for the joint is desired. Welding tubes to tubesheets does not guarantee that a leak will not occur, but it does decrease the incidence of joint leaks. Special test procedures may be initiated to improve the reliability of welded tube-to-tubesheet joints at fairly nominal costs.

DOUBLE TUBESHEETS

No known method of joining tubes to tubesheets entirely eliminates the possibility of leakage. Joints that are tight under shop-testing conditions may leak after a startup, operating upset, or shutdown. Double tubesheets do not eliminate leakage, but they do reduce the probability of leaking fluid on one side into fluid on the other side. However, double tubesheets are an exotic solution that have caused considerable maintenance problems and should not be specified unless the mixing of the two fluids must absolutely be avoided.

Conventional Design

Conventional double tubesheets incorporate two separate tubesheets at each end of the exchanger; one tube sheet contacts the tubeside fluid, the other contacts the shellside fluid. A space is provided between the two tubesheets; this space is normally open to the atmosphere but can be sealed in order to be purged or monitored for leaks.

There are some problems in the manufacture of conventional double tubesheets. Alignment of tube holes may not be correct with resultant lateral shearing and bending forces imposed on the tubes after assembly. It may then be difficult to achieve a tight seal. Because the two tubesheets may be subjected to considerably different process temperatures, there can be differential expansion between the tubesheets that may bend the tubes. The resultant shearing and bending may cause leakage. If temperature changes are cyclical and frequent, tubes may fail because of metal fatigue. In some cases, double tubesheets may need to be spaced 12 to 24 inches apart to avoid overstressing the tubes, especially those in the outer rows. Double tubesheets in horizontal units may present unusual support problems.

If leaks develop at the inboard tubesheet, they cannot be easily stopped by plugging the tube. If rerolling the tubes does not stop the leaks, it may be necessary to retube the bundle. The tubes cannot be feasibly welded to the inboard tubesheet.

Integral Design

An integral double tubesheet can be constructed to overcome many of the problems associated with conventional designs. A single tubesheet can be machined with inter-connecting passages between tube holes leaving metal lands to provide strength and maintain more uniform temperature profiles throughout the tubesheet. The grooves straddle the neutral plane of the tubesheet so there is little reduction in the bending strength of the tubesheet.

The tubesheet passages can be pressurized, purged, vented, or monitored to detect leaks. Separate vents and drains are required in the gap in each pass of multipass units. The integral design construction requires fairly expensive tooling and careful attention to tool speeds and coolants. The method is usually limited to relatively free-machining materials.

Economics favor the integral tubesheet when exchangers are larger than approximately 15 inches in diameter. Consequently, when double tubesheets are justified, these special integral types are more reliable and less costly. Again, leaks are not easily plugged at the inboard face; retubing may be required if rerolling fails to stop and leaks.

Summary

Double tubesheets are seldom justified; more reliability can be achieved by welding tubes to tubesheets and conducting tests to improve the reliability of the welds. Fluids can still mix; although they mix in the space between tubesheets. Appropriate purging or pressurization may be required to avoid any reactions in the space as a result of simultaneous leaks of fluids from both sides. However, if double tubesheets can be justified, the integral type should be used.

INSPECTION TECHNIQUES

The efficiency is a welded joint is expressed as a numerical (decimal) quantity and is used in the design of a joint as a multiplier of the appropriate allowable stress value for the material. The joint efficiency depends upon the type of joint and the extent of examination of the joint.

Radiography is the process of passing electronic radiation through an object to obtain a record of its soundness upon sensitized film.

Spot Radiography is a random sampling of welding used to aid quality control. About one percent of the weld is examined.

Partial Radiography is an examination at 6-inches long of any section of the weld picked at random, plus a similar examination of certain intersections of the welds as required by the ASME Code. Acceptance standards for partial radiography are more stringent than for spot radiography.

Full Radiography is frequently used to indicate that the main seams of a vessel are to be radiographed. The use of the term in this manner should be avoided; it is much more precise to specify the extent of radiographic examination required by specifying radiographing for each of the various categories of joints established in the ASME Code. It is proper to say that a particular weld is to be fully radiographed, meaning that the entire weld is to be so examined.

Magnetic Particle Inspection is a method of detecting cracks and similar discontinues at or near the surface in iron and magnetic alloys of steel. It consists of properly magnetizing the materials and applying finely divided magnetic particles which form patterns indicating discontinuities.

Liquid-Penetrant Inspection is a method of testing for surface defects or sub-surface defect with surface openings by relying upon a penetrant seeping into the defect. It is normally used for non-magnetic materials.

Ultrasonic Inspection is made by bounding a beam of ultrasonic energy into a specimen and measuring the reflected energy. Defects in the specimen affect wave reflection.

Stress Relief is the uniform heating of a structure or portion thereof to a sufficient temperature below the critical range to relieve the major portion of the residual stresses, followed by uniform cooling.

36
Safety

Safety can be approached and resolved through adequate attention to detail, knowledge of basic principles and design techniques, and proper identification or definition of process safety hazards. Recognized hazardous situations can preferably be eliminated by proper design techniques. If elimination is not possible, the hazard can be controlled, or isolated. The unrecognized conditions are usually those that cause injury or property damage.

Incidents involving fires, explosions and accidents can be initiated by many different acts. Most incidents can be characterized as resulting from the following:

(1) spills or leaks of flammable fluids

(2) uncontrolled reactions

(3) process upsets

(4) operating errors

(5) mechanical equipment failure

(6) improper equipment venting or pressure relief

(7) poor housekeeping

In general, the designer of an evaporation system is responsible for designing the process to minimize the potential for a failure that would result in personal injury, property damage, or lost production. Specific responsibilities include:

(1) Process safety should be considered from the very beginning of preliminary design. This involves the selection of the process configuration, temperatures, pressures and vessel volumes with a view of what might go wrong.

(2) The basic principles and design techniques of safety engineering should be understood.
(3) Hazardous situations should be identified, defined and eliminated. Control or isolation of hazardous situations are less desirable safety solutions than eliminating such situations.
(4) The thinking that resulted in a solution of safety problems should be well documented.
(5) All safety procedures should be met.
(6) All safety calculations and relief device sizings should become a permanent part of the project file.
(7) Safety problems should not be intentionally left for solution by any safety reviewers.

COMMON ERRORS

Several errors are commonly made during process design:

(1) Changing service in a piece of equipment without checking the safety considerations for chemical reactivity, materials of construction, relief requirements, control systems, etc.
(2) Modifying heat input into a system without recalculation of relief requirements.
(3) Discharging an additional safety valve to a vent header without checking the header's capacity or compatibility with other vented materials that might be present in the header.
(4) Failing to recognize significance of pressure drop in inlet and discharge piping of safety valves.
(5) Failing to recalculate relief requirements after trim in control or pressure reducing valve has been revised.
(6) Taking full credit for water spray protection or insulation for fire exposure in relief calculations when only partial protection is being installed.
(7) Revising a process without considering increased hazard to personnel facilities, such as control room, operator's shelter, maintenance shop, etc.
(8) Adding a new chemical to an area without considering effect on the existing electrical hazard classification.
(9) Adding a shutoff valve in a process line between a safety valve and the vessel it protects.
(10) Failing to provide explosion venting for solids handling or processing equipment.

SAFETY RELIEF

The ASME Code requires all equipment to be protected by pressure-relieving devices to prevent the pressure from rising above the maximum allowable working pressure. In some cases, the pressure can be permitted to rise to a value of ten percent higher than the maximum allowable working pressure.

The required capacity of a pressure relieving system for a given installation should generally be determined by evaluation of the "worst, credible, single condition" that can occur. If the "worst condition" introduces or activates additional requirements, the control system must unfailingly override them or additional capacity must be provided in the pressure relieving system. The relief requirements must be based on an evaluation of all credible conditions.

The "worst, credible, single condition" is usually determined from one of the following circumstances:

(1) fire exposure
(2) abnormal operation
(3) hydraulic accumulation
(4) thermal expansion
(5) runaway reactions
(6) changed service
(7) tube failures
(8) subatmospheric pressures

Evaporation systems incorporate virtually all of the relief considerations previously discussed.

Pressure Relief

Overpressure in an evaporator may result from:

(1) loss of cooling water
(2) loss of electric power
(3) loss of mechanical drive on fans
(4) failure of steam, steam controller, or other heat source controller
(5) failure or heat exchanger tube
(6) chemical reaction, which may be initiated from improper feed rates, overtemperature, contamination, loss of vacuum, etc.
(7) failure of temperature controls
(8) fire exposure
(9) loss of vacuum source

Each of the possibilities should be examined to determine the condition which would result in the greatest relief requirement. A study should determine which of the contingencies under consideration would result in the occurrence of a second contingency. The possibility of precluding the simultaneous occurrence of the second contingency should be considered.

After all the evaluations, a decision as to which condition or conditions will dictate the emergency relief requirements. In some instances, the calculated relief will be impractical. When this happens, a reasonable means of reducing the maximum contingency, redesigning the evaporation system, or providing a rupture section may be required.

Hazards created when changes are made in service, equipment, or controls which affect adequate relief capacity must be recognized. Such conditions include:

(1) removal of the cascade controls on a steam-heated calandria
(2) placing the system in manual control
(3) replacing the trim in a control valve with trim of higher capacity
(4) addition of shutoff valves

37
Materials of Construction

Evaporators can be fabricated from most metals as well as from some nonmetallic materials, such as impervious graphite, glass, plastics, and ceramics. The choice of materials of construction for an evaporator system depends upon the fluids being handled, the process conditions, the type of equipment used, and a balance of initial capital investment against expected life and maintenance costs. The corrosion properties of the fluid being handled are often the determining factor in selecting the materials of construction. Some components of evaporators are contacted by more than one fluid and must therefore be compatible with all fluids. Other components are contacted by only one of the fluids and need only materials compatible with the fluid in question. When using tubular evaporators, it is generally more economical to place the fluid requiring more expensive materials of construction on the tube side. Clad materials are often used to reduce cost and to provide different materials on either side of a component.

BASIC QUESTIONS

The designer of an evaporator system must have a fundamental knowledge of the properties and applications of common materials of construction and of the corrosive effects of common chemicals. He must also be familiar with the basic standards and practices for using materials of construction. This basic knowledge should enable the designer to:

(1) make preliminary selection of materials, except for unusual problems

(2) avoid common problems with materials in common situations

(3) recognize unusual and untried or severe materials situations which require consultation

(4) communicate with a materials engineer
(5) assemble all information required for proper selection of materials of construction.

Materials of construction can be determined only by understanding the equipment required for the process conditions. Several properties of process fluids are important for proper selection of materials of construction:

(1) compositions and concentrations including any possible impurities, inhibitors, reaction products, dissolved gases, and solids
(2) pH
(3) temperature
(4) pressure
(5) flow velocities, agitation, and turbulence
(6) range of all these variables in normal operation, startup, shutdown, and possible abnormal conditions. Also to be considered:
   (a) minor components
   (b) concentration changes within equipment
   (c) components in remote parts of the system which may have a serious effect if they are inadvertently introduced into equipment
   (d) source of cooling water
   (e) effects of leaks and spills on the exterior of equipment and piping
   (f) effects of any cleaning procedure.

Several questions should be addressed when considering materials of construction for a particular application:

(1) What are the effects of failure of a particular part or equipment item? Can failure result in a hazardous situation or cause serious disruption of operation?
(2) What is the desired service life for the equipment or part?
(3) Will corrosion products degrade product quality? Will it affect operation of the process (reaction, catalyst, etc.)?
(4) Will corrosion effects on dimensions or appearance be a problem?
(5) What materials have been used in identical or similar situations? With what results? Are inspection reports available for existing equipment?
(6) Is plant corrosion test data available?
(7) Has experience with materials of construction been gained in pilot plant work?

(8) Have laboratory corrosion tests been conducted?

(9) Is literature information available for materials of construction for this application?

SELECTION

Selection of materials of construction must be made only after considering safety and economics of the process as well as its workability. Factors that must be considered include:

(1) investment cost, including any necessary corrosion allowance

(2) maintenance costs

(3) reliability

(4) expected life of the equipment

(5) effects on project schedules that any long delivery materials exert.

38
Testing Evaporators

Performance testing is a means to enable understanding of an evaporator system. Tests permit detection of unsatisfactory performance and often indicate methods to improve operation. Performance tests may also be required to establish that a new evaporator system has met performance guaranteed by the supplier. Tests may also serve to plan for maintenance or cleaning. Tests may be used to determine evaporator capacity under different operating conditions or to obtain data for designing a new evaporator system. Tests may also be necessary to establish base performance for evaluation of possible upgrading methods. The American Institute of Chemical Engineers has published a procedure entitled "AIChE Equipment Testing Procedure: Evaporators". This procedure covers methods for such performance tests and discusses several factors influencing performance and accuracy of test results.

Tests are conducted to determine capacity, heat transfer rates, steam economy, product losses, and cleaning cycles. Practically all the criteria of evaporator performance are obtained from differences between test measurements. Errors can result when measuring flow rates, temperatures and pressures, concentrations, and steam quality. Factors which can have a great effect on performance include: dilution, vent losses, heat losses, and physical properties of fluids.

PLANNING THE TEST

In setting up any performance test, the following principal steps should be taken:

(1) establish the test objectives
(2) plan the test and the test data log
(3) check all instruments and measurement methods

(4) check for leaks
(5) take all steps necessary to ensure smooth operation, especially during the tests
(6) conduct a preliminary test and evaluate the results to spot any possible discrepancies
(7) select data for evaluation from a period of smooth and steady operation.

It is of great advantage that the person who will analyze plant data participate in the test program. Many factors should be evaluated during the test to determine the validity of the test data. Evaporator tests can be costly, especially if the data obtained are not sufficient to meet the test objectives. Adequate test personnel must be available to gather all necessary data over a relatively short period. Testing a sextuple-effect evaporator can require several hours for one data set at one operating point.

Diagnosis of plant data depends heavily upon actual physical inspection of the equipment, checking on conditions not apparent from any panel board or instrument readings. Included are:

(1) vent temperatures (may be measured in some cases)
(2) complete removal of condensate from heating elements
(3) pump speeds and degree of wear
(4) presence of scale on the inside or outside of tubes
(5) obstructed circulating piping
(6) obstructions in separator or vapor piping.

The test team must be completely familiar with the evaporator system. Flow sheets and instrument diagrams must be understood. Physical location of measuring devices must be known and evaluated.

Accuracy of the data required must be established. Errors in the interpretation of results may arise either from errors of measurement or from factors that are not normally subject to measurement. The effect of these errors depends to a large extent on the intended use of the test results. If test data are used to predict performance of the same evaporator under slightly different conditions, most of the normal errors are self-compensating. If test data for one effect of an evaporator are to be used in the design of a new evaporator, even minor errors can be of great importance. Accurate data is necessary when comparing actual peformance to that expected.

CAUSES OF POOR PERFORMANCE

Evaporator system problems usually are evidenced by one or all of the following symptoms:

(1) reduced evaporator capacity

(2) reduced steam economy

(3) loss of product

(4) frequent cleaning cycles.

Frequent causes of poor performance of an evaporator system include the following:

(1) low steam economy

(2) low rates of heat transfer

(3) excessive entrainment

(4) short cleaning cycles.

Low Steam Economy

Steam economy with a fixed feed arrangement can be calculated from heat and material balances. Steam economies lower than that calculated during the design of the unit may be the result of one or more of the following:

(1) leakage of pump gland seal water

(2) excessive rinsing

(3) excessive venting

(4) flooded barometric condensers

(5) dilution from condensate leakage

(6) steam leakage

(7) leaking across supposedly closed or sealed valves.

Low Rates of Heat Transfer

Poor heat transfer may occur for the following reasons:

(1) salted, scaled, or fouled surfaces (both process and steam side)

(2) inadequate venting

(3) condensate flooding

(4) inadequate circulation

(5) inadequate liquid distribution

(6) temperature differences that are too high or too low

Excessive Entrainment

Product losses through excessive entrainment may result from:

(1) air leakage (especially below liquid surface)

(2) excessive flashing

(3) sudden pressure changes

(4) inadequate liquid levels

(5) inadequate pressure levels

(6) operation at increased capacity

(7) plugged drains

Short Cleaning Cycles

Downtime required for cleaning may not agree with the expected frequency of cleaning. Short cleaning cycles may be caused by:

(1) sudden changes in operating conditions (such as pressure or liquid levels)

(2) improper liquid level control

(3) low velocities

(4) introduction of hard water or other foulants during cleaning, rinsing, or from seal leaks

(5) high temperature differences

(6) improper cleaning procedures.

39
Troubleshooting

In spite of precautions taken during the design of evaporator systems, problems do arise during startup and operation. Parameters which cannot always be precisely determined make it nearly impossible to define all the problems during the design stage. These parameters include:

(1) composition changes

(2) fouling

(3) validity of heat transfer data

(4) changes in product specification

(5) changes in raw materials

(6) changes in utilities.

Occasionally it becomes necessary to investigate the performance of an evaporator in order to evaluate its performance at other operating conditions or to determine why the system does not perform as expected. Problems encountered during the operation of evaporator systems can often be simply explained. Troubleshooting, therefore, often means checking for small details which have a great effect on the performance of the evaporator system. Of course, it is possible that a type of evaporator has been misapplied, the heat transfer surface that has been provided in not adequate for the intended service, or fouling is occurring. Often, however, the trouble can be traced to one or more of the conditions to be discussed later. A troubleshooter's axiom is to always look for the simple, almost obvious things before raising an alarm. Don't take anything for granted.

Discrepancies in performance may be caused by deviations in physical properties of fluids, flow rates, inlet temperatures, mechanical construction of the equipment. The troubleshooter should first check to see that compositions, flows, temperatures, and physical properties agree with those specified for design. He should then examine the equipment drawings to determine if the problem could be in the manner in which the equipment was constructed. The adequacy of the equipment installation should also be examined.

Problems encountered in evaporators fall into four major categories:

(1) general operations
(2) vacuum
(3) steam economy
(4) relationship of evaporator performance to other parts of the plant.

Typical operation problems involve:

(1) scale formation
(2) poor performance
(3) excessive entrainment losses
(4) mechanical failure of vessel internals.

These difficulties can often be traced to faulty operation, mechanical wear, or improper design.

Problems arising in one or more of the system components will result in problems in other areas as well.

If a detailed evaluation indicates that the basic design and installation is appropriate, then the task becomes to check specific things. Some of these will now be discussed.

It should be emphasized that routine periodic inspections of evaporator systems be scheduled. Such inspections will uncover minor changes which appear harmless and are made for ease of cleaning and operation but which have seriously detrimental effects on the evaporator performance. Some items to be wary of:

(1) lack of venting
(2) open bypass valves
(3) miscalibrated instruments
(4) lack of seal water on pumps
(5) leaking valves
(6) control valves not free to move.

CALANDRIAS

Specific things to be checked for calandrias include:

(1) Has the steam side been vented to remove air or other entrapped gases?
(2) Has the steam control valve been adequately sized? What is the actual steam pressure in the steam chamber?

(3) Has the steam trap been properly selected, sized, and installed?

(4) Are the control valve and steam trap functioning correctly?

(5) Is steam condensate flooding part of the surface? What is the temperature of the steam condensate? Is the condensate nozzle large enough? Is steam trap piping adequately sized?

(6) Is the process liquid level maintained at the proper place? Are liquid level instruments calibrated? Are instrument leads plugged?

(7) Is the liquid holdup adequate to prevent surging?

(8) Are process compositions and temperatures equal to those used for design? Does the process material contain enough volatile to provide adequate boiling?

(9) What is the temperature of the top head for natural circulation calandrias? A temperature higher than the liquid temperature may indicate inadequate circulation from some reason.

(10) Is the available steam pressure equal to that used for design?

(11) Are process nozzles adequate?

(12) Is the process side adequately blown down on purged?

(13) Were all debris and other foreign objects removed from equipment and piping prior to startup? What is the appearance of the equipment before cleaning? How often is the unit cleaned? Is the cleaning adequate?

(14) If a pump is provided, does the pump and the system match? Is the pump cavitating?

(15) Has enough back pressure been provided to prevent boiling of the process fluid when the evaporator operation requires?

(16) Is entrainment occurring? Are entrainment separators properly sized and installed? Are they plugged?

(17) Is dilution important?

(18) Are flows adequate to maintain flow regimes used in design? Is the pressure drop out of line?

(19) For falling film units, is the unit plumb? Is the inlet channel vented to remove and flashed vapors? Are flows adequate to ensure that a film is formed? Is the outlet flowrate sufficient to prevent the film from breaking?

CONDENSERS

Specific things to be checked for condensers include:

(1) Has a constant pressure vent system been provided? Has it been properly installed? Inerts should be injected downstream of the condenser, not upstream.

346 Handbook of Evaporation Technology

(2) Is the vent system adequate?

(3) Are condensate connections properly sized? Is liquid being entrained into the condenser? If horizontal, are the tubes level (or sloped toward the outlet)?

(4) Is all piping adequate?

(5) Is the water side operating under a vacuum?

(6) Are temperatures and composition equal to those used for design?

(7) Is the water flow adequate? Properly vented?

(8) Was all debris or other foreign objects removed from equipment and piping prior to startup?

(9) If air-cooled, is the inlet piping adequate to effect good distribution? Are fan blades properly pitched? Are motors delivering rated power? Are fan belts slipping? Is there noticeable recirculation of hot exhaust air?

VACUUM FAILS TO BUILD

If vacuum fails to build in the evaporator, the following guidelines will help to determine the cause.

(1) manhole cover and inspection ports not properly sealed

(2) check for other leaks

(3) liquid inlet or outlet valve not closed or seated properly

(4) packing glands not tight. No seal water flow.

(5) pump discharge check valves leaking

(6) safety valves not sealed

(7) barometric leg not sealed

(8) sight glasses leaking

(9) fouled condensers

(10) tube failure.

Steam Jet Vacuum Sytems

(1) steam pressure too low

    (a) steam valves not fully open

    (b) strainer may be plugged

    (c) boiler pressure too low

(2) wet steam

(3) restriction in exhaust line from second-stage ejector

(4) if two single-stage units are used, valves may be closed

(5) barometric condenser flooded
   (a) water flows too high
   (b) leak in barometric leg or barometric leg plugged
   (c) discharge pump not adequate
(6) not enough water to intercondensers to condense first-stage steam exhaust
(7) ejector steam nozzle plugged
(8) steam nozzles of different stages reversed by mistake

Mechanical Pumps

(1) check pump speed
(2) check rotation direction
(3) make sure enough hurling fluid is provided at the correct temperature
(4) check back pressure
(5) check for excessive pressure drop between pump and condenser
(6) check maintenance records

NO VACUUM IN STEAM CHEST

(1) vent valves not open
(2) steam valves open or not seated
(3) condensate pump seal leak or no seal water
(4) vent valve leak
(5) condensate pump check valve leak
(6) safety valve leak

VACUUM BUILDS SLOWLY

(1) air leaks
   (a) fluid lines not tight
   (b) covers and inspection ports leaking
(2) improper operation of first stage ejector
(3) wet steam
(4) steam pressure too low
(5) nozzle wear or plugging
(6) tube leak

FOAMING

(1) air leaks in feed piping
(2) air leaks in sight glasses
(3) steam pressure increase rapidly, perhaps as a result of faulty control valve or regulator
(4) air mixed into product in feed tank or hotwell
(5) air leaks below liquid level
(6) air leak in pump
(7) unsteady vacuum during operation

INADEQUATE CIRCULATION

(1) inadequate steam
(2) cold feed
(3) loss of pressure or drop in vacuum
(4) inadequate condensate removal
(5) fouled tubes
(6) pump problems

SUDDEN LOSS OF VACUUM

(1) inadequate cooling water to condenser or intercondensers
(2) air leak through water supply
(3) loss of steam pressure to ejector
(4) wet steam
(5) empty feed tank with resultant air leak
(6) sudden system leak
(7) fouled condenser

VACUUM FLUCTUATES

(1) wet steam
(2) flooded barometric condenser
(3) inadequate tail pipe installation
(4) fluctuating steam supply
(5) fluctuating water supply

WATER SURGE IN TAIL PIPE

(1) no seal water to pump
(2) pump leak
(3) pump check valve leak
(4) pump speed not constant
(5) water flow too great
(6) pump loses prime
(7) fluctuating vacuum

BAROMETRIC CONDENSER FLOODING

(1) water flow too great
(2) pump speed inadequate
(3) tail pipe plugged
(4) leak in tail pipe
(5) pump failure

40
Upgrading Existing Evaporators

Evaporation is one of the most energy-intensive unit operations used by industry. A recent estimate places the nation's equivalent energy use for evaporation above 110 million barrels of fuel oil per year. Because of the importance of energy to evaporation, it is likely that even the oldest evaporators in a manufacturing plant were at some time subjected to an economic analysis comparing the energy cost and consumption to the capital cost for additional hardware that would conserve energy. However, because energy was relatively cheap, the economics and increased operating convenience of relatively simple units favored the installation of less energy-efficient systems.

The radical increases in energy costs in relation to capital equipment costs have made all past analyses obsolete. The engineering evaluation of more energy-efficient equipment may have many constraints and in most cases will be more complicated than describing a new plant. The changes in the process conditions with each modification must be evaluated not only to determine the energy savings, but also to meet the demands of maintaining product quality and capacity. The evaluation of performance and the applicability of the technology must also consider physical layout and retrofit problems.

The economic analysis for upgrading existing evaporators also is more complex than are cost estimates for new systems. Many individual evaluations must be made and then compared to provide the best alternative. The analysis may need to consider the effects that upgrading the evaporator will have on the remainder of the manufacturing facility. The installed costs for retrofitting are not easy to calculate. Schedules must be set up to minimize loss of production during the changeover, and the economic evaluation must include estimates of loss of production and delays that will measurably increase the costs for the modification. However, it is certain that in most plants, high returns on both investment and evaluation expenses can be achieved.

In the past, the major criteria for the specification of the final evaporator design have been reliability of operation, minimization of capital costs, and the reduction of utility use, in that order. There were sound reasons for managerial decisions that led to the design of the present evaporators, which now are lacking in energy-conservation technology. In view of the energy costs of those times, the managerial reasoning behind this philosophy was not faulty. The bulk of the highly energy-efficient evaporators installed in the past was therefore the result of special circumstances in which energy use was especially critical or costly.

The restriction has now changed radically. The economics of energy conservation are now sufficiently favorable to override much of the past design and cost reasoning. Although a greater economic incentive now exists, it is important to understand the limiting factors that led to past designs because many of these pressures are still present and can still restrict the number of alternatives or decrease the operating efficiencies of the evaporator.

Any evaporators that have not been reevaluated against the new energy costs are likely to be much more expensive to run than is necessary.

AREAS FOR UPGRADING EXISTING EVAPORATORS

The capital investment required to implement reduced energy consumption in existing evaporators varies considerably. In most cases, the energy cost savings will increase directly with the amount of capital invested. In some cases, such as improved procedures for operation and maintenance of an existing evaporator, sizable cost and energy reductions can be achieved at low capital costs with very little hardware modification.

The risks of lowering evaporator reliability and product quality by upgrading the system are minimal if the appropriate technical evaluation is made. Basically, there are three general types of prospects for modifying existing evaporators:

(1) Fine tuning existing evaporator: These low investment changes do not change the basic evaporator layout or operation. Examples: increased attention to preventing water leakage into the process, improved maintenance, or additional insulation.

(2) Modifying auxiliary hardware: These moderate-investment changes normally can be authorized at the plant level. Example: improved heat recovery from condensates and product streams.

(3) Major hardware modifications: These high-investment changes normally require approval at the corporate level. Examples: installing additional effects or adding a vapor-compression system.

Any of these changes can show high returns on investment and expanded technical efforts.

Fine Tuning Existing Systems

Unless the performance of the existing system is known, the savings obtainable through modifications cannot be accurately evaluated. Studying the existing plant normally can lead to immediate short-term reductions and improved operating efficiencies. The starting point for the upgrading analysis is an accurate determination of the theoretical and actual heat and mass balances, together with an assessment of the proper operation of all internal controls. A discussion of the major points to check during this analysis follows.

Venting Rates: Noncondensibles enter the system with the feed or by system leaks. These noncondensibles must be vented at a rate that is sufficient to maintain pressure control and heat transfer. If venting rates are too low, the loss of heat transfer must be recovered by additional steam at a higher pressure to increase the temperature differences for heat transfer. However, excessive venting of the noncondensibles does not improve operation; high venting rates will require additional heating steam to offset the losses from vapors vented to the atmosphere or condenser.

Water Leakage: When water from external sources enters the process side of the evaporator, thermal efficiency decreases since such leaks dilute the product and increase the amount of water to be evaporated. Possible sources of water leakage into the process are corroded heat-exchange surfaces, pump seal flushes, barometric condenser water flooding, condenser flushes, and leaking of the water valves used to clean out the evaporator for startup and shutdown.

Operating Pressures: The proper operating pressures within single-effect and multi-effect evaporators must be maintained to enhance thermal efficiency and capacity. The pressures can be affected by loss of vacuum and line restrictions. As frictional pressure drops or vapor pressure within each effect increases, the boiling point of the liquor is higher and additional steam at a higher pressure is needed to maintain capacity and the achievable heat transfer ΔT's of each effect.

Poor steam quality can reduce the performance of the steam jets used to vent the noncondensibles from the evaporator system. Also, steam quality can cause erroneous readings of steam rates during the evaluation of the system and can indicate poorer efficiencies than are actually experienced. Air leaks will overload the noncondensible venting system; this reduces the vacuum capability of both mechanical pumps and steam jets and increases the system pressures. In every case an increase in pressure will reduce the ΔT. The ΔT loss is overcome by increasing the steam pressure and rate to maintain capacity.

Fouling: Fouling of the evaporator heat-transfer surfaces on the process or condensing vapor side decreases heat-transfer rates and increases pressure drops. Reduced heat-transfer rates cause loss of capacity that can be overcome only at the expense of higher steam rates at higher pressure. This increased pressure causes poorer steam economy by increasing condensate heat losses. Systems exhibiting a high fouling frequency also will require additional energy to "boil out" scale from the heat-transfer surface. A higher constant power requirement for auxiliary operation per unit of product also occurs as the capacity drops.

Separator Efficiencies: If vapor-liquid separation is poor, the product will be carried over into the condensate of the following effect. If solids are carried up into the wire mesh of demister-type separators, a high pressure drop and loss in ΔT result. Recovered product yield and the capacity are then reduced, and the pollution load increases. The reduction in yield wastes all the energy previously expended to produce the lost product and increases the cost of raw material. If scaling materials are entrained, fouling is accelerated.

Heat Losses: Heat losses to the surroundings are a common source of reduced efficiency in energy use. These losses can be decreased by further insulating the system or by repairing existing insulation. The effect of the losses show up as a reduction of intermediate vapor and liquor temperatures and separator or vapor-line reflux. In operation this problem directly affects steam consumption and is overcome by increasing steam rates and pressure in the first effect.

Improper Cleaning: Cleaning may be required to minimize the effects of fouling, scaling, or salting. If proper cleaning procedures are not followed, the system may be placed back in service in only a partially-cleaned condition with resultant loss of efficiency. Cleaning cycle time needs to be determined and the correct intervals observed between appropriate cleaning cycles. The cleaning technique should not introduce water into the system which is not removed in order to avoid resultant lowered dfficiency or capacity.

Modifying Auxiliary Hardware

These hardware modifications normally can be applied without changing the basic flow scheme or operation of the existing evaporator, but they require some capital expenditures. The most important of these are:

(1) Heat recovery

(2) Condensate reuse

(3) Improved instrumentation and control.

Heat Recovery: An evaporator system normally loses some energy through the hot concentrated product, condensates from the evaporator steam chests, final-effect vapor, or steam-jet exhaust streams. The decision on whether to use heat recovery is usually made by a straightforward balancing of energy cost savings and the capital and increasing operation costs for the heat-exchange equipment.

The arrangement of the optimum recovery scheme requires the investigation of several options. Most heat-recovery applications will result in higher feed temperature. The amount of thermal energy required in a single-effect or multi-effect evaporator to raise the feed to its boiling point is therefore reduced. The material balance, vaporization rates, liquid concentrations, heat-transfer efficiencies, and temperature differentials remain unchanged. The steam rate to the first effect is decreased since less energy is required to bring the cold feed to its initial boiling point.

A good heat-recovery scheme will allow the recycling low-availability waste heat to replace a portion of valuable high-energy steam. Any areas in the existing system which reduce temperature or condenses vapor with cooling water, such as barometric condensers and product coolers, have heat-recovery potentials. The proper arrangements of the feed-liquid sequence can sometimes enhance heat recovery with additional heat exchangers.

Condensate Reuse: Steam can be generated by flashing a high-pressure condensate to a lower pressure to preheat feed, to supply steam to other effects, or to be used in another section of the plant. Flash tanks are used for this purpose. The applicability of this approach depends primarily on the existing condensate return conditions to the steam boiler and on an economic comparison with heat-recovery methods described earlier.

Instrumentation and Control: Controllers reduce the problems associated with operational errors and maintain process conditions for optimum performance. An evaporator must be instrumented adequately to provide the data necessary for evaluating the system's performance and to provide indicators for identifying problems affecting thermal efficiencies. Pressure, temperature, and flow indicators and possibly alarms must be incorporated at critical areas, such the first-stage condensate discharge line and evaporator vapor lines.

Major Hardware Modifications

The techniques discussed earlier for upgrading evaporators do not involve significant retrofitting or operating changes to the existing evaporator system. The temperature and pressures within each effect remain nearly constant. For significant reductions in steam consumption, more extensive modifications usually must be made, and these technologies will change the process conditions. The selections of the optimum approach for these methods becomes somewhat more complex. The options available for radically reducing the steam consumption involve either the compression of the overhead vapors for reuse or the use of these vapors in additional effects. Often a combination of these two approaches becomes the best option to minimize capital costs. Frequently, the method may be selected because the space in the plant available for the modifications is limited or because retrofitting costs are much higher than the cost of the hardware. Therefore, the selection of the major hardware modifications must include an evaluation of the changes in the process conditions and a determination of both the retrofit costs and feasibility.

Vapor Compression: The vapor exiting the evaporator contains significant heating value as latent heat of vaporization. This energy can be recovered by heat-recovery techniques or multi-effect evaporation. The vapor can also be reused in the same unit if the temperature and pressure of the vapor stream are increased to the conditions existing at the time in the steam chests. The methods for reusing a fraction of this vapor are:

(1) thermal compression

(2) mechanical compression

Thermal Compression — Of the two vapor-compression methods, thermal compression requires less capital but yields lower heat recovery than does mechanical compression. A steam-jet booster is used to compress a fraction of the vapor leaving the evaporator so that the pressure and temperature are raised. Thermal compression is normally applied to the first effect on existing evaporators or where the conditions are right for the application in single-effect units.

The thermal energy savings depend on the available motive-steam pressure and the compression ratio. From steam-jet performance curves, the ratio of suction vapor to motive steam flow rates is obtained and matched with the total steam requirements for the evaporator. The difference in the low rates between the existing evaporator steam without thermal compression and motive steam is the steam savings. The motive steam rate is the new steam consumption. The condenser water rate is also reduced because the vapor rate to the condenser is decreased.

A steam-jet thermocompressor is employed when low-pressure steam is used for the evaporation and the available steam pressure is high. The jet compressor essentially replaces the existing pressure regulator or reducing valve that controls the steam pressure (and temperature) for evaporation and can be applied easily to single-effect and multi-effect units under the right conditions.

The variability in energy savings and capital costs still depends on the existing evaporator conditions. For existing multi-effect units in which the total amount of first-effect vapor is now used in the second effect, the choice and location of the steam-jet thermocompressor are more complicated because the heat balances and heat-transfer rates are affected for each evaporator reaction. The economics may still be favorable to justify the technical evaluation and modification. Since the limitations of steam-jet thermocompressors are a compression ratio below 1.8 and new heat-transfer rates, the first effect normally becomes the best location.

Mechanical Compression—In most mechanical compression operations, all the vapor leaving the evaporator is compressed The compressors include positive-displacement, axial-flow, or radial flow (centrifugal) types that can be driven by a electric motor, steam turbine, or gas engine. Under the proper conditions, the mechanical energy required by the compressor is low compared to the recoverable energy in the vapor. Under optimum conditions, the equivalent of up to 20 effects can be achieved by mechanical compressions. This high energy recovery must be balanced against the capital, maintenance, and electrical-energy costs for the compressor. The vapor leaving the compressor is superheated, but this usually has no effect on heat transfer rates.

If high-pressure steam is available and a demand exists for low-pressure exhaust steam, as is the normal case for multi-effect units, a steam-turbine-driven compressor drastically reduces the operating costs of the compressor as compared to those for electrically driven units. In this case, the turbine exhaust can be passed on to the next effect of an existing multi-effect unit. If electrical costs are low, the choice based on operating costs would be an electrically driven compressor. The total energy consumption is nearly identical with either drive.

The use of mechanical compression becomes more costly as higher compression ratios are required because both the capital costs and the operating costs increase rapidly. Since a vapor-compression evaporator normally operates with a low ΔT between the condensing compressed vapor and the boiling liquid in the vapor body, a large heat-transfer surface is required. Most older evaporators, particularly single-effect units, do not have sufficient heat-transfer surface to allow this modification unless the existing compression ratio (steam chest to vapor pressure) is below 1.5 to 2.0. The existing vapor capacity of the unit is also a prime factor to consider in evaluating the economic benefits of mechanical compressors.

The need for expensive construction materials also will limit the application of mechanical compression. A large rise in boiling point, corrosive vapor, or a tendency by the liquor to foam also work against vapor compression. For concentration of dilute streams, the initial preconcentration step is a likely candidate for mechanical compression because boiling point rise is usually low in this first stage. If boiling point rise is not significant, a multi-effect evaporator converted to parallel operation with mechanical compression also becomes a possibility.

The thorough evaluation of vapor-compression options is an important part of any upgrading program.

Installing Additional Effects: When most existing evaporation systems were selected, one of the major design evaluations was determining the optimum number of effects. Much of the data may still be available, and these can be updated and used. The initial choice was probably based on economics and capital availability and in some cases, the physical size of the system also may have influenced the decision. The addition of effects sometimes can be used to increase production capacity as well as to reduce energy consumption, and thus there is a dual purpose for studying this option.

The addition of effects also eventually runs into economic limitations, even at high energy costs. If the initial steam pressures and final concentrator pressure are set by other processing considerations, the allowable ΔT between condensing vapor and boiling liquid decreases in proportion to the number of effects. As the average ΔT decreases, the surface and therefore capital cost increases. This economic balance then must be met.

The best location within the evaporation scheme for adding effects will depend on the capacity objectives, the existing heat-transfer surfaces, and the projected heat-transfer surface required for the additional effects. When the existing evaporator heat-transfer surfaces are equal, the location of the additional effects is still important. When the evaporator surfaces are not equal or when the boiling point and viscosities of solution are important factors in design, the projected heat transfer coefficients for each effect in the revised system must be calculated according to the new concentration gradients and temperature differentials. Except in cases of heat-sensitive materials or physical-space restrictions, the additional stages can be designed into the system from the beginning to the end of the existing layout. The consideration of additional effects may require extensive calculations to determine the optimum system.

Combinations: A combination of additional effects and vapor compression often will give a optimum configuration. Thermal compression and one additional effect, for example, can be combined to achieve energy savings nearly equivalent to the addition of two effects but at a much lower capital cost. The usefulness of combination schemes depends on the properties of the liquid being concentrated, the expected energy costs, and the level of energy available at the plant site.

ECONOMIC EFFECTS OF IMPROVEMENTS

The actual energy savings and operating cost reductions will depend on the operating conditions of the existing evaporator, as well as on the upgrading techniques. Each upgrading technique also has a range of energy-savings potentials. Some typical energy savings that can normally be achieved are shown below.

Typical Energy Savings

Method	Capital Requirement	Achievable Energy Savings
Venting and thermal insulation	low	5%
Improved maintenance	low	5%
Heat-recovery exchangers	low	10%
Condensate reuse	low	5%
Thermal compression	medium	45%
Mechanical compression	high	70-90%
Additional effects*	high	$[1-(\frac{N}{N+n})]$ 100%

*N = original number of effects
n = number of added effects

GUIDELINES FOR UPGRADING PROGRAM

It is important to realize at the beginning that it is more difficult and more time consuming to upgrade existing evaporators than to install new evaporators in new plants. Quite often the retrofit construction and plant modifications may result in production loss, with the prospect of further losses if the construction schedule slips.

Evaluating Existng Evaporators

The first and most important part of any upgrading program for energy conservation is a thorough evaluation of the existing equipment. There are two important reasons for this:

(1) to determine the problem areas, if any, and to provide a base point from which to measure potential savings. Very often this step will discover savings that can be achieved by minor changes in operating procedures and can be implemented readily.

(2) to provide operating date, which will be of great importance in later phases of the upgrading program.

The evaluation should include heat and material balances and should examine major consumers of the thermal energy outside the evaporator area. The most efficient heat-recovery system is usually one that can be used as a heat sink to receive waste heat, and the best location for this may be in another area which can readily receive a high-temperature stream for the evaporator.

A thorough analysis of the plant will:

(1) determine the best operating procedures and conditions for the existing plant.
(2) identify the areas for which corrections are needed.
(3) point out short-term savings and cost reductions.
(4) develop a firm economic and technical basis for planning the program and for selecting system modifications.

Outside assistance may be justified for conducting the upgrading program.

Determining Appropriate Methods

It is generally better to evaluate each upgrading method separately and then to determine how the individual techniques could be combined obtain the maximum savings and return on investment.

Economic Analysis

The final steps in the program start with a detailed economic analysis of each feasible method. The installed cost for each alternative must be established and compared to the resultant advantages as a result of savings is energy.

Implementation

After the economic analysis is completed, the decision is made concerning the approach to use. Then the detailed engineering and design of the installation begins. Construction schedules are developed in order to coordinate completion of any necessary plant modifications. Proper scheduling will reduce the risks of additional costs associated with interruption of production. After the new equipment has been installed and is operating, the performance should be evaluated to determine the new process conditions and actual savings. Periodic performance evaluations will insure that the system is operating at the optimum conditions and maximum reductions in energy consumption are being acheived.

41
Energy Conservation

The designer of an evaporator system has the responsibility of ensuring that each system is as energy-efficient as he can make it, subject to the budgetary, scheduling, and manpower constraints under which he must work. The following questions must be addressed:

(1) Have the energy consuming, heat recovery, and other energy conservation facilities been economically optimized for energy cost appropriate for the years in which the system will be operational?

(2) Have these appropriate values for energy been communicated to those who need them for proper design of energy consuming equipment and facilities?

(3) What energy saving/conservation measures have been taken to reduce energy usage? Is there any new or improved technology that reduces energy usage?

(4) What will be the energy required per unit product? How does this compare with existing operations or with those of other producers? Significant differences should be investigated.

42
Specifying Evaporators

Evaporator systems are major pieces of process equipment and are often purchased on a total responsibility basis. This is especially true of vapor compression and highly heat integrated systems. Specific design information and fabrication are often proprietary to vendors. Evaporator manufacturers generally are rather specialized. Few offer a complete range of evaporator types; some specialize in one type only.

The purchaser's task often is to define the process and mechanical limitations accurately to enable vendors to engineer economical systems and offer fair process guarantees. Accurate costs for various energy forms must be established. Overall plant energy balances must be determined and any limitations described to the vendor. Performance guarantees come in a variety of forms and may not be advantageous to the purchaser.

Pilot tests may be required to evaluate new processes or products. Most evaporator manufacturers maintain test facilities and the purchser is encouraged to avail himself of this opportunity, especially if little experience is available for his specific product. Often a vendor's broad experience with all types of evaporators may contribute as much to the solution of new problems as intensive experience in the application of a specific design. Pilot tests may need to be treated confidentially and should not be used indiscriminately.

Many intangible factors may influence the selection of evaporator fabricator; vendor experience with the process or product, his reputation and responsibility, value of the offered guarantee, prior experience with a vendor, confidence in the vendor. These factors are more difficult to evaluate.

The process engineer should carefully evaluate the total system proposed by any vendor. Heat transfer rates, steam economy and frequency of cleaning are the obvious factors to be evaluated. But other, smaller details are often almost as important. Minor equipment must be given proper engineering. Mechanical equipment, instrumentation, condenser type, and piping must all be given careful evaluation.

Vendors should be given an opportunity to make the best proposal. However, if certain features will not be acceptable, these should be specified. Such action should only be taken for valid reasons; not having prior operating experience with a particular evaporator type is not necessarily a valid reason.

In many cases, the purchaser may elect to design his own evaporation system. There are many advantages to this approach; the primary advantage is that the purchaser is in a better position to evaluate tradeoffs necessary in design. The purchaser is certainly more familiar with his requirements than can often be translated into a set of detailed specifications.

COMPARING VENDORS' OFFERINGS

All equipment vendors strive to provide equipment to meet specifications. However, all vendors employ human beings who make mistakes just as do purchasers and operators of evaporator systems. Few vendors can offer every type of evaporator; consequently his proposal may reflect the types he does provide or the types he has the most confidence with. In many cases, the proposed equipment may have been selected with an eye on initial investment only; specifications should specifically spell out any penalties to be assessed for energy usuage and the manner in which they will be applied in any evaluation.

In many cases several evaporator types may be adequate for the intended service. The task then becomes to weigh all factors specified to arrive at the proper selection. Following is a procedure to assist in such evaluations.

(1) Compare cost
(2) Check to see that specifications have been met. If not, what specific exceptions have been raised?
   (a) product quality
   (b) capacity
   (c) residence time
   (d) cooling water temperatures
   (e) condenser type
   (f) steam levels
   (g) materials of construction
   (h) type of vacuum system
   (i) safety considerations
   (j) any special requirements.
(3) Compare operating costs
   (a) steam economy
   (b) pumping power
   (c) water usuage
   (d) effective steam economy for mechanical compressors
   (e) vent rates
   (f) operating personnel required
   (g) probability of upsets
(4) Maintenance considerations
   (a) velocities

(b) temperature levels
(c) controls
(d) type vacuum system
(e) type of entrainment separator
(f) once-through or recirculated
(g) natural or forced circulation
(h) condenser type
(i) spare parts required
(j) corrosion

(5) Check engineering
   (a) heat transfer
   (b) pressure drops
   (c) line sizes
   (d) vapor body sizes
   (e) pumping rates
   (f) operating parameters
   (g) control systems
   (h) surface temperatures
   (i) steam trap sizes
   (j) heat losses
   (k) compression ratios
   (l) safety requirements

(6) Compare guarantees

(7) Service after sale
   (a) start up assistance
   (b) spare parts availability
   (c) engineering expertise
   (d) test facilities
   (e) personnel training

(8) Previous experience with vendors
   (a) good or bad?
   (b) response to problems
   (c) willingness to help
   (d) confidence level

(9) Intangibles
   (a) is system similar to existing one?
   (b) will extensive operator training be required?
   (c) vendor reputation

Information required to make the evaluation above should be provided with the vendors' proposals, and specifications should make plain that such information is required. However, this information is to be treated discriminately. Ethics should preclude revealing one vendor's design to a competitor. Consequently, engineering by the purchaser prior to receiving any proposals will enable a more effective evaluation to be made.

43
New Technology

Evaporation is one of the oldest industrial technologies, and it has undergone an extensive evolution. The high cost of energy, and the anticipation of further price increases, can be expected to lead to additional technical developments and application to industrial evaporation processes.

One of the major new developments that can be expected is the use of more sophisiticated design technology to provide evaporator configuration for optimum energy use. Recent improvements in computer technology now allow the designer to investigate a multitude of options. Advanced initial techniques are expected to be applied to evaporators.

Another major area in which improvements are certain is in the application of enhanced heat transfer surfaces. Enhancement will enable more efficient operation and also permit the use of lower grade heat. Consequently, streams at temperatures not normally considered usable will be used to provide energy for evaporation.

There are other methods for concentrating solutions with a theoretical energy use lower than that of evaporation, and much funding has been devoted to developing these technologies. Application of some of these will result in future energy savings.

Fouling will become better understood. As this happens, improvements in operation and reduced maintenance will be realized. Cleaning procedures will also become more efficient.

Evaporator systems will be combined with other unit operations such as distillation or extraction. The Carver-Greenfield process, which uses a carrier oil to permit a product to be evaporated to nearly complete moisture removal, will find wider application. Ion-exchange will be integrated with evaporation.

The following trends in evaporator design can be expected:

(1) Evaporation applications that, in the past, used single-effect designs because of low capacity or the need for expensive construction materials will increasingly use vapor compression as a means of

improving efficiency. As a minimum, thermal compression will be used.

(2) An increased number of effects is now economical for multi-effect evaporators. Each evaporator system design will be analyzed more closely to define the most economical number of effects.

(3) Extensive heat exchange between outgoing streams and the incoming feed will be used. Gasketed plate heat exchangers will be used increasingly for this application.

(4) New evaporators will be equipped with the instrumentation and measurementt devices necessary to allow the operators to monitor closely the performance of operating evaporators and to minimize energy use.

(5) Increased automated and computerized control will be used to maintain optimum operation.

(6) There will be less use of evaporation schemes that inhibit the recovery of the latent heat energy of the vaporized water. For example, submerged combustion evaporation will be used only when absolutely necessary.

(7) Mechanical compression, often combined with multi-effects, will gain increasing application for solutions with low boiling point rises.

(8) Each evaporator system will be designed to reduce energy loss.

(9) More attention will be given to the effects of time/temperature on product quality.

44
Nomenclature

a	=	wave speed
a_c	=	crossflow flow area
a_w	=	window area flow area
A	=	heat transfer surface
A_{cc}	=	acceleration loss
A_d	=	dimensionless number defined by Dukler
A_i	=	inside surface of tube
A_t	=	total outside surface of fintube
B	=	factor in Equation 8.40
BHP	=	brake horsepower
c	=	specific heat
C	=	concentration, weight percent
C_{sf}	=	constant in Equation 7.50
d_e	=	effective diameter inches
d_i	=	inside tube diameter, inches
d_{io}	=	tube inside diameter under fin, inches
d_o	=	outside tube diameter, inches
d_o''	=	outside tube diameter, inches
d_r''	=	root diameter, inches
D	=	diameter

D_c = coil diameter
D_e = equivalent diameter, feet
D_g = gap between tubes (Equation 8.28)
D_i = inside diameter
D_j = diameter of agitated vessel
D_o = outside diameter
D_r = root diameter
D_s = shell diameter
D_s'' = shell diameter, inches

E = modulus of elasticity, psi
EFF = efficiency
E_w = weighted fin efficiency

f = friction factor
f_a = acoustic frequency, Hertz
f_c = friction factor for coils
f_n = tube natural frequency, Hertz
f_{fb} = turbulent buffeting frequency, Hertz
f_{vs} = vortex shedding frequency, Hertz
F = fintube condensing heat transfer coefficient correction factor
F = experimentally determined constant for Equation 14.2
F_b = bypass correction factor for shellside heat transfer
F_c = crossflow correction factor for shellside heat transfer
F_c = sensible heat transfer correction for convection boiling
FDA = free dry air equivalent
F_l = leakage correction factor for shellside heat transfer
F_r = flow regime correction factor for shellside heat transfer
F_s = correction factor, Equation 8.16
FV = face velocity, feet per minute

g = gravitional acceleration
g_c = gravitational constant
G = mass velocity
G_G = gas mass velocity
G_L = liquid mass velocity

Nomenclature

G_T = total mass velocity

h = heat transfer coefficient
h_a = air side heat transfer coefficient, Btu/(hr)sq ft) (°F)
h_A = heat transfer coefficient for fluid A
h_B = heat transfer coefficient for fluid B
h_B = boiling heat transfer coefficient
h_c = condensing heat transfer coefficient, stratified flow
h_c = convective heat transfer coefficient
h_{cg} = combined condensing-cooling coefficient
h_f = apparent fouling coefficient
h_{fi} = effective inside heat transfer coefficient for fintube
h_g = gas cooling sensible heat transfer coefficient
h_h = heat transfer coefficient for helical coil
h_L = heat transfer coefficient at length L
h_{LO} = sensible heat transfer coefficient for Equation 7.29
h_o = outside heat transfer coefficient for tube banks
h_o = condensing heat transfer coefficient for horizontal tube
h_s = heat transfer coefficient for straight tube
h_s = heat transfer coefficient for vapor shear condensing
h_2 = heat transfer coefficient (Equation 7.57)
H = total head
H = height of liquid
HD = head, feet of fluid

k = thermal conductivity
K = constant

l = brink depth
L = tube length
L = diameter of agitator
L_e = equivalent length, feet
L_o = straight length of U-bend, feet

m = mode number, integer
M = molecular weight
M_L = molecular weight of load gas

M_m = molecular weight of motive gas

n = number of tubes
N = rotational speed
N_c = number of condensers
N_b = number of baffles
N_I = interstage number
N_p = number of passes
N_r = number of vertical rows
N_s = total number of stages
N_w = number of tube rows in baffle window

p_s = vapor pressure at Ts
p_w = vapor pressure at T_w
P_a = vapor pressure at actual conditions
P_b = baffle spacing
P_b'' = baffle spacing, inches
P_D = discharge pressure
P_I = interstage pressure
P_L = load gas pressure
P_m = motive gas pressure
P_s = suction pressure
P_{st} = static pressure, inches of water column
P_{60} = vapor pressure at 60°F
P = pressure

ΔP = pressure drop, psi
ΔP_a = air side presure drop, inches of water column
ΔP_c = condensing pressure drop
ΔP_e = entrance, exit pressure drop, psi
ΔP_f = frictional pressure drop, psi
ΔP_{fi} = frictional pressure drop for inlet and outlet baffles, psi
ΔP_G = gas pressure drop
ΔP_{GO} = pressure drop for total mass flowing as gas
ΔP_L = liquid pressure drop
ΔP_{LO} = pressure drop for total mass flowing as liquid

ΔP_n	=	nozzle pressure drop, psi
ΔP_t	=	total pressure drop, psi
ΔP_{TP}	=	two-phase flow pressure drop
ΔP_w	=	window pressure drop, psi
ΔP_1	=	pressure drop based on inlet conditions
Q	=	heat load
Q	=	volumetric flow rate
Q_F	=	volumetric flow rate, gallons per minute
Q_g	=	gas sensible heat load
Q_G	=	gas volumetric flow rate
Q_L	=	liquid volumetric flow rate
Q_T	=	total heat load
r	=	ratio of holding volume to feed volumetric rate
r_o	=	outside fouling factor
R	=	gas constant
R	=	compression ratio
R	=	bend radius, feet
R_b	=	bypass correct factor for pressure drop
Re	=	Reynolds number
Re_c	=	critical Reynolds number for turbulent flow
Re_1	=	liquid Reynolds number
Re_{ns}	=	no-slip Reynolds number
R_g	=	volume fraction gas
R_l	=	leakage correction factor for pressure drop
R_m	=	ratio of motive to load gas
s_l	=	specific gravity of liquid
s_t''	=	tube center-to-center spacing, inches
S	=	velocity of sound
S_c	=	nucleate boiling suppression factor
S_l	=	tube center-to-center spacing parallel to flow
S_n	=	axial stess, psi
S_t	=	tube center-to-center spacing perpendicular to flow
St	=	Strouhal number

t	=	time
T	=	temperature
T_s	=	saturation temperature
T_w	=	wall or surface temperature
ΔT	=	temperature difference
U	=	overall heat transfer coefficient
v	=	velocity, feet per second
v_c	=	crossflow velocity, feet per second
v_{gf}	=	flooding velocity, feet per second
v_l	=	liquid velocity
v_m	=	maximum vapor velocity, feet per second
v_n	=	nozzle velocity, feet per second
v_{ns}	=	no-slip velocity, feet per second
v_t	=	tube velocity, feet per second
v_v	=	vapor velocity
v_w	=	velocity in baffle window, feet per second
v_1	=	inlet velocity
v_2	=	outlet velocity
V_a	=	actual volumetric flow
V_A	=	volume fraction of liquid A
V_{60}	=	volumetric flow at $60°F$
W	=	mass flow rate
W_l	=	mass flow rate of liquid condensed
W_s	=	mass flow rate of steam, pound per hour
W_v	=	mass flow rate of vapor
x	=	mass fraction
x	=	fraction of feed removed
X_{tt}	=	two-phase parameter
Y	=	characteristic length
z	=	compressibility factor
Z	=	elevation
ΔZ	=	change in elevation

GREEK

α	=	angle from horizontal, degrees
β	=	coefficient of expansion
Γ	=	tube loading
Γ_{min}	=	minimum tube loading, pound per hour per foot
Γ_T	=	terminal tube loading, pound per hour per foot
δ	=	log decrament
θ	=	time
γ	=	specific heat ratio
σ	=	surface tension
σ_w	=	surface tension of water
ρ	=	density
λ	=	latent heat of vaporization
λ_L	=	volume fraction liquid
μ	=	viscosity
μ_b	=	bulk viscosity
μ_w	=	wall viscosity
ϕ	=	viscosity correction factor
ϕ	=	surface condition factor

SUBSCRIPTS

a	=	average
l	=	liquid
L	=	liquid
G	=	gas
ns	=	no-slip
s	=	shellside
o	=	initial condition
v	=	vapor
1	=	upstream or inlet
2	=	downstream or outlet

Bibliography

EVAPORATION

(1) Standiford, F. C., "Evaporation", *Chemical Engineering,* Dec. 9, 1963, pp. 157-176.

(2) Coates, J., and Pressburg, B. A., "How Heat Transfer Occurs In Evaporators", *Chemical Engineering,* Feb. 22, 1960, pp. 139-144.

(3) Klaren, D. G. and Holberg, N., "Development of a Multi-Stage Flash/Fluidized Bed Evaporator", *Chemical Engineering Progress,* July 1980, pp. 41-43.

(4) Peak, W. E., "Desalting Seawater by Flash Evaporation", *Chemical Engineering Progress,* July, 1980, pp. 50-53.

(5) Caruana, G., "A Review of Evaporators and Their Application", *British Chemical Engineering,* July 1965, pp. 466-476.

(6) Baker, R. A., "The Flash Evaporator", *Chemical Engineering Progress,* June 1963, pp. 80-83.

(7) "Evaporator Has No Heat Transfer Surface", *Chemical Engineering,* Aug. 15, 1966, p. 180.

(8) Yundt, B., Rinesmith, R., "Horizontal Stray-Film Evaporator", *Chemical Engineering Progress,* Sept. 1981, pp. 69-73.

(9) Farin, W. G., "Low-Cost Evaporation Method Saves Energy By Reusing Heat", *Chemical Engineering,* March 1, 1976, pp. 100-106.

(10) Rosenblad, A. E., "Evaporator Systems For Black Liquor Concentration", *Chemical Engineering Progress,* April, 1976, pp. 53-60.

(11) Rosenblad, A. E., "Selection of Evaporation Cycle", *TAPPI Journal,* Sept. 1973, pp. 86-88.

(12) Fosberg, T. M. and Claussen, H. L., "Falling-Film Evaporators Recover Chemicals Effectively", *TAPPI Journal,* Aug. 1982, pp. 63-66.

(13) Bennett, R. C. and Fakasseles, "The Elbow Separator Evaporator", *Chemical Engineering Progress,* Nov. 1980, pp. 64-67.

(14) Rogers, A. N., et. al., "Treatment of Cooling Tower Blowdown", *Chemical Engineering Progress,* July 1981, pp. 31-38.

(15) Sephton, H. H., "Vertical Tube Foam Evaporator", *Chemical Engineering Progress,* Oct. 1981, pp. 83-86.

(16) Renshaw, T. A., Sopakie, S. F., and Hanson, M. C., "Concentraton Economics In The Food Industry", *Chemical Engineering Progress,* May 1982, pp. 33-40.

(17) Freese, H. L and Glover, W. B., "Mechanically Agitated Thin-Film Evpaporators", *Chemical Engineering Progress,* Jan. 1979, pp. 52-58.

(18) Coates, J. and Pressburg, B. S., "Multiple-Effect Evaporators", *Chemical Engineering,* March 21, 1960, pp. 157-180.

(19) Radovic, L. R. et al., "Computer Design and Analysis of Operation of a Multiple-Effect Evaporation System In the Sugar Industry", *Industrial Engineeing Chemistry Process,* Vol. 18, No. 2, 1979, pp. 318-323.

(20) Kleinman, G., "Double Effect Evaporation of Crude Phosphoric Acid", *Chemical Engineering Progress,* Nov. 1979, pp. 37-40.

(21) Sandfort, P., "Energy Saving Techniques For Liquid Food Concentration", AIChE Paper No. 32a, AIChE Winter Meeting, Orlando, March 1, 1982.

(22) Moore, J. G. and Hesler, W. E., "Evaporation of Heat Sensitive Materials", *Chemical Engineering Progress,* Feb. 1963, pp. 87-92.

(23) Moore, J. G. and Pinkel, E. G., "When to Use Single Pass Evaporators", *Chemical Engineering Progress,* July 1968, pp. 39-44.

(24) Dedert, W. G. and Moore, J. G., "New Trends In Evaporation", *Industrial and Engineering Chemistry,* June 1963, pp. 57-62.

(25) Perry, R. H. and Chilton, C. H., "Chemical Engineers' Handbook", 5th Edition, F. C. Standiford (author), "Evaporation", Section 11-27, McGraw-Hill Book Co., New York (1973).

HEAT TRANSFER

(1) McAdams, W. H., "Heat Transmission", 3rd Edition, McGraw-Hill, New York (1954).

(2) Kern, D. Q., "Process Heat Transfer", McGraw-Hill, New York (1950).

(3) Kreith, F., "Principles of Heat Transfer", 3rd edition, Intext Educational Publishers, New York (1973).

(4) McCabe, W. L. and Smith, J. C., "Unit Operations of Chemical Engineering", 2nd edition "Evaporation Section", pp. 439-477, McGraw-Hill, New York (1967).

(5) Perry, J. H. (editor), "Chemical Engineers Handbook" 4th edition, C. H. Gilmour (author), "Heat Transmission", Section 10, McGraw-Hill, New York (1963).

(6) Perry, R. H. and Chilton, C. H., "Chemical Engineers Handbook 5th edition, pp. 10-22, McGraw-Hill, New York (1973).

(7) Minton, P. E. Lord, R. C., Slusser, R. P., "Design of Heat Exchangers", *Chemical Engineering,* Jan. 23, 1970, pp. 76-118.

(8) Minton, P. E., Lord R. C., Slusser, R. P., "Design Parameters For Condensers and Reboilers", *Chemical Engineering,* March 23, 1970, pp. 127-134.

(9) Minton, P. E., Lord, R. C., Slusser, R. P., "Guide to Trouble Free Heat Exchangers", *Chemical Engineering,* June 1, 1970, pp. 153-160.

(10) Minton, P. E., "Design Spiral-Plate Heat Exchangers", *Chemical Engineering,* May 4, 1970, pp. 103-112.

(11) Minton, P. E., "Designing Spiral-Tube Heat Exchangers", *Chemical Engineering,* May 18, 1970, pp. 145-152.

(12) Butterworth, D. "Introduction to Heat Transfer", Oxford University Press, London (1977).
(13) Bergles, A. E. et al., "Two Phase Flow and Heat Transfer In The Power and Process Industries", McGraw-Hill Book Co., New York (1981).
(14) "Advances In Enhanced Heat Transfer", 1981, ASME HTD- Vol. 18, American Society Of Mechanical Engineers, New York.
(15) LoPinto, L., "Fog Formation In Low-Temperature Condensers", *Chemical Engineering,* May 17, 1982, pp. 111-113.
(16) Foxall, D. H., and Chappell, H. R., "Superheated Vapor Condensation in Heat Exchanger Design", *Chemical Engineering,* Dec. 29, 1980, pp. 41-50.

BOILING

(1) Collier, J. G., "Convective Boiling and Condensation", 2nd edition, McGraw-Hill Internation Book Co., New York (1981).
(2) Brisbane, T. W. C., Grant, I. D. C., Whalley, P B., "A Prediction Method for Kettle Reboiler Performance", Paper 80-HT-42, American Society of Mechanical Engineers, New York.
(3) Palen, J. W., Shih, C. C., Taborek, J., "Mist Flow in Thermosiphon Reboilers", *Chemical Engineering Progress,* July, 1982, pp. 59-61.
(4) Westwater, J. W., "Nucleate Pool Boiling", *Petro/Chem Engineer,* Sept. 1961, pp. 53-60.
(5) Johnson, D. L. and Yukawa, Y., "Verical Thermosiphon Reboilers in Vacuum Service", *Chemical Engineering Progress,* July, 1979, pp. 47-52.
(6) Fair, J. R., "What You Need to Design Thermosiphon Reboilers", *Petroleum Refiner,* Feb. 1960, pp. 105-123.
(7) Collins, G. K., "Horizontal-Thermosiphon-Reboiler Design", *Chemical Engineering,* July 19, 1976, pp. 149-152.
(8) Webb, R. L., "The Evolution Of Enhanced Surface Geometries For Nucleate Boiling", *Heat Transfer Engineering,* Jan.-June, 1981, pp. 46-69.
(9) Webb, R. L., "Nucleate Boiling On Porous Coated Surfaces", *Heat Transfer Engineering,* July-Dec., 1983, pp. 71-82.

HEAT EXCHANGERS

(1) Scaccia, C. and Theoclitus, G., "Heat Exchangers: Types, Performance, and Application", *Chemical Engineering,* Oct. 6, 1980, pp. 121-132.
(2) Devore, A., Vargo, G. J., Picozzi, G. J., "Heat Exchangers: Specifying and Selecting", *Chemical Engineering,* Oct. 6, 1980, pp. 133-148.
(3) Fanarites, J. P., and Benevino, J. W., "Designing Shell-and-Tube Heat Exchangers", *Chemical Engineering,* July 5, 1976, pp. 62-71.
(4) Howarth, F., et al., "Symposium on Special Heat Exchange Equipment", *The Chemical Engineer,* June 1962, pp. A83-A114.
(5) Raju, K. S. N. and Chand, J., "Consider the Plate Exchanger", *Chemical Engineering,* Aug. 11, 1980, pp. 133-144.
(6) Patel, R. K., Shende, B. W., and Ghosh, P. K., "Designing a Helical- Coil Heat Exchanger", *Chemical Engineering,* Dec. 13, 1982, pp. 85-88.

(7) Shah, R. K., "What's New In Heat Exchanger Design?", *Mechanical Engineering,* May 1984, pp. 50-59.

FLOW-INDUCED VIBRATION

(1) "Flow Induced Heat Exchanger Tube Vibration- 1980", (G00182), HTD-Vol. 4, American Society of Mechanical Engineers, New York.
(2) Schwarz, G. W., Jr., "Preventing Vibration In Shell-and-Tube Heat Exchangers", *Chemical Engineering,* July 19, 1976, pp. 134-140.
(3) Barrington, E. A., "Acoustic Vibrations In Tubular Exchangers", *Chemical Engineering Progress,* July 1973, pp. 62-68.
(4) Chenoweth, J. M. and Kissell, J. H., "Flow Induced Vibration In Shell-and-Tube Heat Exchangers". AIChE Today Series, American Institute of Chemical Engineers, New York.
(5) Blevins, R. D., "Flow-Induced Vibration", Van Nostrand Reinhold Co., New York (1977).

FOULING

(1) Somerscales, E. G. C. and Knudsen, J. G., "Fouling of Heat Transfer Equipment", Hemisphere Publishing Co., Washington (1978).
(2) Helzner, A. E., "Operating Performance of Steam-Heated Reboiler", *Chemical Engineering,* Feb. 14, 1977, pp. 73-76.
(3) "Fouling In Heat Transfer Equipment", ASME-HTD-Vol. 17, American Society of Mechanical Engineers, New York (1981).
(4) Knudsen, J. G., "Fouling of Heat Exchangers: Are We Solving The Problem?", *Chemical Engineering Progress,* Feb. 1984, pp. 63-69.

DIRECT CONTACT HEAT TRANSFER

(1) Vener, R. E., "Liquid-Gas Contacting", *Chemical Engineering,* Aug. 1956, pp. 175-206.
(2) How, H., "How to Design Barometric Condensers", *Chemical Engineering,* Feb. 1956, pp. 174-182.
(3) Huckaba, C. E., Master, N., Santoleri, J J., "Performance of Novel Sub-X Heat Exchanger", *Chemical Engineering Progress,* July, 1967, pp. 74-80.
(4) Cronan, C. S., "Submerged Combustion Flares Anew", *Chemical Engineering,* Feb. 1956, pp. 163-167.
(5) Pick, A. E., "Consider Direct Steam Injection For Heating Liquids", *Chemical Engineering,* June 28, 1982, pp. 87-89.
(6) Tate, R. W., "Sprays and Spraying For Process Use", Part I, *Chemical Engineering,* July 19, 1965, pp. 157-162, Part II, *Chemical Engineering,* Aug. 2, 1965, pp. 111-116.
(7) Fair, J. R., "Designing Direct-Contact Coolers/Condensers", *Chemical Engineering,* June 12, 1972, pp. 91-100.

ENERGY CONSERVATION

(1) "Upgrading Existing Evaporators to Reduce Energy Consumption", (COO/2870-2) National Technical Information Service, U. S. Department of Commerce, Springfield, VA, 22161 (1977).

(2) "Energy Conservation In Distillation" (DOE/CS/4431-T2) National Technical Information Service, U. S. Department Of Commerce, Springfield, VA, 22161 (1980).

(3) Robnett, J. D., "Engineering Approaches to Energy Conservation", *Chemical Engineering Process,* March 1979, pp. 59-67.

(4) Barber, R. E., "Rankine-Cycle Systems for Waste Heat Recovery", *Chemical Engineering,* Nov. 25, 1974, pp. 101-106.

(5) Crozier, R. A., Jr., "Designing a 'Near-Optimum' Cooling-Water System", *Chemical Engineering,* April 21, 1980, pp. 118-127.

(6) Holiday, A. D., "Conserving and Reusing Water", *Chemical Engineering,* April 19, 1982, pp. 118-137.

(7) Linnhoff, B. and Vredeveld, D. R., "Pinch Technology Has Come Of Age", *Chemical Engineering Progress,* July 1984, pp. 33-40.

(8) Boland, D. and Hindmarsh, E., "Heat Exchanger Network Improvements", *Chemical Engineering Progress,* July 1984, pp. 47-54.

(9) Aegerter, R., "Energy Conservation In Process Plants", *Chemical Engineering,* Sept. 3, 1984, pp. 93-96.

VAPOR COMPRESSION EVAPORATION

(1) Casten, J. W., "Mechanical Recompression Evaporation", *Chemical Engineering Progress,* July 1978, pp. 61-67.

(2) Bennett, R. C., "Recompression Evaporation", *Chemical Engineering Progress,* July, 1978, pp. 67-70.

(3) Hough, G. W., "Pre-Evaporation of Kraft Black Liquor by Vapor Recompression Evaporation", *TAPPI Journal,* July, 1978, pp. 23-25.

(4) Beagle, M. J., "Recompression Evaporation", *Chemical Engineering Progress,* Oct. 1962, pp. 79-82.

(5) Malleson, J. H., "Chemical Process Application for Compression Evaporation", *Chemical Engineering,* Sept. 2, 1963, pp. 75-82.

(6) Beasley, A. H. and Rhinesmith, R. D., "Energy Conservation by Vapor Compression Evaporation," *Chemical Engineering Progress,* Aug. 1980, pp. 37-41.

(7) Zimmer, A., "Developments in Energy-Efficient Evapoation", *Chemical Engineering Progress,* Aug. 1980, pp. 50-56.

(8) Weimer, L. D., Dolf, H. R., Austin, D. A., "A Systems Engineering Approach to Vapor Recompression Evaporation", *Chemical Engineering Progress,* Nov. 1980, pp. 70-77.

(9) Meo, D. III, "Effective Thermocompressor For Solvent Stripper", *Chemical Engineering Progress,* Oct. 1981, pp. 33-36.

(10) Dev, L., "Heat Recovery in the Forest Products Industry", *Chemical Engineering Progress,* Dec. 1979, pp. 25-29.

(11) Hughes, C. H. and Emmermann, D. K., "VTE.VC For Sea Water", *Chemical Engineering Progress,* July 1981, pp. 72-73.

(12) Dev, L. and Kelso, R. W., "Steam Stripping of Kraft Mill Condensates", *Chemical Engineering Progress,* Jan. 1978, pp. 72-75.

(13) Dansinger, R. S., "Distillation Columns With Vapor Recompression", *Chemical Engineering Progress,* Sept. 1979, pp. 58-64.

(14) "Compressor Handbook", Hydrocarbon Processing, P.O. Box 2608, Houston, TX 77001.

(15) Flores, J., Castells, F. and Ferre, J. A., "Recompression Saves Energy", *Hydrocarbon-Processing,* July 1984, pp. 59-62.

(16) Davis, H., "Evaluating Multistage Centrifugal Compressors", *Chemical Engineering,* Dec. 26, 1983, pp. 35-38.

VACUUM SYSTEMS

(1) Power, R. B., "Steam Jet Ejectors", *Oil and Gas Equipment,* October 1965 through July 1966.

(2) Huff, G. A., Jr., "Selecting a Vacuum Producer", *Chemical Engineering,* Mar. 15, 1976, pp. 83-86.

(3) Ryans, J. L. and Croll, S., "Selecting Vacuum Systems," *Chemical Engineering,* Dec. 4, 1981, pp. 72-90.

(4) "ASME Performance Test Code; Ejectors" (C00010), American Society of Mechanical Engineers, New York.

(5) "Standards for Direct Contact Barometric and Low Level Condensers", 5th edition, 1970, Heat Exchange Institute, 1230 Keith Building, Cleveland, Ohio 44115.

(6) "Standards for Steam Jet Ejectors", 3rd edition, 1956, Heat Exchange Institute.

(7) "Standard for Field Testing: Addition to Standard for Steam Jet Ejectors", 3rd edittion, 1975, Heat Exchanger Institute.

(8) "Construction Standards for Surface Type Condensers" For Ejector Service", 1972, Heat Exchange Institute.

(9) "Vacuum Systems Catalog", Croll-Reynolds Co., Inc., 751 Central Ave., Westfield, NJ 07091.

(10) "Kinney Vacuum Catalog", Kinney Vacuum Co., 495 Turnpike, Canton, MA 02021.

(11) Dobrowolski, Z. C., "Mechanical Vacuum Pumps In The Process Industry", *Chemical Engineering Progress,* July 1984, PP. 75-83.

(12) Ryans, J. L., "Advantages Of Integrated Vacuum Pumping Systems", *Chemical Engineering Progress,* June, 1984, pp. 59-62.

(13) Patton, P. W., Vacuum Systems In The CPI", *Chemical Engineering Progress,* Dec. 1983, pp. 56-61.

STEAM TRAPS

(1) Mathur, J., "Steam Traps", *Chemical Engineering Deskbook Issue,* Feb. 26, 1973, pp. 47-52.

(2) Vallery, S. J., "Are Your Steam Traps Wasting Energy?", *Chemical Engineering,* Feb. 9, 1981, pp. 84-98.

(3) Monroe, E. J., Jr., "Effects of CO_2 In Steam Systems, *Chemical Engineering,* Mar. 23, 1981, pp. 209-212.

378 Handbook of Evaporation Technology

(4) Monroe, E. J., Jr., "Select the Right Steam Trap", *Chemical Engineering,* Jan. 5, 1976, pp. 129-134.

(5) Monroe, E. J., Jr., "Instill Steam Traps Correctly", *Chemical Engineering,* May 10, 1976, pp. 121-126.

(6) "Steam Trap Selection and Application Guide", Sarco Co., P. O. Box 119, Allentown, PA 18105

(7) "Hook-Up Design For Steam And Fluid Systems", Sarco Co.

(8) "Industrial Steam Trapping", Yarway Corp., Blue Bell, PA 19422.

(9) "Handbook of Steam Trapping", TLV Co., LTD., Tokyo.

(10) Industrial Steam Trapping Handbook", Yarway Corporation, Blue Bell, Pennsyvlania.

(11) "Steam Utilization Course", Spirax-Sarco, Inc., Allentown, Pennsylvania.

(12) Condensate Manual", Gestra AG, Bremen, West Germany.

CONTROL

(1) Hepp, P. S., "Internal Column Reboilers-Liquid Level Measurement", *Chemical Engineering Progress,* Feb. 1963, pp. 66-69.

(2) Baker, D. F., "Surge Control for Multistage Centrifugal Compressors", *Chemical Engineering,* May 31, 1982, pp. 117-122.

(3) Shinskey, F. G., "Control Systems Can Save Energy", *Chemical Engineering Progress,* May 1978, pp. 43-46.

(4) Sanders, C. W., "Better Control of Heat Exchangers", *Chemical Engineering,* Sept. 21, 1959, pp. 145-148.

(5) Renard, I. H, "A Roadmap to Control-System Design", *Chemical Engineering,* Nov. 29, 1982, pp. 46-58.

(6) J. W. Hutchison, "ISA Handbook of Control Valves", Instrument Society of America, Pittsburgh.

(7) Shinskey, F. G., "Distillation Control", McGraw Hill Book Co., New York (1977).

(8) Badavas, P. C., "Feedforward Methods For Process Control Systems", *Chemical Engineering,* Oct. 15, 1984, pp. 103-108.

(9) Dealy, J. M., "Viscometers For Online Measurement And Control", *Chemical Engineering,* Oct. 1, 1984, pp. 62-70.

(10) Hoeppner, C. H., "Online Measurement of Liquid Density", *Chemical Engineering,* Oct. 1, 1984, pp. 71-78.

PUMPS

(1) McLeon, M. G., "How to Select and Apply Flexible-Impeller Pumps", *Chemical Engineering,* Sept. 20, 1982, pp. 101-106.

(2) Karassik, I. J., "Centrifugal Pumps and System Hydraulics", *Chemical Engineering,* Oct. 4, 1982, pp. 84-106.

(3) Doll, T. R., "Making The Proper Choice of Adjustable-Speed Drives", *Chemical Engineering,* Aug. 9, 1982, pp. 46-60.

(4) "Hydraulic Institute Standards", 13th edition, Hydraulic Institute, Cleveland (1975).

(5) Stepanoff, A. J., "Centrifugal And Axial Flow Pumps", 2nd edition, John Wiley and Sons, Inc., New York (1957).

(6) Kristal, F. A. and Annett, F. A., "Pumps", 2nd edition, McGraw-Hill, New York (1953).

(7) Karassik, I. J. and Wright, E. F., "Pump Questions and Answers", McGraw-Hill, New York (1949).

(8) Church, A. H., "Centrifugal Pumps and Blowers", John Wiley and Sons, New York (1944).

PROCESS PIPING AND FLUID FLOW

(1) Simpson, L. L. and Weirick, M. L., "Designing Plant Piping", *Chemical Engineering Deskbook Issue,* April 3, 1978, pp. 35-48.

(2) Kistler, H. Z., "Outlets and Internal Devices for Distillation Columns", *Chemical Engineering,* July 28, 1980, pp. 79-83.

(3) Simpson, L. L., "Sizing Piping for Process Plants," *Chemical Engineering,* June 17, 1968, pp. 192-214.

(4) Govier, G. W. and Aziz, K., "The Flow of Complex Mixtures in Pipes", Van Nostrand Reinhold, New York (1972).

SEPARATORS

(1) Gerunda, A., "How to Size Liquid-Vapor Separators", *Chemical Engineering,* May 4, 1981, pp. 81-84.

(2) Droz, N. A. R., "Urea Evaporator Entrainment Separator", *Chemical Engineering Progress,* March 1982, pp. 62-65.

(3) Calvert, S., "Guidelines For Selecting Mist Eliminators", *Chemical Engineering,* Feb. 27, 1978, pp. 109-112.

(4) Holmes, T. L. and Chen, G. K., "Design and Selection Of Spray/Mist Elimination Equipment", *Chemical Engineering,* Oct. 15, 1984, pp. 82-89.

(5) Wu, F. H., "Drum Separator Design: A New Approach", *Chemical Engineering,* April 2, 1984, pp. 74-80.

THERMAL INSULATION

(1) Neal, J. E. and Clark, R. S., "Saving Heat Energy In Re-Fractory-Lined Equipment", *Chemical Engineering,* May 4, 1981, pp. 56-70.

(2) Chapmen, F. S. and Holland, F. A., "Keeping Piping Hot, Part I, By Insulation", *Chemical Engineering,* Dec. 20, 1965, pp. 79-90.

(3) Barnhart, J. M., "Insulation Saves Heat, Saves Money," *Chemical Engineering,* June 11, 1962, pp. 164-169.

(4) Hughes, R. and Deumaga, V., "Insulation Saves Energy", Chemical Engineering, May 27, 1974, pp. 95-100.

(5) Turner, W. C., "Criteria For Installing Insulation System in Petrochemical Plants", *Chemical Engineering Progress,* Aug. 1974, pp. 41-45.

- (6) Marks, J. B. and Holton, K. D., "Protection of Thermal Insulation", *Chemical Engineering* Progress, Aug. 1974, pp. 46-49.
- (7) McChesney, M. and McChesney, P., "Insulation Without Economics", *Chemical Engineering,* May 3, 1982, pp. 70-79.
- (8) McChesney, M. and McChesney, P., "Preventing Burns From Insulated Pipes", *Chemical Engineering,* July 27, 1981, pp. 58-64.
- (9) Turner, W. C. and Malloy, J. F., "Thermal Insulation Handbook", McGraw-Hill, New York (1981).
- (10) Turner, W. C., and Malloy, J. F., "Handbook of Thermal Insuation Design Economics for Pipes and Equipment", McGraw-Hill, New York (1980).

TROUBLESHOOTING

- (1) Swartz, A., "A Guide for Troubleshooting Multiple-Effect Evaporators", *Chemical Engineering,* May 8, 1978, pp. 175-182.
- (2) Vargas, K. J., "Troubleshooting Compression Refrigeration Systems", *Chemical Engineering,* Mar. 22, 1982, pp. 137-143.
- (3) Shah, G. C., "Troubleshooting Distillation Columns, *Chemical Engineering,* July, 31, 1978, pp. 70-78.
- (4) Shah, G. C., "Troubleshooting Reboiler Systems", *Chemical Engineering Progress,* July 1979, pp. 53-58.
- (5) Gilmour, C. H., "Troubleshooting Heat-Exchanger Designs", *Chemical Engineering,* June 19, 1967, pp. 221-228.
- (6) Yundt, B., "Troubleshooting VC Evaporators", *Chemical Engineering,* December 24, 1984, pp. 46-55.

VENTING

- (1) Khan, R. A., "Effect of Noncondensables in Sea Water Evaporators", *Chemical Engineering Progress,* July, 1972, pp. 79-80.
- (2) Standiford, F. C., "Effect of Non-Condensibles On Condenser Design and Heat Transfer", *Chemical Engineering Progress,* July 1979, pp. 59-62.

AIR-COOLED HEAT EXCHANGERS

- (1) Monroe, R. C., "Consider Variable Pitch Fans", *Hydrocarbon Processing,* Dec. 1980.
- (2) Franklin, G. M. and Munn, W. B., "Problems With Heat Exchangers In Low Temperature Environments", *Chemical Engineering Progress,* July 1974, pp. 63-67.
- (3) Shipes, K. V., "Air-Cooled Exchangers in Cold Climates", *Chemical Engineering Progress,* July 1974, pp. 53-58.
- (4) Rothernberg, D. H. and Nicholson, R. L., "Interacting Controls for Air Coolers", *Chemical Engineering Progress,* Jan. 1981, pp. 80-82.
- (5) Gunter, A. Y. and Shipes, K. V., "Hot Air Recirculation By Air Coolers", AIChE Preprint 8—presented at 12th National Heat Transfer Conference, August 15-18, 1971, American Institute of Chemical Engineers, New York.

(6) Kals, W., "Wet-Surface Aircoolers", *Chemical Engineering,* July 26, 1971, pp. 90-94.

HEAT TRANSFER FLUIDS

(1) Minton, P. E. and Plants, C. A., "Heat Transfer Media Other Than Water", Volume 12, pp. 171-191, Kirk-Othmer Encyclopedia of Chemical Technology, John Wiley and Sons, Inc., New York (1980).
(2) Fried, J. R., "Heat-Transfer Agents for High-Temperature Systems", *Chemical Engineering,* May 28, 1973, pp. 89-98.
(3) Singh, J., "Selecting Heat-Transfer Fluids for High-Temperature Service", *Chemical Engineering,* June 1, 1981, pp. 53-58.

TESTING

(1) Standiford, F. C., "Testing Evaporators", *Chemical Engineering Progress,* Nov. 1962, pp. 80-83.
(2) Newman, H. H., "How to Test Evaporators," *Chemical Engineering Progress,* July 1968, pp. 33-38.
(3) "AIChE Equipment Testing Procedure: Evaporators", 2nd edition, American Institute of Chemical Engineers, New York (1979).

ELECTRICAL HEATING

(1) Ando, M., and Othmer, D. R., "Heating Pipelines With Electrical Skin Current," *Chemical Engineering,* Mar. 9, 1970, pp. 154-158.
(2) Silverman, D., "Electrical Heating", *Chemical Engineering,* July 20, 1964, pp. 161-164.
(3) Brown, C. W., "Electric Pipe Tracing", *Chemical Engineering,* June 23, 1975, pp. 172-178.
(4) Yurkanin, R. M., "HPI Applications of Electric Process Heating", *Petro/Chem Engineer,* Aug. 1967, pp. 32-35.
(5) Woollen, K. N., "Electrical Energy: A Versatile Heat Source", *Chemical Engineering,* June 11, 1962, pp. 145-151.
(6) Lansdale, J. T. and Mundy, J. E., "Estimating Pipe Heat Tracing Costs", *Chemical Engineering,* Nov. 29, 1982, pp. 89-93.

STEAM TRACING

(1) Bertran, C. G., Desai, V. J., and Interess, E., "Designing Steam Tracing", *Chemical Engineering,* April 13, 1972, pp. 74-80.
(2) Nouse, F. F., "Pipe Tracing and Insulation," *Chemical Engineering,* June 17, 1968, pp. 243-246.
(3) Chapman, F. S. and Holland F. A., "Keeping Piping Hot, Part III, By Heating", *Chemical Engineering,* Jan. 17, 1968, pp. 133-144.

(4) Fisch, E. "Winterizing Process Plants", *Chemical Engineering,* Aug. 20, 1981, pp. 128–143.

JACKETED VESSELS

(1) Bollinger, D. H., "Assessing Heat Transfer In Process Vessel Jackets", *Chemical Engineering,* Sept. 20, 1982, pp. 95–100.
(2) Fogg, R. M. and Uhl, V. W., "Resistance to Heat Transfer In The Half-Tube Jacket", Paper 87e, 74th National Meeting, AIChE, March 14, 1973, American Institute of Chemical Engineers, New York.

TURBINES

(1) Swearingen, J. S., "Turboexpanders and Processes That Use Them", *Chemical Engineering Progress,* July 1972, pp. 95–102.
(2) Chadha, N., "Use Hydraulic Turbines to Recover Energy", *Chemical Engineering,* July 23, 1984, pp. 57–61.

MECHANICAL DESIGN

(1) Yokell, S., "Heat-Exchanger Tube-to-Tubesheet Connections", *Chemical Engineering,* Feb. 8, 1982, pp. 78–94.
(2) Crozier, R. A., Jr., "Pressure Relief to Prevent Heat-Exchanger Failure", *Chemical Engineering,* Dec. 15, 1980, pp. 79–83.

MATERIALS OF CONSTRUCTION

(1) Kirby, G. N., "How to Select Materials", *Chemical Engineering,* Nov. 3, 1980, pp. 86–131.
(2) Redmond, J. D. and Miska, K. H., "The Basics of Stainless Steels", *Chemical Engineering,* Oct. 18, 1982, pp. 79–93.

DESALINATION

(1) Wagner, W. M. and Finnegan, D. R., "Select A Seawater Desalting Process", *Chemical Engineering,* Feb. 7, 1983, pp. 71–75.
(2) Peak, W. E., "Desalting Seawater In Flash Evaporators", *Chemical Engineering Progress,* July, 1980, pp. 50–53.
(3) Gooding, C. H., "Reverse Osmosis and Ultrafiltration Solve Separation Problems", *Chemical Engineering,* Jan. 7, 1985, pp. 56–62.
(4) Applegate, L. E., "Membrane Separation Processes", *Chemical Engineering,* June 11, 1984, pp. 64–89.
(5) Hoffman, D., "Low Temperature Evaporation Plants", *Chemical Engineering Progress,* Oct. 1981, pp. 59–62.

(6) "Desalination Momentum Picks Up", *Compressed Air Magazine,* Sept. 1983, pp. 6-11.
(7) "Process Selection Guide to Seawater Desalting", Technical Presentation 750-3550, Aqua-Chem, Inc., Milwaukee, Wisconsin.

EVAPORATORS

(1) Parker, N. H., "How to Specify Evaporators", *Chemical Engineering,* July 22, 1963, pp. 135-140.
(2) Kerridge, A. E., "How to Evaluate Bids For Major Equipment", *Hydrocarbon Processing,* May 1984, pp. 141-154.
(3) "Guide To Dryer and Evaporator Equipment", *Chemical Processing,* Nov. 1984, pp. 53-69.

Index

Accessories - 210
Acoustic vibration - 53
   Detuning baffles - 53
Agitated vessels - 17, 90
Air-cooled heat exchangers - 145
   Condensers - 145
   Control - 308
   Heat transfer - 17
   Pressure drop, air side - 46

Ball mill evaporators - 108
Barometric legs - 228
Basket evaporators - 78
Bibliography - 372
Biofouling - 124
Boiling - 29
   Convective - 34
   Enhanced surfaces - 37
   Film - 31, 34
   Natural circulation - 33
   Nucleate - 33
Boiling point rise - 305
   Control - 305

Calandrias - 4, 60
   Control - 310
   Feed location - 69
   Flow instabilities - 65
   Forced circulation - 84

Internal - 67
Liquid level - 60, 68
Natural circulation - 60
Propeller - 80
Surging - 64
Cascade evaporator - 221
Cavitation - 272
Centrifugal pumps - 273
Centrifugal separators - 159
Chemical reaction fouling - 121
Circulating pumps - 86
Cleaning cycles - 126
Climbing film evaporators - 82
Coils - 73
Compression evaporation - 175
   Mechanical vapor compression - 176
   Thermocompression - 177
Compressors
   Axial flow centrifugal - 189
   Capacity limitation - 193
   Characteristics - 192
   Constant speed - 192
   Control - 195
   Drive systems - 192
   Positive displacement - 189
   Radial flow centrifugal - 190
   Reliability - 203
   Selection - 189

Compressors (cont'd)
    Variable speed - 195
Concentration - 6, 304
Condensate connections - 221
Condensate removal - 259
    Liquid level control - 259
    Steam traps - 261
Condensation - 18
    Desuperheating - 26
    Fog formation - 26
    Enhanced surfaces - 24
    Immiscible condensates - 24
    Inside horizontal tubes - 21
    Outside horizontal tubes - 21
    Sloped bundles - 23
    Subcooling - 24
    Vapors with noncondensables - 25
    Vertical tubes - 19
Condensers - 211
    Air-cooled - 145
    Condensate connections - 221
    Control - 306
    Direct contact - 211
    Evaporative cooled - 143
    Flooding in horizontal tubes - 217
    Integral condensers - 218
    Shell-and-tube - 214
    Shellside flooding - 218
    Surface - 213
    Updraft versus downdraft - 215
Control - 297
    Auto-select system - 304
    Base section - 312
    Boiling point rise - 305
    Buoyancy float - 306
    Calandria - 310
    Compressor - 195
    Condenser - 306
    Conductivity - 305
    Densitometer - 306
    Differential pressure - 306
    Evaporator - 302
    Evaporator system - 298
    Feedback - 299
    Feed forward - 300
    Gamma gage - 306
    Liquid level - 259, 312
    Manual - 297
    Mechanical vapor compression - 199
    Multielement - 299
    Product concentration - 304
    Thermocompressors - 184
    Vacuum systems - 252
Convective boiling - 34
Cooling water
    Air injection - 326
    Backwashing - 325
    Biofouling - 124
    Energy economy - 141
    Fouling - 119
    Siphons - 322
    Tempered water systems - 141
    Tubewall temperatures - 131
    Venting - 141
Corrosion - 7
Corrosion fouling - 123
Costs
    Vacuum systems - 256
Cyclones - 159

Desalination - 206
    Capital cost - 208
    Complexity - 206
    Energy efficiency - 207
    Fouling - 119
    Maintenance - 207
    Operating temperature - 208
    Startup and operability - 206
Design considerations
    Condensers - 214
    Extended surfaces - 319
    Flow distribution - 321
    Fouling - 127
    Impingement protection - 319
    Maintenance - 324
    Pumping systems - 272, 277
    Shellside design - 130
    Steam trapping - 135, 264
    Tubeside velocities - 129
    Tube size and arrangement - 318
    Venting - 134
Desuperheating of vapors - 26
Detuning baffles - 53

Direct contact
    Condensers - 211
    Evaporators - 107
Disk evaporator - 107
Downcomers - 68
Drain piping - 280

Electrical heaters - 108
Energy conservation - 357
    Vacuum systems - 258
Energy economy
    Air-cooled condensers - 145
    Cooling water - 141
    Evaporative cooled condensers - 143
    Heat exchange - 142
    Heat pumps - 148
    High temperature media - 147
    Multiple-effect evaporators - 148, 166
    Pumping systems - 147
    Steam - 141
    Steam condensate recovery systems - 150
    Tempered water systems - 141
    Thermal engine cycles - 149
    Vacuum systems - 246, 258
Enhanced surfaces
    Boiling - 37
    Condensing - 24
Entrainment - 153
Evaporation
    Definition - 2
    Fouling - 119, 120, 125
Evaporative cooled condensers - 143
Evaporators
    Accessories - 210
    Agitated - 90
    Ball mills - 108
    Basket - 78
    Batch - 70
    Cascade - 107
    Climbing film - 82
    Coils - 73
    Compression - 175
    Continuous - 70
    Control - 297

    Direct contact - 107
    Disk - 107
    Electrically heated - 108
    Elements - 5
    Falling film - 83
    Fired heaters - 108
    Flash - 103
    Fluidized bed - 105
    Foaming - 106
    Forced circulation - 84
    Function - 3
    Grainer - 107
    Heat transfer - 9
    Horizontal tube - 74
    Improvements - 6
    Inclined tube - 79
    Jacketed - 70
    Long tube vertical - 81
    Mechanical compression - 203
    Mechanically aided - 90
    Multiple effect - 148
    Once-through - 70
    Performance - 133
    Plate - 87
    Propeller - 80
    Recirculated - 70
    Refrigerant heated - 108
    Rising film - 82
    Scraped surface - 90
    Short tube vertical - 77
    Solar - 70
    Special types - 106
    Spray film - 76
    Submerged combustion - 100
    Thermal compression - 176
    Thin-film - 92
    Types - 70
    Upgrading - 350

Falling film
    Distributors - 29
    Evaporators - 83
    Heat transfer - 27
    Spray film - 76
Feed location - 69
Film boiling - 31, 34
Fin efficiency - 15
Fired heated evaporators - 108

Index 387

Flash evaporators - 103
Flashing entrainment - 164
Flash tanks - 155
Flow distribution - 321
Flow induced vibration - 48
Flow instabilities - 65
Fluid-elastic whirling - 52
Fluidized bed evaporators - 105
Fluted tubes - 24
Foaming - 6, 164
Foaming evaporators - 106
Fog formation - 26
Foodstuffs fouling - 121
Forced circulation
    Evaporators - 84
    Pumps - 86
Fouling - 7, 113
    Biofouling - 124
    Chemical reaction - 121
    Classification - 114
    Corrosion - 123
    Cost of - 114
    Design considerations - 127
    Evaporation - 125
    Optimum cleaning cycles - 126
    Particulate - 120
    Philosophy of design - 127
    Precipitation - 118
    Sequential events - 116
    Shellside design - 130
    Shellside velocity - 131
    Solidification - 125
    Tubeside velocity - 129
    Tubewall temperature - 131
Free dry air equivalent - 244

Grainer - 107

Heat exchangers - 142
Heat pumps - 172
Heat tracing - 290
Heat transfer - 9
    Agitated vessels - 17
    Air-cooled heat exchangers - 17
    Boiling - 29
    Condensation - 18
    Falling film - 27
    Modes - 10

    Sensible heat transfer inside
        tubes - 11
        Helical coils - 12
    Sensible heat transfer outside
        tubes - 13
        Banks of tubes - 13
        Lowfin tubes - 14
    Types of operations - 10
High flux tubes - 37
High temperature media - 147
Horizontal tube evaporators - 74

Immiscible condensates - 24
Impingement protection - 319
Inclined tube evaporators - 79
Installation - 322
    Equipment layout - 323
    Piping - 323
Instrument guidelines - 313
Insulation - 288
Internal calandrias - 67

Jacketed vessels - 70
Jet ejectors - 222

Liquid characteristics - 6
Liquid ring pumps - 237
Long tube vertical evaporators - 81
Lowfin tubes - 14
    Condensation - 23
    Pressure drop - 45

Maintenance - 324
    Air injection - 326
    Backwashing - 325
    Cleaning - 325
    Design practices - 324
    Repair - 325
    Standard practices - 325
    Steam traps - 269
Materials of construction - 336
Mechanical design - 327
    Double tubesheets - 329
    Inspection techniques - 331
    Thermal expansion - 328
    Tube-to-tubesheet joints - 329
    Upset conditions - 328

Mechanically-aided evaporators - 90
   Thin-film - 92
Mechanical vapor compression - 186
   Application - 204
   Compressor selection - 189
   Drive systems - 192
   Economics - 204
   Evaporator design - 203
   Factors affecting costs - 188
   Factors influencing design - 190
   Reliability - 203
   System characteristics - 199
   Thermodynamics - 187
Multiple-effect evaporators - 148, 166
   Backward feed - 168
   Calculations - 170
   Forward feed - 167
   Heat recovery systems - 170
   Mixed feed - 169
   Optimization - 170
   Parallel feed - 169
   Staging - 169

Natural circulation calandrias - 60
Natural convection heat transfer - 33
New technology - 363
Nomenclature - 365
Non-condensable gases
   Condensation of vapors containing - 25
   Effect on heat transfer - 134
Nucleate boiling - 33

Operation
   Energy economy - 140
   Pressure versus vacuum - 140
   Thermal compressors - 177
Optimization
   Cleaning cycles - 126
   Multiple-effect evaporators - 170

Particulate fouling - 120
Performance of evaporators - 133
Physical properties - 38

Plate evaporators - 87
   Gasketed plate - 89
   Patterned plate - 89
   Spiral plate - 87
Porous boiling surfaces - 37
Precipitation fouling - 118
Pressure drop - 39
   Across tube banks - 42
   Air-cooled heat exchangers - 46
   Condensing vapors - 41, 46
   Helical coils - 41
   Inside tubes - 39
   Lowfin tubes - 45
   Two-phase flow - 46
   U-bend tubes - 40
Process computers - 316
Process piping - 279
   Layout - 285
Process pumps - 270
Process vessels - 292
Product quality - 7
Propeller calandrias - 80
Propeller pumps - 275
Pumping systems - 270
   Cavitation - 272
   Centrifugal pumps - 273
   Common problems - 277
   Energy economy - 147
   NPSH - 271
   Principles - 272
   Propeller pumps - 275
   Safety factors - 274
   Types of pumps - 270

Refrigerant heated evaporators - 108
Refrigeration - 294
   Absorption - 296
   Mechanical - 295
   Steam jet - 295
Reverse osmosis - 206
Rising film evaporators - 82, 84

Safety - 332
Salting - 7
Scaling - 7
Scraped-surface evaporators - 90
Sensible heat transfer
   Inside tubes - 11

Sensible heat transfer (cont'd)
　　Outside tubes - 13
Shellside design - 130
Slurries - 284
Solidification fouling - 125
Solids deposition - 162
Special evaporator types - 106
Specifying evaporators - 360
　　Comparing vendor's offerings - 361
Splashing - 164
Spray-film evaporators - 76
Steam
　　Condensate recovery - 150
　　Desuperheating - 26
　　Energy economy - 141
Steam jet ejectors - 222
　　Aftercondensers - 225
　　Barometric legs - 228
　　Basic performance curve - 231
　　Efficiencies - 230
　　Intercondensers - 225
　　Multistage - 225
　　Multistage ejector characteristics - 233
　　Precondensers - 225
　　Stage characteristics - 230
　　Troubleshooting - 234
　　Unstable ejectors - 232
Steam traps - 261
　　Common problems - 264
　　Design considerations - 261
　　Effects of carbon dioxide - 268
　　Installation - 265
　　Maintenance - 269
　　Mechanical traps - 263
　　Operating principles - 262
　　Selection - 263
　　Specification - 263
　　Thermodynamic traps - 263
　　Thermostatic traps - 263
Subcooling of condensate - 24
Submerged combustion evaporators - 100
Surface condensers - 213
Surging in calandrias - 64

Temperature sensitivity - 6

Testing evaporators - 339
　　Causes for poor performance - 340
　　Planning - 339
Thermal compression - 176
　　Application - 184
　　Control - 184
Thermal engine cycles - 147
Thermal insulation - 288
Thermocompressors - 177
　　Characteristics - 178
　　Control - 180
　　Estimating data - 180
　　Fixed nozzle - 179
　　Multiple nozzle - 180
　　Operation - 177
　　Spindle operated - 180
　　Types - 179
Thin-film evaporators - 92
　　Application - 92
　　Maintenance - 180
　　Process considerations - 97
Time/temperature relation - 139
Troubleshooting - 343
　　Calandrias - 344
　　Condensers - 345
　　Steam jet ejectors - 234
　　Vacuum systems - 346
Tubewall temperatures - 131
Turbulent buffeting - 51
Two-phase flow - 281
　　Flow instability - 65
　　Pressure drop - 46

Upgrading existing evaporators - 350
　　Areas to consider - 351
　　Economic effect - 356
　　Guidelines - 357

Vacuum producing equipment - 222
　　Air leakage - 242
　　Control - 252
　　Costs - 256
　　Estimating energy requirements - 246
　　Initial system evacuation - 251
　　Jet ejectors - 222
　　Liquid ring pump - 237

Vacuum producing equipment (cont'd)
    Load gas - 253
    Maintenance - 240
    Mechanical pumps - 236
    Multistage combinations - 240
    Precondensers - 225
    Reliability - 240
    Safety factors - 245
    Sizing information - 241
Vane impingement separators - 157
Vapor-liquid separators - 153
    Centrifugal - 159
    Cyclones - 159
    Entrainment - 153
    Falling film evaporators - 164
    Flashing - 164
    Flash tanks - 155
    Foaming - 164
    Other types - 162
    Solid deposition - 162
    Splashing - 164

Vane impingement - 157
Wire mesh - 155
Velocity
    Shellside - 131
    Tubeside - 129
Venting - 133
Vertical evaporators - 77
    Long tube - 81
    Short tube - 77
Vibration - 48
    Acoustic - 53
    Design criteria - 56
    Fixing field problems - 58
    Fluid-elastic whirling - 52
    Mechanisms - 49
    Proprietary designs - 59
    Turbulent buffeting - 51
    Vortex shedding - 50
Vortex shedding - 50

Wire mesh separators - 155

RETURN TO ➡ **CHEMISTRY LIBRARY**
100 Hildebrand Hall